For M. F. S.

STUDY GUIDE

JOSEPH BOYLE

PHYSICS

PRINCIPLES WITH APPLICATIONS

Fifth Edition

GIANCOLI

PRENTICE HALL, Upper Saddle River, NJ 07458

Assistant Editor: *Wendy Rivers*
Production Editor: *Mindy DePalma*
Special Projects Manager: *Barbara A. Murray*
Supplement Cover Manager: *Paul Gourhan*
Production Coordinator: *Ben Smith*

Printed in the United States of America

10 9 8 7 6 5 4 3 2

ISBN 0-13-627944-9

Prentice-Hall International (UK) Limited, *London*
Prentice-Hall of Australia Pty. Limited, *Sydney*
Prentice-Hall Canada, Inc., *Toronto*
Prentice-Hall Hispanoamericana, S.A., *Mexico*
Prentice-Hall of India Private Limited, *New Delhi*
Prentice-Hall of Japan, Inc., *Tokyo*
Simon & Schuster Asia Pte. Ltd., *Singapore*
Editora Prentice-Hall do Brasil, Ltda., *Rio de Janeiro*

CONTENTS

PREFACE

CHAPTER

PREFACE

This study guide was written to accompany PHYSICS: PRINCIPLES WITH APPLICATIONS, Fifth Edition by Douglas C. Giancoli. It is intended to provide additional help in understanding the basic principles covered in the text and add to the student's problem solving skills.

The Concept Summary section of the study guide summarizes the main topics covered in the corresponding chapter of the text. Each chapter begins with a list of key terms and phrases discussed in the Concept Summary section. A summary of the basic mathematical equations used in the Concept Summary section follows the Key Terms and Phrases section and the each symbol used in the equations is defined.

The end of each chapter in the text includes questions as well as problems. The study guide contains a section entitled Selected Questions with Answers where three or more questions from the question section of the text are discussed.

Because beginning physics courses emphasize problem solving, hints on problem solving skills are placed just before the section of programmed problems. A suggestion for the proper used of the programmed problems is included in chapter 1. The section of the text to which the programmed problem corresponds is included as part of the solution. Every effort has been made to present the solution to each problem in a step by step manner. Each chapter concludes with four practice problems.

The student should be aware that the study guide is meant to complement the text, not to replace the text as a learning tool. Because of this, it is suggested that the student carefully read the chapter in the text before using the study guide.

I wish to acknowledge the cooperation and assistance given by Wendy Rivers of Prentice-Hall, Inc. and a special thanks is extended to Peggy Boyle for her help in typing and editing the manuscript.

Every effort has been made to avoid errors; however, I alone have responsibility for any errors which remain and corrections and comments are most welcome.

Joseph J. Boyle
Miami-Dade Community College

CHAPTER 1

INTRODUCTION

KEY TERMS AND PHRASES

science attempts to explain natural phenomena that can be detected with our senses or by instruments designed to extend our senses, e.g., a telescope.

physics is the branch of science that deals with natural laws and processes, and the states and properties of matter and energy.

matter refers to any object that has substance and occupies space.

energy is the ability to do useful work. Energy takes a number of forms, such as mechanical energy, electromagnetic energy, heat energy, and nuclear energy.

model, as used in physics, is an analogy or mental image used to explain a physical phenomena.

scientific theory, as used in physics, is a plausible principle offered to explain a physical phenomena. A theory leads to predictions that can then be tested by experiment to see if there is agreement with the phenomenon.

scientific law is applied to certain statements that are found to be valid over a large range of observed phenomena. An example of a scientific law is the law of universal gravitation.

significant figures in a measurement include the figures that are certain plus the first doubtful digit.

SI system of measurement is the system of measurement established by the French Academy of Science. For example, in SI units, the unit of length is the meter, time is second, and mass is kilogram.

order-of-magnitude is a rough estimate of the value of a quantity. This estimate usually contains one significant figure and the associated power of ten. For example, one often hears the world population given to the nearest billion.

CONCEPT SUMMARY

Science and Creativity

Science attempts to explain natural phenomena that can be detected with our senses or by instruments designed to extend our senses, e.g., a telescope. Science is a creative endeavor that resembles other creative activities of the human mind, e.g., art and music. **Physics** is the branch of science that deals with natural laws and processes, and the states and properties of **matter** and **energy**.

To explain a particular natural phenomenon, a scientist constructs a **model** that leads to a **theory** designed to explain the phenomenon. The theory leads to predictions that can then be tested by experiment to see if there is agreement with the phenomenon.

In science, the term **law** is applied to certain statements that are found to be valid over a large range of observed phenomena. An example of a scientific law is the law of universal gravitation. Scientific laws are descriptive in that they "describe how nature does behave" as compared to a traffic law which tells us how we should behave.

Measurement and Uncertainty

There is limitation in terms of accuracy in every measurement. This limitation is usually associated with the measuring instrument and human "inability to read the instrument beyond some fraction of the smallest division shown." Because of this, it is common to include the estimated uncertainty associated with a scientific measurement. For example, the width of a table might be 85.10 ± 0.01 inches. The ±0.01 inches is the uncertainty in the measurement. The percent uncertainty is the ratio of the uncertainty to the measured value, for example, 0.01/85.10 x 100% = 0.01%.

The number of significant figures in a measurement includes the figures that are certain and the first doubtful digit. In the example of the table, which is 85.10 ± 0.01 inches long, there are four significant figures. The 8, 5, and 1 are certain and the 0 is the doubtful digit. Calculations must follow the rules for significant figures. The final answer must have the same number of significant figures as the least significant factor used in the calculation.

SI System of Measurement

The measurement of any quantity is made relative to a particular standard or unit. The system used almost exclusively in this book is the Systeme International (SI system of measurement). In SI units, the unit of length is the meter, time is second, and mass is kilogram. Length, time, mass, electric current, temperature, amount of substance, and luminous intensity are base quantities. The base unit associated with electric current is the ampere, temperature is degrees kelvin, amount of substance is the mole, and luminous intensity is candela. All other quantities can be derived from the base quantities. The units associated with these derived quantities are called derived units. An example of a derived quantity is force and the unit of force is the newton (N) where 1 N = 1 kg m/s^2.

Order of Magnitude: Rapid Estimating

It is sometimes useful to give a rough estimate of the value of a quantity. This estimate is called the **order-of-magnitude** and this number contains one significant figure and the associated power of ten. For example, one often hears the world population given to the nearest billion.

An example of rapid estimating would be the number of people attending a parade. Suppose the parade route is 1 mile long with people lined up 5 deep on both sides of the street, with each person taking up an average of two feet of space. A rough estimate of the crowd could be made as follows:

(1 person/2 feet) x (5280 feet/1 mile) x 2 sides of street x (5 people/1 space) ≈ 30,000 people

SELECTED TEXT QUESTIONS WITH ANSWERS

QUESTION 1. It is advantageous that base standards (such as for length and time) be accessible (easy to compare to), invariable (do not change), indestructible, and reproducible. Discuss why these are advantages and whether any of these criteria can be incompatible with others.

ANSWER: For a standard to be useful it must meet the above criteria everywhere. The problem that is now faced is that while the criteria invariable and indestructible might be met, the current standards are not accessible or easily understood by the average person.

For example, from 1893 to 1960 the United States standard for length was the distance between two scratches on a metal bar made of a platinum-iridium alloy. This standard meets the criteria plus has the advantage of being easily understandable to the common person. Since 1960, the meter has been defined as 1,650,763.73 times the wavelength of red-orange spectral line of the krypton-86 isotope. This standard meets the criteria but is less accessible and certainly less understandable to the average person.

QUESTION 2. What are the merits and drawbacks of using a person's foot as a standard? Discuss in terms of the criteria mentioned in question 4. Consider both a) a particular person's foot and b) any person's foot.

ANSWER: Whether the standard is the foot of a particular person or any person's foot, the major drawback is that the length of the foot tends to change with time and is not reproducible. Also, the standard is lost and therefore destructible when the person dies. The only advantage is that a foot is accessible and can be easily used to give a rough estimate of short distances by stepping off the distance.

QUESTION 4. Suggest a way to measure the distance from the Earth to the Sun.

ANSWER: The distance from the Earth to the Sun can be measured by triangulation, as shown in the diagram. The Sun is simultaneously observed by two people (A and B) who are standing on the equator at directly opposite points on the Earth. The angle θ is measured by each observer and the radius (R) of the Earth is known. The distance (L) to the Sun can be found from tan θ = L/R and L = R tan θ.

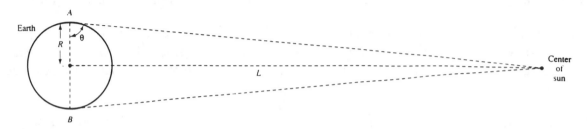

PROBLEM SOLVING SKILLS

For problems involving percent uncertainty in area or volume and the linear dimensions are given:

1. Determine the maximum and minimum possible values for the area or volume.
2. Determine the uncertainty in the measurement.
3. Divide the uncertainty by the measurement and multiply by 100%. If the problem involves a linear measurement, then steps 1 and 2 are not necessary to solve the problem.

For problems involving conversion of English units to SI units or vice versa.

1. List the quantity given and, if necessary, express it in one unit. For example, 1 min and 10 seconds = 70 seconds.
2. Use the factor-label method to solve the problem.
3. Where appropriate, use power of ten notation. Make sure to follow the rules for multiplication, division, addition, and subtraction of quantities which are expressed as powers of ten.
4. Apply the rules of significant figures in solving the problem. Remember that the answer must have the same number of significant figures as the least significant factor.

HOW TO USE THE PROGRAMMED PROBLEM SECTION

The programmed problem section involves solving problems similar to but not the same as the end-of-chapter problems found in your textbook. The problems are arranged in the same step-by-step process that should be used to solve the end-of-chapter problems found in the text.

Each part of the problem is broken down into individual steps. The steps are represented by individual frames which are divided into a left side where a question is posed and a right side where the answer is located. Each step is designated with a letter and number. For example, b. 2 indicates that this frame is the second step to the solution of part b of the problem. The right side should be covered with a blank sheet of paper on which you should attempt to answer the question. After completing your answer, uncover the right frame and check your work.

The individual frames are separated by a line that extends across the page and the end of the problem is indicated by a thick black line.

PROGRAMMED PROBLEMS

PROBLEM 1. What is the percent uncertainty in the measurement 2.54 ± 0.02 cm?

a. 1	Solution: (Section 1-4)
Determine the percent uncertainty.	(0.02 cm)/(2.54 cm) x 100% = 0.79%

PROBLEM 2. Determine the percent uncertainty in the area of a square that is 2.54 ± 0.02 m on a side.

a. 1	Solution: (Section 1-4)
Determine the area if each side is 2.54 m in length.	area = 2.54 m x 2.54 m = 6.45 m² Note: using a calculator, the product is 6.4516. However, there are only three significant figures in each factor and the answer must be rounded off to three significant figures.
a. 2 Determine the maximum possible area.	maximum length = 2.54 + 0.02 m = 2.56 m maximum area = 2.56 m x 2.56 m = 6.55 m²
a. 3 Determine the minimum possible area.	minimum length = 2.54 - 0.02 m = 2.52 m minimum area = 2.52 m x 2.52 m = 6.35 m²
a. 4 Write the area with the associated uncertainty.	The maximum area is greater than the area by 0.10 m² and the minimum area is less by 0.10 m². The area can now be written as area = 6.45 ± 0.10 m²
a. 5 Determine the % uncertainty.	% uncertainty = (0.10 m²)/(6.45 m²) x 100% Percent uncertainty = 1.6 %

PROBLEM 3. Express the height of a person 5'10" tall in a) centimeters and b) meters. Express your answer to the correct number of significant figures.

a. 1	Solution: (Section 1-6)
Convert 5 feet to inches.	(5 feet)(12 inches/1 feet) = 60 inches

a. 2	60 inches + 10 inches = 70 inches
Express the person's height in centimeters.	(70 inches)(2.54 cm/1 inch) = 177.8 cm
	70 inches has only two significant figures; therefore, 177.8 cm must be reduced to two significant figures. The person's height to the correct number of significant figures is 180 cm.
b. 1	(180 cm)(1.00 m/100 cm) = 1.80 m
Express the height in meters.	The answer must be reduced to two significant figures. The person's height is 1.8 m.

Problem 4. The equatorial diameter of the Earth given to three significant figures is 7930 miles. Express the diameter in a) meters and b) kilometers and use power of ten notation in your final answer.

a. 1	Solution: (Section 1-6)
Express the diameter in meters. Note: 1 mile = 1609 m	Using a calculator, the diameter is
	(7930 mi)(1609 m/1 mi) = 12,759,370 m
	The diameter of the Earth is given to three significant figures. It is necessary to round off the answer to three significant figures.
	Therefore, the diameter = 12,800,000 m
	The diameter in power of ten notation = 1.28×10^7 m
b. 1	1.000 km = 1000 m
Determine the diameter in km.	$(1.28 \times 10^7 \text{ m})(1.000 \text{ km}/1000 \text{ m}) = 1.28 \times 10^4$ km

PROBLEM 5. The radius of the Sun is roughly 100 times the diameter of the Earth. Use the process of rapid estimating to give an order of magnitude estimate of the number of planets the size of the Earth that could be contained within the volume of the Sun.

a. 1	Solution: (Section 1-7)
Write down the formula for the volume of both the Earth and the Sun Note: assume that both are perfect spheres.	$V_s = 4/3 \; \pi \; R_s^3$ $V_e = 4/3 \; \pi \; R_e^3$

a. 2

Take the ratio of the Sun's radius to the Earth's volume. This ratio is a rough estimate of the number planets that could contained within the Sun.

$V_s / V_e = (4/3 \ \pi \ R_s^3)/(4/3 \ \pi \ R_e^3)$ Note: both 4/3 and π cancel.

$V_s / V_e = R_s^3/R_e^3$

$V_s / V_e = 100^3/1^3 = 1.0 \ x \ 10^6$

An order of magnitude estimate would be that approximately 1 million planets the size of the Earth could be contained within be the volume of the Sun.

Note: the correct value is closer to 1.3 million planets the size of the Earth because the radius of the Sun is approximately 109 times greater than the Earth's radius.

PRACTICE PROBLEMS

PROBLEM 1. Determine the percent uncertainty in the area of a circle of radius 2.54 ± 0.01 cm.

ANS. 0.7%

PROBLEM 2. Alpha Centauri is approximately 4.2 light-years from the Earth. Express this distance in a) miles and b) kilometers. Note: l light year is the distance that light travels in one year. The speed of light is 186,000 miles per second or 300,000 km/s.

ANS. a) 2.5×10^{13} miles,   b) 4.0×10^{13} km

PROBLEM 3. The speed limit along a certain road is 55 miles per hour. Express this speed in meters per second to the correct number of significant figures.

ANS. 25 m/s

PROBLEM 4. Estimate the number of styrofoam balls 2.5 cm in diameter it would take to fill a box of volume 1.0 cubic meter.

ANS. 15,000 balls

CHAPTER 2

DESCRIBING MOTION: KINEMATICS IN ONE DIMENSION

KEY TERMS AND PHRASES

kinematics is the study of the motion of objects and involves the study of distance, speed, acceleration, and time.

average speed of an object is determined by dividing the distance that the object travels by the time required to travel that distance.

instantaneous speed is the speed of an object at a particular point in time.

acceleration is the rate of change of speed in time.

average acceleration is the change of velocity divided by the time required for the change.

instantaneous acceleration is the change of speed that occurs in a very small interval of time.

uniformly accelerated motion occurs when the rate of acceleration does not change, i.e., the rate of acceleration is constant.

free fall occurs when air resistance on an object is negligible and the only force acting on it is gravity.

gravitational acceleration for all objects in free fall is approximately 9.8 meters per second per second or 9.8 m/s^2.

vector is a quantity that has both magnitude and direction. Examples of vector quantities are velocity, acceleration, displacement, and force.

scalar is a quantity that has magnitude but has no direction associated with it. Examples of scalar quantities include speed, distance, mass, and time.

SUMMARY OF MATHEMATICAL FORMULAS

average acceleration	$\bar{a} = \Delta v / \Delta t$	The average acceleration equals the change in speed divided by the change in time
kinematics equations for uniformly accelerated motion	$v = v_0 + a t$	Speed as related to initial speed, acceleration and time
	$x = x_0 + v_0 t + \frac{1}{2} a t^2$	Distance as related to initial distance, initial speed, acceleration, and time
	$v^2 = v_0^2 + 2 a (x - x_0)$	Speed as related to initial speed, acceleration, and distance.
	$x - x_0 = \bar{v} t$	Distance traveled equals the product of the average speed and the time.
	$\bar{v} = \frac{1}{2}(v + v_0)$	average speed as related to the initial speed and final speed

CONCEPT SUMMARY

Kinematics

Kinematics is the study of the motion of objects and involves the study of the following concepts: distance, speed, acceleration, and time. This chapter is restricted to the motion of an object along a straight line. This is known as one-dimensional or linear motion.

The **average speed** of an object is determined by dividing the distance that the object travels by the time required to travel that distance. The **instantaneous speed** refers to the speed of an object at a particular point in time. In this study guide the SI system of units will be used. The SI unit of speed is meters per second (m/s) or kilometers per hour (km/h). SI is an abbreviation of the French words Systeme International. This system was formerly referred to as the MKS (meter-kilogram-second) system.

Acceleration refers to the rate of change of speed in time. The SI unit of acceleration is meters per second per second or m/s^2. The **average acceleration** is defined as the change of velocity divided by the time required for the change. The **instantaneous acceleration** refers to the change of speed that occurs in a very small interval of time ($\Delta t \rightarrow 0$).

Uniformly Accelerated Motion

Uniformly accelerated motion occurs when the rate of acceleration does not change, i.e. the rate of acceleration is constant. The following equations apply to this type of motion:

$v = v_o + a\,t$ $x = x_o + v_o\,t + \frac{1}{2}\,a\,t^2$

$v^2 = v_o^2 + 2\,a\,(x - x_o)$ $x - x_o = \overline{v}\,t$ where $\overline{v} = \frac{1}{2}(v + v_o)$

v_o = initial speed of the object (at t = 0 s) v = speed of the object after time t

a = rate of acceleration x = position of the object after time t

$\overline{v}$ = average speed x_o = initial position of the object
                                                 Note: $x_o = 0$ unless otherwise specified.

t = time interval during which the motion has occurred; unless otherwise specified, the time at the start of the motion will be zero seconds.

Free Fall

One application of uniformly accelerated motion is to the problem of objects in **free fall.** When an object is in free fall, we assume that air resistance is negligible and that the only force acting on it is gravity. Assuming air resistance is negligible, the rate of acceleration (g) of all objects in free fall is approximately 9.8 meters per second per second or 9.8 m/s^2.

The equations for uniformly accelerated motion can be applied to free fall. Since the motion is vertical, y replaces x and y_o replaces x_o. Also, v_y replaces v and v_{yo} replaces v_o while g replaces the symbol a.

Frames of Reference

The description of motion of any object must always be given relative to a **frame of reference** or **reference frame.** The reference frame is usually specified by using **Cartesian Coordinates.** The x, y, and z axes shown in the figures below can be used to locate the position of an object with respect to a fixed point o. At times, the x and y directions will be used to represent the direction of cardinal points: north (N), south (S), east (E), and west (W), with "up" above the plane of the paper and "down" below the plane of the paper. In certain problems, it will be convenient to use the x axis to represent the horizontal direction while the y axis represents the vertical direction.

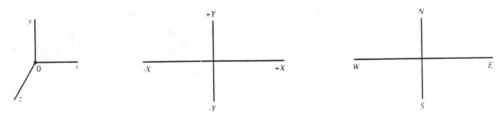

Vectors and Scalars

A **vector** is a quantity that has both magnitude and direction. An example of a vector quantity is velocity. A car traveling at 20 m/s must be traveling in a specified direction, e.g., due north. The vector quantities studied in the first three chapters include displacement, velocity, acceleration, and force. A vector is designated by the symbol for the vector quantity in bold face type as in the text. For example, the velocity vector will be represented by a bold face v, i. e., **v**.

A **scalar** is a quantity that has magnitude but has no direction associated with it. A scalar is specified by giving its magnitude and units (if any). An example of a scalar quantity is speed. Speed refers to the magnitude of an object's motion but not the direction in which it is traveling. Scalar quantities introduced in the first three chapters of the text include speed, distance, mass, and time.

Graphical Analysis

Graphs can be used to analyze the straight line motion of objects. Although **graphical analysis** can be used for uniformly accelerated motion, the method is especially useful when dealing with the motion of an object that is not undergoing uniform acceleration.

In a distance versus time graph the instantaneous velocity at any point can be determined from the slope of a tangent line drawn to the point in question. In a velocity versus time graph the slope of the tangent line represents the instantaneous acceleration while the area under the curve represents the distance traveled.

SELECTED TEXT QUESTIONS WITH ANSWERS

QUESTION 1. Does a car speedometer measure speed, velocity, or both?

ANSWER: A car's speedometer indicates the magnitude of the car's motion. It does not indicate the car's direction. Since velocity is both magnitude and direction, the speedometer measures speed but does NOT measure velocity.

QUESTION 2. Can an object have a varying velocity if its speed is constant? If yes, give examples.

ANSWER: An object's velocity can vary even though its speed is constant. Velocity is a vector quantity while speed is a scalar quantity. A car traveling along a curve at 55 miles per hour is traveling at a constant speed. However, its direction of motion changes as it moves along the curve. The direction of the velocity vector is tangent to the curve and is changing from one moment to the next. As a result, while the speed is constant the velocity is not.

QUESTION 9. Can an object have a northward velocity and a southward acceleration? Explain.

ANSWER: If the object is traveling northward, then the direction of the velocity vector is northward. If the object slows, but is still traveling northward, then the velocity vector is northward but the car is decelerating. Acceleration (or deceleration) is a vector quantity. If the object is accelerating, then the acceleration vector is in the same direction as the velocity vector. In the instance where the object is decelerating, the acceleration vector is directed opposite from the velocity vector. Therefore, for an object traveling northward but decelerating, the velocity vector is northward but the acceleration vector is southward.

PROBLEM SOLVING SKILLS

For problems involving uniformly accelerated motion:

1. Obtain a mental picture by drawing a diagram that reflects the motion of the object in question. This is especially useful in the free fall problems where the initial motion may be vertically upward or downward.
2. Complete a data table using information both given and implied in the wording of the problem.
3. Use the proper sign for the quantity represented by the symbol in the data table. For example, if a car is slowing down, then the rate of acceleration is negative. If an object in free fall was initially thrown downward, then the downward direction is taken to be positive and both the initial velocity and the rate of acceleration are positive. If the object in free fall was given an initial upward motion, then the upward direction is taken to be positive. This means that the initial upward velocity is positive but the rate of acceleration is negative because it is slowing down as it travels upward.
4. Memorize the formulas for uniformly accelerated motion. It is also necessary to memorize the meaning of each symbol in each formula. Using the data from the completed data table, determine which formula or combination of formulas can be used to solve the problem.

For problems related to graphical analysis where velocity is a function of time:

1. Determine the area of one block. This area represents the distance represented by a single block.
2. Count the number of blocks in the time interval being considered. Multiply the total number of blocks by the distance represented by one block. The product is the total distance traveled during the time interval. This technique is known as graphical integration.
3. The instantaneous acceleration at a particular moment of time is determined as follows: a) draw a tangent line to the graph at the point in question, and b) determine the magnitude of the slope of the tangent line. The magnitude of the slope of the line represents the instantaneous value of the acceleration at that moment in time. If the slope is positive, the object is accelerating. If the slope is negative, the object is decelerating. If the slope is zero, the object is traveling at constant speed.

For problems related to graphical analysis where distance is a function of time:

1. The instantaneous velocity at a particular moment in time equals the slope of the tangent line drawn to the curve at the point in question. If the slope is positive, the object's speed is positive. If the slope is negative, the object's speed is negative. If the slope is zero, the object is not moving.

PROGRAMMED PROBLEMS

PROBLEM 1. A car accelerates from 2.0 m/s to a speed of 10.0 m/s in 4.0 s. If the rate of acceleration is uniform, determine the a) rate of acceleration and b) distance traveled during the 4.0 s of motion.

a. 1	Solution: (Sections 2-5 and 2-6)
List each symbol and complete a data table based on the information given in the problem.	v_o = 2.0 m/s t = 4.0 s v = 10 m/s a = ? x = ? x_o = 0

a. 2	
Determine the rate of acceleration.	$v = v_o + a\,t$ 10 m/s = 2.0 m/s + a (4.0 s) a = 2.0 m/s^2

b. 1	
Determine the distance traveled in 4.0 seconds.	$x = x_o + v_o\,t + \frac{1}{2}\,a\,t^2$ x = 0 m + (2.0 m/s)(4.0 s)+ ½(2.0 m/s^2)(4.0 s)^2 x = 24 m

PROBLEM 2. A car is traveling at the posted speed limit of 15.0 miles per hour (6.70 m/s) in a school zone. The car is passing a school bus when a child darts out from in front of the bus. The car is 10.0 m from the child when the driver of the car applies the brakes. The car decelerates at a constant rate of 4.00 m/s^2 until coming to a halt. Calculate the a) distance the car travels while decelerating and b) distance required to stop if the car is initially traveling at 30.0 miles per hour (13.4 m/s).

a. 1	Solution: (Sections 2-5 and 2-6)
Complete a data table for the deceleration.	x = ? x_o = 0 m a = - 4.00 m/s^2 v_o = 6.70 m/s v = 0 m/s (car comes to a halt) t = ?

a. 2	
Calculate the distance the car would travel during the deceleration.	$2\,a\,(x - x_o) = v^2 - v_o^2$ 2 (-4.00 m/s^2)(x - x_o) = (0 m/s)^2 - (6.70 m/s)^2 (-8.00 m/s^2)(x - x_o) = -44.9 m^2/s^2 x - x_o = 5.61 m The car stops approximately 4.39 m (14 feet) in front of the child.

b. 1	$x = ?$	$x_o = 0$ m
Complete a data table for the deceleration.	$a = -4.00$ m/s^2	$v_o = 13.4$ m/s
	$v = 0$ m/s (car comes to a halt)	$t = ?$

b. 2	$2 a (x - x_o) = v^2 - v_o^2$
Calculate the distance the car would travel during the deceleration.	$2 (-4.00$ m/s$^2)(x - x_o) = (0$ m/s$)^2 - (13.4$ m/s$)^2$
	$(-8.00$ m/s$^2)(x - x_o) = -180$ m^2/s^2
	$x - x_o = 22.4$ m
	At 30.0 mph, the car requires approximately four times more distance to come to a complete stop as compared to 15.0 mph. Therefore, in order to protect children, the various States set a rather low speed limit when traveling through a school zone.

PROBLEM 3. A physics student is driving home after class. The car is traveling at 33.0 miles per hour (14.7 m/s) when it approaches an intersection. The student estimates that he is 66.0 feet (20.0 m) from the entrance to the intersection when the traffic light changes from green to amber (yellow) and the intersection is 32.8 feet (10.0 m) wide. The light will change from amber to red in 3.00 seconds. The maximum safe deceleration of the car is 4.00 m/s^2 while the maximum acceleration of the car is 2.00 m/s^2. Should the physics student a) decelerate and stop or b) accelerate and travel through the intersection? Note: there is a police car directly behind the student.

a. 1	Solution: (Sections 2-5 and 2-6)	
Complete a data table for the deceleration.	$x = ?$	$x_o = 0$ m
	$a = -4.0$ m/s^2	$v_o = 14.7$ m/s
	$v = 0$ m/s (car comes to a halt)	$t = 3.0$ s

a. 2	$x = x_o + v_o t + \frac{1}{2} a t^2$
Calculate the distance the car would travel during the deceleration.	$x = 0$ m $+ (14.7$ m/s$)(3.0$ s$) + \frac{1}{2} (-4.0$ m/s$^2)(3.0$ s$)^2$
	$x = 44.1$ m $+ -18.0$ m $= 26.1$ m
	The start of the intersection is 20 m from the car's initial position and the car must travel 30 m to be beyond the intersection. Based on the calculation, the student should not stop because the car would come to a halt 6.1 m (appoximately 20 feet) into the intersection.

b. 1 Complete a data table for the acceleration.	$x = ?$ $x_o = 0$ m $a = +2.0$ m/s^2 $v_o = 14.7$ m/s $t = 3.0$ s $v = ?$
b. 2 Determine the distance that the accelerating car would travel in 3.0 s.	$x = x_o + v_o t + \frac{1}{2} a t^2$ $x = 0$ m $+ (14.7$ m/s$)(3.0$ s$) + \frac{1}{2}(+2.0$ m/s$^2)(3.0$ s$)^2$ $x = $ 44.1 m + 9.0 m = 53.1 m Based on the calculation, the student should accelerate because the car would be approximately 23 m beyond the intersection after 3.0 s.

PROBLEM 4. A car traveling at a constant speed of 30 m/s (approximately 67 miles per hour) passes a highway patrol police car which is at rest. The police officer accelerates at a constant rate of 3.0 m/s^2 and maintains this rate of acceleration until he pulls next to the speeding car. Assume that the police car starts to move at the moment the speeder passes his car. Determine the (a) time required for the police officer to catch the speeder and (b) distance traveled during the chase.

a. 1 Complete a data table for both vehicles using information both given and implied in the statement of the problem.	Solution: (Sections 2-5 and 2-6) motorist's car police car $v_o = 30$ m/s $v = 0$ m/s $v = 30$ m/s $v = ?$ $a = 0$ m/s $a = 3.0$ m/s^2 $t = ?$ $t = ?$ $x = ?$ $x_o = ?$ $x_o = 0$ m $x_o = 0$ m
a. 2 Write an equation for the motorist's car as a function of time.	motorist $x - x_o = \bar{v} t$ but $x_o = 0$, therefore $x = \frac{1}{2}(30$ m/s $+ 30$ m/s$)t$ and $x = (30$ m/s$)t$

a. 3 Write an equation for the position of the police car as a function of time.	police car $x = v_o t + \frac{1}{2} a t^2 + x_o$ $x = (0 \text{ m/s}) t + \frac{1}{2} (3.0 \text{ m/s}^2) t^2 + 0 \text{ m}$ $x = (1.5 \text{ m/s}^2) t^2$
a. 4 Determine the time required for the police to catch the motorist. Note: when the police car pulls alongside the motorist, it is traveling faster than the motorist's car. However, both cars have traveled the same distance.	Since $x_{motorist} = x_{police\ car}$ $(30 \text{ m/s}) t = (1.5 \text{ m/s}^2) t^2$ writing the expression as an algebraic equation and solving for t, $30 t = 1.5 t^2$ $0 = 1.5 t^2 - 30 t$ $0 = (t - 20)(1.5 t)$ Either $0 = t - 20$ or $0 = 1.5 t$ $t = 20 \text{ s}$ or $t = 0 \text{ s}$ The equation is a quadratic equation and two values are obtained for the time of motion. However, $t = 0$ s merely corresponds with our initial assumption that the two cars were at the same position at $t = 0$. The correct solution is $t = 20$ s.
b. 1 Determine the distance each vehicle travels during the time interval.	Add $t = 20$ s to the data table and solve for the distance traveled during the chase. $x - x_o = \overline{v} t$ $x - 0 \text{ m} = (30 \text{ m/s})(20 \text{ s})$ $x = 600 \text{ m}$

PROBLEM 5. A stone is thrown vertically downward with an initial speed of 4.90 m/s from the top of a building 19.6 m high. Determine the a) velocity of the stone just before it strikes the ground and b) time that the stone is in the air.

a. 1	Solution: (Section 2-7)
Because the motion is downward, let the downward direction be positive. Complete a data table based on the information given.	v_o = 4.90 m/s $\qquad$ t = ? v = ? $\qquad$ y = 19.6 m a = 9.80 m/s² $\qquad$ y_o = 0 m
a. 2 Determine the velocity of the stone just before it strikes the ground. Note: after it leaves the person's hand it is in free fall.	$v^2 = v_o^2 + 2 a (y - y_o)$ v^2 = (4.90 m/s)² + 2(9.80 m/s²)(19.6 m - 0 m) v^2 = 24 m²/s² + 384 m²/s² v^2 = 408 m²/s² v = 20.2 m/s
b. 1 Determine the time that the stone is in the air.	The velocity of the stone just before it strikes the ground can now be included in the data table. Select the appropriate equation and solve for the time of flight. $v = v_o + a t$ 20.2 m/s = 4.9 m/s + (9.80 m/s²) t t = 1.56 s

PROBLEM 6. A stone is thrown vertically upward from the edge of a building 19.6 m high with an initial velocity of 14.7 m/s. The stone just misses the building on the way down and strikes the street below. Determine the a) time of flight and b) velocity of the stone just before it strikes the ground.

a. 1	Solution: (Section 2-7)
Complete a data table based on information both given and implied in the problem.	v_o = +14.7 m/s $\qquad$ t = ? v = ? $\qquad$ y_o = 0 a = - g = - 9.80 m/s² $\qquad$ y = - 19.6 m The stone's initial velocity is 14.7 m/s upward but it lands 19.6 m below its starting position. The initial direction of motion is upward, so let the stone's initial velocity be positive. Since it is decelerating, then a = - g. Also, since the stone lands below its starting point its displacement from its starting point is -19.6 m.

a. 2 Select the appropriate equation and solve for the time of flight.	$y = v_o t + \frac{1}{2} a t^2$ $-19.6 \text{ m} = (14.7 \text{ m/s})t + \frac{1}{2}(-9.80 \text{ m/s}^2)t^2$ Rearranging and dividing by -4.90 gives $0 = 1.0 t^2 - 3.0 t - 4.0$ and factoring $0 = (t - 4.0)(t + 1.0)$ Either $0 = t - 4.0$ and $t = 4.0$ s or $0 = t + 1.0$ and $t = -1.0$ s. Since t cannot be negative, the answer must be $t = 4.0$ s.
b. 1 The time of flight can now be added to the data table. Solve for the velocity of the stone just before it strikes the ground.	$v = v_o + a t$ $\quad = 14.7 \text{ m/s} + (-9.80 \text{ m/s})(4.0 \text{ s})$ $v = -24.5 \text{ m/s}$ The negative value for the velocity indicates that the stone is traveling downward. As you recall, the upward direction was arbitrarily selected as the positive direction.

PROBLEM 7. The graph shown below represents the motion of a car over a period of 10.0 s. Use graphical analysis to determine the a) rate of acceleration of the car at 2.0 s, 5.0 s and 8.0 s and b) distance traveled during the 10.0 s of motion.

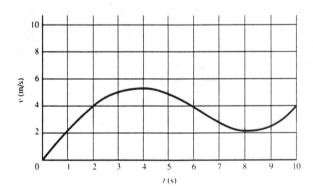

a. 1 Determine the acceleration of the car at 2.0 s, 5.0 s, and 8.0 s.	Solution: (Section 2-8) To determine the rate of acceleration, a tangent line is drawn at each point in question. Two data points are selected from each tangent line and the rate of acceleration is determined by using the following formula: $a = (v_2 - v_1)/(t_2 - t_1)$

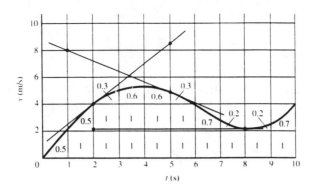

at t = 2.0 s: a = (8.5 m/s - 4.0 m/s)/(5.0 s - 2.0 s) = +1.5 m/s²

at t = 5.0 s: a = (8.0 m/s - 5.0 m/s)/(1.0 s - 5.0 s) = - 0.75 m/s²

at t = 8.0 s: a = (2.2 m/s - 2.2 m/s)/(8.0 s - 2.0 s) = 0 m/s²

There is a certain amount of judgment required in drawing the tangent lines. The actual value of the acceleration at each point may be different from the values obtained above. Your knowledge of algebra will help in checking whether or not the answers are reasonable. The tangent line at t = 2.0 s has a positive slope and therefore the car should be accelerating. Since the car's speed is increasing at this point it is accelerating. At t = 5.0 s, the car is slowing down and is therefore decelerating. The slope of the tangent line at t = 5.0 s is negative and therefore agrees with observation. The slope of the tangent line at t = 8.0 s is zero. This indicates that the car is neither accelerating nor decelerating. Looking at the graph, we can see that the car has stopped slowing down and has not yet begun to increase in speed; thus its rate of acceleration is indeed zero.

b. 1

Use graphical integration to determine the distance traveled during the 10.0 s of motion.

The distance traveled can be determined by calculating the area under the graph. This can be done by determining the distance represented by the area of one block and then multiplying this value by the total number of full and partial blocks that lie between the curve and the time axis. Note: a major source of error in determining the distance traveled is the judgment required in estimating the value of a partial partial block. A more accurate value can be obtained by using graph paper that contains a fine grid.

2.0 m/s ⬚ 1.0 s   distance represented by one block = (2.0 m/s)(1.0 s)

distance = 2.0 m

sum of complete blocks = 13.0

sum of partial blocks = 0.5 + 0.5 + 0.3 + 0.6 + 0.6 + 0.3
                         + 0.7 + 0.2 + 0.2 + 0.7 = 4.6

total number of blocks = 13.0 + 4.6 = 17.6

total distance = (17.6 blocks)(2.0 m/1 block) = 35 m

PRACTICE PROBLEMS

PROBLEM 1. A driver maintains a steady speed of 20.0 m/s for 45.0 min and then slows to 10.0 m/s for the next 15.0 min. Determine the a) total distance traveled and b) average speed for the trip.

ANS. a) 63000 m, b) 17.5 m/s

PROBLEM 2. A truck traveling at 20.0 m/s decelerates uniformly to 10.0 m/s over a distance of 100 m. Determine the a) rate of deceleration and b) time required for the truck to travel the 100 m.

ANS. a) - 1.50 m/s², b) 6.67 s

PROBLEM 3. A ball is thrown vertically upward with an initial speed of 4.90 m/s from the top of a building 19.6 m high. Determine the a) time to reach maximum height and b) maximum height above the top of the building reached by the ball.

ANS. a) 0.50 s, b) 1.2 m

PROBLEM 4. The graph shown below represents the motion of a car over a period of 10.0 s. Use graphical analysis to determine the a) acceleration of the car at 0 s, 4.5 s, 7.0 s, and 9.0 s and b) distance traveled during the 10.0 s of motion.

ANS. a) - 1.2 m/s²; 0 m/s²; 1.3 m/s²; - 0.7 m/s²   b) 50 m

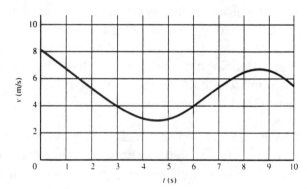

CHAPTER 3

KINEMATICS IN TWO DIMENSIONS; VECTORS

KEY TERMS AND PHRASES

resultant vector is the arithmetic sum (or difference) of the magnitudes and the directions of two or more vectors.

tip to tail method is a graphical method used to determine the vector sum of two or more vectors. The two vectors are drawn to scale and then moved parallel to their original direction until the tail of one vector is at the tip of the next vector. Once all of the vectors are joined in this manner the resultant vector can be determined. The resultant is drawn from the tail of the first vector to the tip of the last vector. The angle of the resultant above (or below) the x axis is determined by using a protractor.

parallelogram method is a graphical method useful if two vectors are to be added. The two vectors are drawn to scale and joined at the tails. Dotted lines are then drawn from the tip of each vector parallel to the other vector. The finished diagram is a parallelogram. The resultant is along the diagonal of the parallelogram and extends from the point where the tails of the original vectors touch to the point where the dotted lines cross. The angle of the resultant above (or below) the x axis is determined by using a protractor.

vector component method is used to replace each vector with components in the x and y directions. The arithmetic sum of the x components (ΣX) and y components (ΣY) are then determined. Since ΣX and ΣY are at right angles, the Pythagorean theorem can be used to determine the magnitude of the resultant. The definition of the tangent of an angle can be used to determine the angle of the resultant above (or below) the x axis.

relative velocity refers to the velocity of an object with respect to a particular frame of reference.

projectile motion is the motion of an object fired (or thrown) at an angle θ with the horizontal. The only force acting on the object during its motion is gravity.

SUMMARY OF MATHEMATICAL FORMULAS

projectile motion equations	$v_{yo} = v_o \sin \theta$	Initial vertical component of a projectile's velocity
	$v_{xo} = v_o \cos \theta$	Initial horizontal component of a projectile's velocity
	$y = v_{yo} t - \frac{1}{2} g t^2$	Vertical component of an object's position (y) as related to its initial speed (v_{yo}), gravitational acceleration (g), and time of motion (t)
	$v_y^2 = v_{yo}^2 - 2 g y$	Vertical component of an object's velocity (v_y) as related to its initial speed (v_{yo}), gravitational acceleration (g), and position (y)
	$v_y = v_{yo} - g t$	vertical component of an object's velocity (v_y) as related to its initial speed (v_{yo}), gravitational acceleration (g), and time (t)
	$v_x = v_{xo}$	horizontal component of velocity (v_x) remains constant during the projectile's flight
	$x = x_o + v_{xo} t$	horizontal component of an object's position (x) as related to the initial horizontal speed (v_{xo}), and time (t)

CONCEPT SUMMARY

Addition of Vectors - Graphical Methods

If two vectors are in the same direction and along the same line, the **resultant vector** will be the arithmetic sum of the magnitudes and the direction will be in the direction of the original vectors. If the vectors are in opposite directions and along the same line, the resultant vector will have a magnitude equal to the difference in the magnitudes of the original vectors and the direction of the resultant will be in the direction of the vector which has the greater magnitude.

If the vectors are at some angle to each other than 0° or 180°, then special methods must be used to determine the magnitude and direction of the resultant. Two graphical methods used are the 1) tip to tail method and 2) parallelogram method.

The graphical methods involve the use of a ruler and a protractor. An appropriate scale factor must be used to represent the vectors. For example, a velocity vector which has a magnitude of 3.0 m/s and is directed due east can be represented by a line 3.0 cm long directed along the +x axis. $\underline{\quad 3.0 \text{ cm} \quad +x}$ ➔

Tip to Tail Method

The tip to tail method can be conveniently used for two or more vectors. The method consists of moving the vectors parallel to their original direction until the tail of one vector is at the tip of the next vector. Once all of the vectors are joined in this manner the resultant vector can be determined. The resultant is drawn from the tail of the first vector to the head of the last vector. For example, the resultant velocity (v_R) relative to the bank of a river for a person swimming at 3.0 m/s downstream in a river with a current of 1.0 m/s can be determined as shown in the figure. The magnitude of the resultant can be determined by measuring the length of the vector and multiplying by the scale factor. The scale used in the diagram is 1.0 cm = 1.0 m/s.

Using a ruler it can be determined that the resultant has a length of 4.0 cm. Therefore, the magnitude is (4.0 cm) x (1.0 m/s)/(1.0 cm) = 4.0 m/s. The direction of the resultant vector is downstream.

Parallelogram Method

The parallelogram method is useful if two vectors are to be added. If more than two vectors are involved, the method becomes cumbersome and an alternate method should be used.

The two vectors are drawn to scale and joined at the tails. Dotted lines are then drawn from the tip of each vector parallel to the other vector. A protractor can be used to ensure that the lines are drawn parallel. The finished diagram is a parallelogram. The resultant is along the diagonal of the parallelogram and extends from the point where the tails of the original vectors touch to the point where the dotted lines cross. As in the tip to tail method, the resultant is determined by measuring the length of the resultant and multiplying by the scale factor used to represent the vectors. The angle of the resultant above (or below) the x axis is determined by using a protractor.

For example, determine the sum of the following vectors: v_1 = 3.0 m/s due east, v_2 = 4.0 m/s due north.

step 1. Draw a diagram representing each with the tail of each joined at a point. Scale: let 1.0 cm = 1.0 m/s.

step 2. Complete the parallelogram. Draw the resultant vector across the diagonal. Determine v_R.

Using a ruler, it can be determined that the resultant has a length of 5.0 cm. Therefore, the magnitude of the resultant is (5.0 cm) x (1.0 m/s)/(1.0 cm) = 5.0 m/s. Using a protractor, the angle θ can be determined to be 53° north of east. Therefore, v_R = 5.0 m/s $\angle$53° N of E.

Subtraction of Vectors

The negative of a vector is a vector of the same magnitude but in the opposite direction. Thus, if vector **v** = 5 m/s due east, then -**v** = 5 m/s due west.

In order to subtract one vector from another, rewrite the problem so that the rules of vector addition can be applied. For example, A - B can be rewritten as A + (-B). Determine the magnitude and direction of -B and apply the rules of vector addition to solve for the resultant vector.

Multiplication of a Vector by a Scalar

The product of a vector times a scalar has the same direction as the vector and a magnitude equal to the product of the magnitude of the scalar times the magnitude of the vector. For example, if c is a scalar while **V** is a vector, then the product has a magnitude cV and the same direction as **V**. If c is a negative scalar, the magnitude of the resultant is still cV but the direction of the resultant is directly opposite that of **V**.

Analytic Method for Adding Vectors

In the trigonometric component method, each of the original vectors is expressed as the vector sum of two other vectors. The two vectors are chosen to be in directions which are perpendicular to another one. At first, this method may seem long and tedious. However, with some practice this method is by far the most useful for this course.

For problems involving the addition of two or more vectors lying in the x-y plane, the method reduces to the following steps:

step 1. Resolve each vector into x and y components.

If the angle is measured from the x axis to the vector, then the x component is equal to the product of the magnitude of the vector and the cosine of the angle. The y component is equal to the product of the magnitude of the vector and the sine of the angle. For example, for
v = 5.0 m/s $\angle$30° N of E

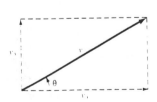

$$v_x = v \cos \theta = (5.0 \text{ m/s})(\cos 30°)$$

then $v_x = (5.0 \text{ m/s})(0.87) = 4.3 \text{ m/s}$

$$v_y = v \sin \theta = (5.0 \text{ m/s})(\sin 30°)$$

then $v_y = (5.0 \text{ m/s})(0.50) = 2.5 \text{ m/s}$

step 2. Sign Convention

Assign a positive value to the magnitude if the component is in the +x or +y direction and a negative value if the component is in the -x or -y direction. Thus, for the vector used as an example in step 1,

$$v_x = + 4.3 \text{ m/s} \quad \text{and} \quad v_y = + 2.5 \text{ m/s}$$

step 3. Reduce the problem to the sum of two vectors.

Determine the sum of the x components (ΣX) and the sum of the sum of the y components (ΣY), where Σ is the upper case Greek letter sigma which is designated to mean "the sum of." Since the x components are along the same line, their magnitudes are added arithmetically. This is also true for the y components.

step 4. Determine the magnitude and direction of the resultant.

Since ΣX and ΣY are at right angles, the Pythagorean theorem can be used to determine the magnitude of the resultant. The definition of the tangent of an angle can be used to determine the direction of the resultant.

Relative Velocity

Relative velocity refers to the velocity of an object with respect to a particular frame of reference. As in chapter 2, the reference frame is usually specified by using Cartesian coordinates, i.e., x, y and z axes, relative to which the position and/or motion of an object can be determined. As stated in the text, the velocity of an object relative to one frame of reference can be found by vector addition if its velocity relative to a second frame of reference and the relative velocity of the two reference frames are known.

Projectile Motion

Projectile motion is the motion of an object fired at an angle θ with the horizontal. This motion can be discussed by analyzing the horizontal component of the object's motion independently of the vertical component of motion. If air resistance is negligible, then the horizontal component of motion does not change; thus $a_x = 0$ and $v_x = v_{xo} = $ constant.

The vertical component of motion is affected by gravity and is described by the equations for an object in free fall as discussed in chapter 2. The following equations are used to describe the motion of a projectile.

vertical component of motion   horizontal component of motion

$$v_y = v_{yo} - gt \qquad\qquad\qquad\qquad x = v_{xo} t$$

$$y = v_{yo} t - \tfrac{1}{2} g t^2 \qquad\qquad\qquad \text{where } v_x = v_{xo}$$

$$v_y^2 = v_{yo}^2 - 2 g y$$

v_y is the vertical component of velocity at time t.

v_{yo} is the initial vertical component of velocity, $v_{yo} = v_o \sin \theta$.

g   is the acceleration due to gravity, $g = 9.8$ m/s^2.

t   is the time interval of the motion, $t_o = 0$ s.

y   is the vertical displacement from the starting point. The starting point is the reference point and the vertical displacement at this point is arbitrarily taken to be zero. Since $y_o = 0$, y_o does not appear in the equations.

x   is the horizontal displacement from the starting point. x_o does not appear in the equation for horizontal motion because $x_o = 0$ at the starting point.

v_{xo} is the initial horizontal component of velocity. Assuming air resistance to be negligible, the horizontal component of velocity does not change during the motion. Therefore, $v_x = v_{xo} = v_o \cos \theta$.

SELECTED TEXT QUESTIONS WITH ANSWERS

QUESTION 1. Does the odometer of a car measure a scalar or a vector quantity? What about the speedometer?

ANSWER: Distance traveled is measured on the odometer while speed is measured on the speedometer. Distance and speed are scalar quantities while displacement and velocity are vector quantities. The odometer will record the distance traveled but does not record the displacement from the starting point. For example, if a student drives one mile to a convenience store and then back to his starting point, the odometer will record a trip of two miles. However, his displacement from his starting point will be zero. Also, the speedometer records the magnitude of the motion but not the direction.

QUESTION 6. Two vectors have length $V_1 = 3.5$ km and $V_2 = 4.0$ km. What are the maximum and minimum magnitudes of their vector sum?

ANSWER: The maximum magnitude occurs if the two vectors are in the same direction. The resultant velocity is the arithmetic sum of their magnitudes, i.e., $V_R = 3.5$ km + 4.0 km = 7.5 km.

The minimum magnitude occurs if the two vectors are in opposite directions. The resultant velocity is the difference of their magnitudes, i.e., $V_R = 4.0$ km - 3.5 km = 0.5 km in the direction of the vector which has the greater magnitude.

QUESTION 13. A projectile has the least speed at what point in its path?

ANSWER: The motion of a projectile at any point in its motion is the vector sum of the horizontal component and the vertical component of its velocity. In the absence of air resistance, the magnitude of horizontal component does not change $v_x = v_{xo}$. As a projectile travels upward the magnitude of the vertical component decreases reaching zero at maximum height, and on the way down, the vertical component of velocity again increases. Since the velocity is the vector sum of the components, the magnitude of the velocity (i.e., the speed) must have a minimum value at the point where the vertical component is zero. Therefore, the minimum value of the speed occurs at the point where the projectile reaches maximum height.

PROBLEM SOLVING SKILLS

For problems involving vector addition or subtraction:

1. Use a protractor and ruler to accurately represent each vector involved in the problem. Make sure to use an appropriate scale factor in representing the vector.
2. Choose an appropriate method to solve the problem. If a graphical method is used, be sure to measure both the magnitude and direction of the resultant.
3. If the trigonometric component method is used, you must
   1) break each vector into x and y components.
   2) use the sign convention and assign a positive sign or a negative sign to the magnitude.
   3) determine the sum of the x components and repeat for the sum of the y components.
   4) use the Pythagorean theorem and simple trigonometry to solve for the magnitude and direction of the resultant.

For problems involving projectile motion:

1. Draw an accurate diagram showing the trajectory of the projectile.
2. Use the trigonometric component method to determine v_{xo} and v_{yo}.
3. Complete a data table using information both given and implied in the wording of the problem.
4. As in the free fall problems of chapter 2, use the appropriate sign convention depending on whether the object was initially moving upward or downward.
5. Memorize the formulas for projectile motion. It is also necessary to memorize the meaning of each symbol in each formula. Using the data from the completed data table, determine which formula or combination of formulas must be used to solve the problem.

PROGRAMMED PROBLEMS

PROBLEM 1. A man walks 50 meters due east, then 50 meters ∠60° north of east. Determine the magnitude and direction of the resultant displacement using the a) tip to tail method, b) parallelogram method, and c) vector component method.

---

a. 1

Displacement is a vector quantity. Write each vector in vector notation.

Solution: (Sections 3-2 and 3-4)

$\mathbf{D_1}$ = 50 m ∠due east

$\mathbf{D_2}$ = 50 m ∠60° north of east

---

a. 2

Draw an accurate diagram locating each vector. Let the scale factor used in the drawing be 1.0 cm = 20 m. A protractor must be used to measure the angles.

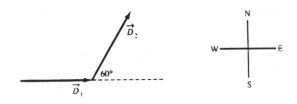

---

a. 3

Determine the magnitude and direction of the resultant vector using the tip to tail method.

This method consists in moving one of the vectors parallel to its original direction until its tail is at the tip of the other vector. The resultant (**R**) is drawn from the unattached tail of the second vector to the unattached head of the first.

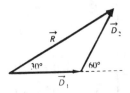

Based on careful measurement with a ruler, the magnitude of the resultant is 4.3 cm. Using the scale factor, (4.3 cm)(20 m/1 cm) = 86 m, the magnitude in meters can be determined. The direction of the resultant is determined by use of a protractor and is 30° N of E. Thus **R** = 86 m ∠30° N of E.

---

b. 1

Draw the vectors and complete the parallelogram.

The two vectors are drawn tail to tail. In order to complete the parallelogram, the dotted lines must be drawn parallel to the opposite sides. A protractor should be used to ensure that this is done properly.

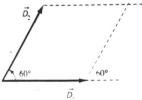

3-8

b . 2	The resultant is along the diagonal of the parallelogram. The direction is from the point where the tails are joined to the point where the dotted lines cross. As in the tip to tail method, the resultant is determined by measuring the length of **R** and multiplying by the scale factor. The direction is determined by using a protractor.
Use a protractor and a ruler to determine the magnitude and direction of the resultant. Note: remember to multiply by the scale factor.	The length of R is measured to be 4.3 cm; therefore, (4.3 cm)(20 m/l cm) = 86 m. Using a protractor, the angle is measured to be 30°. Thus, **R** = 86 m $\angle 30°$ N of E.

Vector Component Method: This method does not require an accurate diagram as part of the solution. However, a diagram reflecting the relative magnitudes and directions of the vectors is helpful.

c. 1	
Make a drawing and resolve each vector into x and y components.	

c. 2	If the angle is measured from the east-west axis (x axis) to the vector, then the x component is equal to the product of the magnitude of the vector and the cosine of the angle. The y component is equal to the product of the magnitude of the vector and the sine of the angle.
Use trigonometry to determine the magnitude of each component.	

$$\text{x components}$$

$$\mathbf{D}_{1x} = (50 \text{ m})(\cos 0°) = + 50 \text{ m}$$

$$\mathbf{D}_{2x} = (50 \text{ m})(\cos 60°) = + 25 \text{ m}$$

$$\text{y components}$$

$$\mathbf{D}_{1y} = (50 \text{ m})(\sin 0°) = 0$$

$$\mathbf{D}_{2y} = (50 \text{ m})(\sin 60°) = +43 \text{ m}$$

c. 3	The x components are in the +x; therefore, both are assigned positive values. The y-component of **D**₂ is in the +y direction; therefore, it has been assigned a positive value.
Determine the sum of the x components (ΣX) and the sum of the of the y components (ΣY).	$\Sigma \mathbf{X} = \mathbf{D}_{1x} + \mathbf{D}_{2x} = 50 \text{ m} + 25 \text{ m} = 75 \text{ m}$ $\Sigma \mathbf{Y} = \mathbf{D}_{1y} + \mathbf{D}_{2y} = 0 \text{ m} + 43 \text{ m} = 43 \text{ m}$

c. 4

Complete the parallelogram using ΣX and ΣY as the vectors to be added. Solve for the magnitude and direction of the resultant vector.

Since ΣX and ΣY are at right angles, the Pythagorean theorem and trigonometry can be used to determine the magnitude and direction of the resultant.

$$R = [(\Sigma X)^2 + (\Sigma Y)^2]^{\frac{1}{2}} = [(75 \text{ m})^2 + (43 \text{ m})^2]^{\frac{1}{2}} = 86 \text{ m}$$

$\tan \theta = (43 \text{ m})/(75 \text{ m}) = 0.57;$ $\theta = 30°$ N of E

Thus, $R = 86 \text{ m} \angle 30°$ N of E.

PROBLEM 2. Use the trigonometric component method to determine the sum of the following displacement vectors: $A = 4.0 \text{ m} \angle 53°$ N of E, $B = 6.0 \text{ m} \angle 45°$ N of W, $C = 2.0 \text{ N} \angle 60°$ S of W.

a. 1

Draw an accurate diagram and locate each vector. Note: the tail of each vector is located at the origin.

Solution: (Section 3-4)

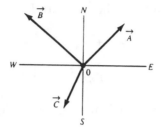

a. 2

Determine the x and y component of each vector and then deter-determine ΣX and ΣY. Remember to apply the sign convention to each component.

$A_x = 4.0 \text{ m} \cos 53° = + 2.4 \text{ m}$ $A_y = 4.0 \text{ m} \sin 53° = + 3.2 \text{ m}$

$B_x = 6.0 \text{ m} \cos 45° = - 4.2 \text{ m}$ $B_y = 6.0 \text{ m} \sin 45° = + 4.2 \text{ m}$

$C_x = 2.0 \text{ m} \cos 60° = - 1.0 \text{ m}$ $C_y = 2.0 \text{ m} \sin 60° = - 1.7 \text{ m}$

$$\overline{}$$
$\Sigma X = - 2.8 \text{ m}$ $\Sigma Y = + 5.7 \text{ m}$

a. 3

Draw a vector diagram showing ΣX and ΣY. Complete the parallelogram and draw in the resultant.

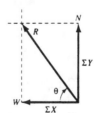

a. 4

Use the Pythagorean theorem and trigonometry to determine the magnitude and direction of the resultant.

$R^2 = (\Sigma X)^2 + (\Sigma y)^2 = (- 2.8 \text{ m})^2 + (+ 5.7 \text{ m})^2 = 40 \text{ m}^2$

$R = 6.4 \text{ m}$

$\tan \theta = (5.7 \text{ m})/(- 2.8 \text{ m}) = - 2.0,$ therefore $\theta = 64°$ N of W

$R = 6.4 \text{ m} \angle 64°$ N of W

PROBLEM 3. The current in a river is 1.0 m/s. A woman swims 300 m downstream and then back to her starting point without stopping. If she can swim 2.0 m/s in still water, determine the time required for the round trip.

a. 1 Determine her velocity relative to the river bank as she swims downstream.	Solution: (Section 3-8) Let $\mathbf{v}_{pw}$ = velocity of the person relative to the river water $\mathbf{v}_{ws}$ = velocity of the water relative to the shore of the river, i.e., the river current $\mathbf{v}_{ps}$ = velocity of the person relative to the shore of the river The tip to tail method can be used to solve for $\mathbf{v}_{ps}$. The vectors are along the same line; therefore, the resultant vector is the arithmetic sum of the magnitudes. $\mathbf{v}_{ps} = \mathbf{v}_{pw} + \mathbf{v}_{ws}$ $\qquad = 2.0 \text{ m/s} + 1.0 \text{ m/s}$ $\mathbf{v}_{ps} = 3.0 \text{ m/s (downstream)}$
a. 2 Determine the time to swim downstream.	time = displacement/resultant velocity $t = \mathbf{D}/\mathbf{v}_{ps} = (300 \text{ m})/(3.0 \text{ m/s}) = 100 \text{ s}$
a. 3 Determine the time required to swim back to the starting point.	The tip to tail method can again be used to solve for $\mathbf{v}_{ps}$. The woman is now swimming against the current. Thus $\mathbf{v}_{ps} = \mathbf{v}_{pw} - \mathbf{v}_{ws}$ $\qquad = \mathbf{v}_{pw} + (-\mathbf{v}_{ws})$ $\qquad = 2.0 \text{ m/s} + (-1/0 \text{ m/s})$ $\mathbf{v}_{ps} = 1.0 \text{ m/s (upstream)}$ time = displacement/resultant velocity $t = \mathbf{D}/\mathbf{v}_{ps} = (300 \text{ m})/(1.0 \text{ m/s}) = 300 \text{ s}$

a. 4	$t_{total} = t_{downstream} + t_{upstream}$
Determine the total time for the round trip.	$t_{total} = 100 \text{ s} + 300 \text{ s} = 400 \text{ s}$

PROBLEM 4. The woman in the previous problem swims across the river to the opposite bank. If the river is 300 m wide, determine the time to swim across the river and back. She swims perpendicular to the river current and as a result she returns to a point downstream from her starting point.

a. 1	Solution: (Section 3-8)
Determine the woman's velocity relative to the riverbank (v_{ps}).	The woman's velocity relative to the water (v_{pw}) is at right angles to the current (v_{ws}). The Pythagorean theorem can be used to solve for the woman's velocity relative to the river bank v_{ps}. $$v_{ps} = [(v_{pw})^2 + (v_{ws})^2]^{1/2}$$ $$= [(2.0 \text{ m/s})^2 + (1.0 \text{ m/s})^2]^{1/2}$$ $v_{ps} = 2.2 \text{ m/s}, \quad \tan \theta = (1.0 \text{ m/s})/(2.0 \text{ m/s}) = 0.50, \quad \theta = 27°$ $\mathbf{v_{ps}} = 2.2 \text{ m/s} \angle 27°$
a. 2	time = displacement/resultant velocity
Determine the time required to cross the river.	$t = D/v_{ps} \quad$ but $\cos \theta = (300 \text{ m})/D$ and $D = (300 \text{ m})/(\cos 27°) = 340 \text{ m}$ $t = D/v_{ps} = (340 \text{ m})/(2.2 \text{ m/s}) = 150 \text{ s}$
a. 3	The return trip can be analyzed in a like manner. The resultant velocity will be 2.2 m/s and the displacement will be 340 m.
Determine the time required to swim back across the river and the total time for the round trip.	$t = D/v_{ps} = (340 \text{ m})/(2.2 \text{ m/s}) = 150 \text{ s}$ The total time for the round trip can now be determined. $t_{total} = t_{across} + t_{back}$ $t_{total} = 150 \text{ s} + 150 \text{ s} = 300 \text{ s}$

Alternate Method: Since the woman swims perpendicular to the current, the current does not affect the time for the round trip. An analogous situation is that of an airplane passenger walking from a window on the left side of the plane to a window on the right side of the plane. The time is the same if the plane is at rest or if it is moving at 450 miles per hour. Thus, time to cross river equals the distance across the river divided by the woman's velocity perpendicular to the river current.

$t = D/v_{pw} = (300 \text{ m})/(2.0 \text{ m/s}) = 150 \text{ s}$

$t_{total} = t_{across} + t_{back} = 150 \text{ s} + 150 \text{ s} = 300 \text{ s}$

PROBLEM 5. A stone is thrown horizontally outward from the top of a bridge. The stone is released 19.6 meters above the street below. The initial velocity of the stone is 5.0 m/s. Determine the a) total time that the stone is in the air and b) magnitude and direction of the velocity of the projectile "just" before it strikes the street.

a. 1	Solution: (Sections 3-5 and 3-6)
Draw an accurate diagram showing the trajectory of the projectile.	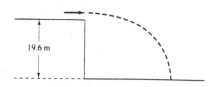

---

a. 2	Since the object is thrown horizontally, the initial vertical component (v_{yo}) is zero. The initial horizontal component (v_{xo}) is 5.0 m/s. Note: the vertical component of motion is independent of the horizontal motion.
Determine the initial horizontal, and vertical components of velocity.	

---

a. 3	Since the vertical component is downward, the downward direction will be designated as the positive direction in the same manner as the free fall problems discussed in chapter 2. Therefore, the acceleration will be + g rather than - g.
Complete a data table using information both given and implied in the problem.	$g = 9.8 \text{ m/s}^2 \qquad v_{yo} = 0 \qquad t = ? \qquad y = 19.6 \text{ m}$
	$v_y = ? \qquad\qquad x = ? \qquad v_{xo} = 5.0 \text{ m/s}$

---

a. 4	$y = \frac{1}{2} g t^2 + v_{yo} t$
Determine the total time that the stone is in the air.	$19.6 \text{ m} = \frac{1}{2} (9.8 \text{ m/s}^2) t^2 + (0 \text{ m/s}) t$
	$t^2 = 2(19.6 \text{ m})/(9.8 \text{ m/s}^2) = 4.0 \text{ s}^2$
	Either t = 2.0 s or - 2.0 s. Since the time of flight cannot be negative, the answer is t = 2.0 s.

---

b. 1	$v_y = v_{yo} + g t$
Determine the magnitude of the vertical component of the stone's velocity just before it hits the street.	$= 0 \text{ m/s} + (9.8 \text{ m/s}^2)(2.0 \text{ s})$
	$v_y = + 19.6 \text{ m/s}$

b. 2	The horizontal velocity remains constant; therefore, both components of motion are now known. Use the Pythagorean theorem and trigonometry to determine the magnitude and direction of the resultant velocity (v_R).

Determine the magnitude and direction of the velocity of the projectile just before it strikes the street.

$$v_R = [(v_y) + (v_{xo})]^{1/2}$$

$$= [(19.6 \text{ m/s})^2 + (5.0 \text{ m/s})^2]^{1/2}$$

$$v_R = 20 \text{ m/s}$$

$$\tan \theta = v_y/v_{xo} = (19.6 \text{ m/s})/(5.0 \text{ m/s}) = 3.9; \quad \theta = 76°$$

$$v_R = 20 \text{ m/s} \angle 76° \text{ below the horizontal}$$

PROBLEM 6. A projectile is fired with an initial speed of 113 m/s at an angle of 60.0° above the horizontal from the top of a cliff 49.0 m high. Determine the a) time to reach maximum height, b) maximum height above the base of the cliff reached by the projectile, c) total time it is in the air, and d) horizontal range of the projectile.

---

a. 1

Draw an accurate diagram showing the trajectory of the projectile.

Solution: (Sections 3-5 and 3-6)

---

a. 2

Determine v_{yo} and v_{xo}.

$v_{yo} = v_o \sin 60.0° = (113 \text{ m/s})(0.866) = 97.9 \text{ m/s}$

$v_{xo} = v_o \cos 60° = (113 \text{ m/s})(0.500) = 56.5 \text{ m/s}$

---

a. 3

Complete a data table using information both given and implied.

$v_{yo} = 97.9 \text{ m/s}$ $t = ?$ (to reach max height)

$y = ?$ (at max height) $v_y = 0$ (at maximum height)

$a = -g = -9.80 \text{ m/s}^2$ (decelerates as it travels upward)

$v_{xo} = 56.5 \text{ m/s}$ $x = ?$

---

a. 4

Determine the time to reach maximum height.

$v_y = v_{yo} - gt$

$0 \text{ m/s} = 97.9 \text{ m/s} - (9.80 \text{ m/s}^2) t$

$t = 9.99 \text{ s}$

b. 1 Determine the maximum height above the cliff reached by the projectile Now determine the maximum height above the ground reached by the projectile.	$y = -\frac{1}{2} g t^2 + v_{yo} t$ $\quad = -\frac{1}{2} (9.80 \text{ m/s}^2)(9.99 \text{ s})^2 + (97.9 \text{ m/s})(9.99 \text{ s})$ $\quad = -489 \text{ m} + 978 \text{ m}$ $y = 489 \text{ m}$ The total height above the base of the cliff can now be determined: $y = 489 \text{ m} + 49 \text{ m} = 538 \text{ m}$
c. 1 Complete a data table and determine the time required for the projectile to fall from maximum height to the base of the cliff.	The downward direction is taken to be positive. Thus +g will be used instead of -g. $v_{yo} = 0$ (at maximum height) $\qquad t = ?$ $v_y = ?$ $\qquad\qquad\qquad\qquad y = 538 \text{ m}$ $g = 9.80 \text{ m/s}^2$ $\qquad\qquad\qquad x = ?$ $v_{xo} = 56.5 \text{ m/s}$ $y = \frac{1}{2} g t^2 + v_{yo} t$ $538 \text{ m} = \frac{1}{2} (9.80 \text{ m/s}^2) t^2 + (0 \text{ m/s}) t$ $t^2 = 2(538 \text{ m})/9.80 \text{ m/s}^2 = 110 \text{ s}^2$ $t = 10.5 \text{ s}$
c. 2 Determine the total time the projectile is in the air.	total time $= 9.99 \text{ s} + 10.5 \text{ s} = 20.5 \text{ s}$
d. 1 Use the equation for horizontal motion to determine the horizontal range.	$x = v_{xo} t = (56.5 \text{ m/s})(20.5 \text{ s})$ $x = 1160 \text{ m}$

PRACTICE PROBLEMS

PROBLEM 1. A man walks 3.0 km due east, then 1.0 km 45° north of east, and finally 2.0 km 30° south of west. Use the tip to tail method, the parallelogram method, and the trigonometric component method to determine his final displacement from his starting point.

ANS. 2.0 km ∠8.4° south of east

PROBLEM 2. Use the trigonometric component method to determine the sum of the following displacement vectors:

A = 5.0 m ∠37° N of E, **B** = 6.0 m ∠45° N of W,
C = 4.0 m ∠30° S of W, **D** = 3.0 m ∠60° S of E.

ANS. 3.4 m ∠50° N of W

PROBLEM 3. In a particular laboratory experiment, a spring gun placed on a table fires a steel ball horizontally outward. A student determines that the ball starts 1.0 m above the floor and travels 2.7 m horizontally before striking the floor. Determine the a) time that the ball is in the air and b) initial velocity of the ball.

ANS. a) 0.45 s, b) 6.0 m/s

PROBLEM 4. The gun described in the previous problem was then arranged at a 45° angle above the horizontal and the student determines that the ball leaves the muzzle of the gun 1.1 m above the floor. Assuming that the initial velocity of the bullet remains the same, determine the a) total time that the ball is in the air and b) horizontal range of the projectile.

ANS. a) 1.1 s,   b) 4.6 m

CHAPTER 4

DYNAMICS

KEY TERMS AND PHRASES

dynamics is the study of the causes of motion.

net force refers to the vector sum of all of the forces acting on an object.

Newton's first law of motion is also known as Galileo's law of inertia, where **inertia** refers to the tendency of an object to resist any change in its state of motion. Inertia is measured by measuring an object's mass.

Newton's second law of motion refers to an object's motion when a net force does not equal zero. The net force (F) will cause an object to accelerate or decelerate. The rate of acceleration (a) is directly proportional to the magnitude of the net force and inversely proportional to the object's mass (m), i.e., $a = F/m$ or $F = m\,a$.

Newton's third law states that whenever one object exerts a force on a second object, the second object exerts an equal but opposite force on the first.

weight is a measure of the force of gravity on an object.

friction is a contact force that opposes the relative motion of two surfaces as they slide past each other. The frictional force depends on the coefficient of friction (μ) and the normal force (F_N).

coefficient of friction (μ) is a pure number without physical units and varies with the types of surfaces that are in contact. When the object is at rest, the **coefficient of static friction** (μ_s) is used to determine the magnitude of the frictional force just before the object starts to move. When the object is moving, the **coefficient of kinetic friction** (μ_k) is used.

normal force is a contact force that acts perpendicular to the common surface of contact.

SUMMARY OF MAHEMATICAL FORMULAS

Newton's first law	Net $\mathbf{F} = 0$ or $\Sigma\mathbf{F} = 0$	If the net force on an object equals zero, then the object's motion will not change, i.e., a = 0.
Newton's second law	$\mathbf{a} = \mathbf{F}/m$ or $\mathbf{F} = m\,\mathbf{a}$	An object's acceleration is related to the net force and the object's mass.
Newton's third law	$\mathbf{F}_{12} = -\mathbf{F}_{21}$	If object 1 exerts a force on object 2, then object 2 exerts an equal but opposite force on object 1.
weight	$w = m\,g$	An object's weight is directly proportional to its mass and gravitational acceleration.
force of friction	$F_{fr} = \mu\,F_N$	Frictional force depends on the coefficient of friction and the normal force.

CONCEPT SUMMARY

Newton's Laws of Motion

Dynamics is the study of the causes of motion. The basic causes of motion can be explained by using **Newton's three laws of motion**. The **first law** explains the motion of an object on which the **net force** equals zero. The net force refers to the vector sum of all of the forces acting on an object. The magnitude and direction of the net force is determined by using the methods of vector addition described in chapter 3. While the tip to tail method and the parallelogram method are useful, the method used most frequently in determining the resultant is the **vector component method**. If the net force equals zero, the object will tend to remain at rest, or if in motion, will remain in motion in a straight line at a constant speed, i.e., constant velocity.

Newton's first law of motion is also known as Galileo's law of inertia, where **inertia** refers to the tendency of an object to resist any change in its state of motion. Inertia is measured by measuring an object's mass.

Newton's second law of motion refers to an object's motion when a net force does not equal zero. The net force (F) will cause an object to accelerate or decelerate. The rate of acceleration (a) is directly proportional to the magnitude of the net force and inversely proportional to the object's **mass** (m).

$\mathbf{a} = \mathbf{F}/m$ or $\mathbf{F} = m\,\mathbf{a}$

The SI unit of force is the **newton** (N) and for mass it is the kilogram (kg). A net force of one newton will cause a 1 kg object to accelerate at 1 m/s²; thus 1 N = 1 kg m/s².

Newton's third law is the "action-reaction" law with which most students are familiar. However, it is necessary to be very careful in interpreting the meaning of this law. The action force and the reaction force are equal and opposite but do not act on the same object. A good way to remember the law is the statement given in the text: "Whenever one object exerts a force on a second object, the second object exerts an equal but opposite force on the first."

Weight

The force of gravity on an object, i.e., the **weight**, varies slightly from place to place on the surface of the Earth. The variation is so small that the weight of an object located close to the Earth's surface is usually considered to be constant.

The formula for an object's weight (w or F_g) near the Earth's surface can be determined as follows:

Net F = m a, but Net F = F_g = w and a = g; therefore,

F_g = w = m g

Mass and weight are often confused. Mass depends on an object's inertia and does not vary with location. Weight is the force of gravity acting on an object and varies from place to place. For example, a person standing on the Earth weighs six times as much as he or she would weigh on the Moon; however, his or her mass would remain the same.

Friction

Friction is a contact force that opposes the relative motion of two surfaces as they slide past each other. The maximum value of the frictional force (F_{fr}) is proportional to the **coefficient of friction** (μ) and also to the **normal force** (F_N) acting on the object.

F_{fr} = μ F_N

The coefficient of friction is a pure number without physical units and varies with the types of surfaces that are in contact with each other. The force to overcome friction when the object is initially at rest is greater than when the object is moving. When the object is at rest, the **coefficient of static friction** (μ_s) is used to determine the magnitude of the frictional force just before the object starts to move. When the object is moving, the **coefficient of kinetic friction** (μ_k) is used.

The normal force is a contact force that acts perpendicular to the common surface of contact. For example, a book at rest on a horizontal table is acted upon by two forces. The book's weight is downward but is opposed by the normal force which is equal in magnitude but directed upward, perpendicular to the surface of the table.

SELECTED TEXT QUESTIONS WITH ANSWERS

QUESTION 3. A stone hangs by a fine thread from the ceiling, and a section of the same thread dangles from the bottom of the stone. If a person gives a sharp pull on the dangling thread, where is the thread likely to break: below the stone or above it? What if the person gives a slow and steady pull? Explain your answers.

ANSWER: The key to the answer lies in the concept of inertia. Inertia is the tendency of an object to resist any change in its state of motion. The stone has inertia and this inertia must be overcome in order to move the stone. If a person gives a SHARP pull to the dangling thread, the inertia of the stone resists any change in its motion. The result is that the stone does not move and the force exerted by the person is not transmitted to the section of thread above the stone. The tension in the section below the stone is much greater than above the stone and the lower section snaps.

If the person exerts a slow and steady pull, the inertia of the stone is overcome and the stone will begin to move slowly. The tension in the section of thread is due to the person. The tension in the section above the stone is due to the sum of the pull exerted by the person and the stone's weight. The tension in the section above the stone is greater than below the stone. As a result the section of thread above the stone breaks.

QUESTION 5. Whiplash sometimes results from an automobile accident when the victim's car is struck violently from the rear. Explain why the head of the victim seems to be thrown backward in this situation. Is it really?

ANSWER: When the car is struck from the rear, the force on the car causes the car to move forward. The seat and seatback are attached to the body of the car and move forward with the car. As the seatback moves forward it pushes the occupant's torso forward. The person's head is above the seatback and the head's inertia causes the head to appear to be "thrown backward" as the car moves forward. In reality, the person's head tends to remain at rest while the car plus victim's torso is pushed forward.

Modern cars are equipped with a headrest which, if properly positioned, will push the head forward along with the torso if the car is struck from the rear.

QUESTION 13. The force of gravity on a 2 kg rock is twice as great as that on a 1 kg rock. Why then doesn't the heavier rock fall faster?

ANSWER: According to Newton's second law of motion, an object's rate of acceleration is directly proportional to the magnitude of the object's weight but inversely proportional to the object's mass, i.e. $a = F/m$. A 2 kg rock has twice as much weight as a 1 kg rock but it also has twice as much mass, i.e. inertia and is twice as hard to accelerate. The ratio of the force to mass is the same for both objects and as a result the acceleration is the same for each.

PROBLEM SOLVING SKILLS

For problems related to Newton's second law of motion:

1. Draw an accurate diagram locating each of the forces acting on the object or system of objects.
2. Draw a free body diagram locating the forces acting on the object(s) in question.
3. If the object is on an incline, the weight is replaced with components acting parallel and perpendicular to the incline.
4. If a frictional force is involved, the magnitude of the force is related to the coefficient of friction and the normal force.
5. Determine the magnitude of the net force acting on the object.
6. Use Newton's second law to write an equation for the motion of each object in the system. Solve for the rate of acceleration of each object in the system.
7. If the problem involves uniform acceleration, then the kinematics equations developed for uniformly accelerated motion can be used to determine the velocity, displacement, time, etc.

PROGRAMMED PROBLEMS

PROBLEM 1. A box of mass 5.0 kg is pulled vertically upward by a force of 68 newtons applied to a rope attached to the box. Determine the a) rate of acceleration of the box and b) vertical velocity of the box after 2.0 s of motion.

a. 1	Solution: (Sections 4-4 and 4-7)
Draw an accurate diagram locating the forces acting on the box, then draw a free body diagram.	

a. 2 the Determine the net force acting on the box.	The box moves upward; therefore, the tension (T) is greater than weight, therefore the net force is $T - mg = 68$ N $- (5.0$ kg$)(9.8$ m/s$^2) = 19$ N

a. 3 Apply Newton's second law of motion and solve for the rate of acceleration.	net $F = ma$ 68 N $- (5.0$ kg$)(9.8$ m/s$^2) = (5.0$ kg$)$ a 19 N $= (5.0$ kg$)$ a $a = 3.8$ m/s^2

b. 1 Determine the velocity the box after 2.0 s.	The acceleration is uniform; therefore, the kinematics equations derived for uniformly accelerated motion can be used to determine the velocity. $v = at + v_o$ $v = (3.8$ m/s$^2)(2.0$ s$) + 0$ m/s $v = 7.6$ m/s upward

PROBLEM 2. A hockey puck of mass 0.50 kg traveling at 10 m/s on a horizontal surface slows to 2.0 m/s over a distance of 80 m. Determine the a) magnitude of the frictional force acting on the puck and b) coefficient of kinetic friction between the puck and the surface.

a. 1	Solution: (Sections 4-4 and 4-7)
Draw an accurate diagram locating the forces acting on the puck, then draw a free body diagram.	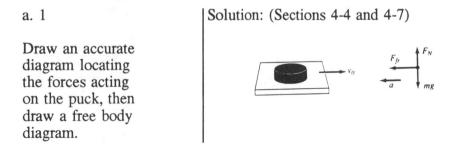

a. 2	m = 0.50 kg, x = 80 m
Complete a data table.	v_0 = 10 m/s, x_0 = 0 m, v = 2.0 m/s

a. 3	$v^2 = v_0^2 + 2 a (x - x_0)$
Use the kinematics equations to determine the acceleration.	$(2.0 \text{ m/s})^2 = (10 \text{ m/s})^2 + 2 a(80 \text{ m} - 0)$
	a = - 0.60 m/s²

a. 4	net F = m a
Use Newton's second law to determine the magnitude of the frictional force.	F_{fr} = (0.50 kg)(- 0.60 m/s²)
	F_{fr} = - 0.30 newtons
	Note: the frictional force is negative because it is in the direction opposite from the direction of motion.

b. 1	Since the puck is on a level surface, the normal force is equal in magnitude but in the opposite direction from the puck's weight.
Determine the magnitude and direction of the normal force.	F_N = mg = (0.50 kg)(9.8 m/s²)
	F_N = 4.9 N

b. 2	$F_{fr} = \mu_k F_N = \mu_k mg$
Determine the coefficient of kinetic friction between the puck and the ice.	0.30 N = μ_k (0.50 kg)(9.8 m/s²)
	μ_k = 0.061

PROBLEM 3. A student of mass 50 kg decides to test Newton's laws of motion by standing on a bathroom scale placed on the floor of an elevator. Assume that the scale reads in newtons. Determine the scale reading when the elevator is a) accelerating upward at 0.50 m/s², b) traveling upward with a constant speed of 3.0 m/s, and c) traveling upward but decelerating at 1.0 m/s².

a. 1	Solution: (Sections 4-4, 4-7)
Draw a diagram locating the forces acting on the student.	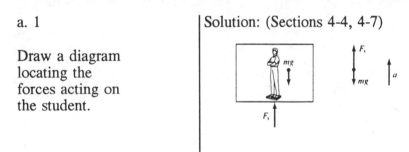

a. 2 Determine the scale reading when the elevator is accelerating upward at 0.50 m/s².	Since the acceleration is upward, then F_s is greater than mg and the net force is F_s - mg, directed upward. The student's apparent weight as determined by the scale is greater than her actual weight. Apply Newton's second law and determine the scale reading (F_s). $F = m \, a$ and $F_s - m \, g = m \, a$ $F_s - (50 \text{ kg})(9.8 \text{ m/s}^2) = (50 \text{ kg})(0.50 \text{ m/s}^2)$ $F_s = 515 \text{ N}$
b. 1 Determine the reading on the scale when the elevator is traveling upward at a constant speed of 3.0 m/s.	The velocity is constant; therefore, the rate of acceleration is zero and the net force must equal zero. The student's apparent weight as read by the scale equals her actual weight. $F = m \, a$ and $F_s - mg = m \, a$ but $a = 0$ $F - (50 \text{ kg})(9.8 \text{ m/s}^2) = (50 \text{ kg}) \, 0$ $F = 490 \text{ N}$
c. 1 Determine the reading on on the scale when it is traveling upward but decelerating at 1.0 m/s².	The elevator is decelerating; therefore, the downward force of her weight is greater than the upward force exerted by the scale. Therefore, the student's apparent weight as determined by the scale is less than her actual weight. $F = m \, a$ and $F_s - mg = m \, a$ $F_s - (50 \text{ kg})(9.8 \text{ m/s}^2) = (50 \text{ kg})(- 1.0 \text{ m/s}^2)$ $F_s = 440 \text{ N}$

PROBLEM 4. A wooden plank is raised at one end until an angle of 30° is achieved. A 2.0 kg box is placed on the incline 1.0 m from the lower end and given a slight tap to overcome static friction. The coefficient of kinetic friction between the box and plank is 0.20. Determine the a) rate of acceleration of the box down the incline and b) speed of the box at the bottom. Assume that the initial speed of the box is zero.

a. 1 Draw a free body diagram locating all of the forces acting on the object.	Solution: (Section 4-8) 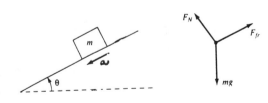

a. 2	$F_{\parallel}/mg = \sin 30°$ and $F_{\parallel} = mg \sin 30°$
Determine the components of the object's weight perpendicular and parallel to the incline.	$F_{\perp}/mg = \cos 30°$ and $F_{\perp} = mg \cos 30°$

a. 3	There is no motion perpendicular to the incline; the magnitude of the normal force equals the perpendicular component of the weight. The frictional force is related to the perpendicular component of the weight by the formula $F_{fr} = \mu_k F_N$. The component of the weight parallel to the incline causes the motion. Since $F_{\parallel}$ is greater than F_{fr}, the net force = $F_{\parallel} - F_{fr}$.
Determine the magnitude and direction of the net force acting on the box.	$F_{\parallel} = mg \sin \theta = (2.0 \text{ kg})(9.8 \text{ m/s}^2) \sin 30° = 9.8 \text{ N}$
	$F_{fr} = \mu_k F_N = \mu_k mg \cos \theta = (0.20)(2.0 \text{ kg})(9.8 \text{ m/s}^2)\cos 30°$
	$F_{fr} = 3.4 \text{ N}$
	net F = $F_{\parallel} - F_{fr}$ = 9.8 N - 3.4 N
	net F = 6.4 N

a. 4	a = F/m
Determine the rate of acceleration of the box.	a = (6.4 N)/(2.0 kg)
	a = 3.2 m/s^2

a. 5	net F = ma
Mathematically show that the rate of acceleration is independent of the object's mass.	$F_{\parallel} - F_{fr}$ = ma
	$mg \sin 30° - \mu_k mg \cos 30° = m\,a$
	Note that the mass of the object appears in each term. Dividing each term by m gives
	$g \sin 30° - \mu_k g \cos 30° = a$
	$(9.8 \text{ m/s}^2)(0.5) - (0.20)(9.8 \text{ m/s}^2)(0.87) = a$
	$4.9 \text{ m/s}^2 - 1.7 \text{ m/s}^2 = a$
	a = 3.2 m/s^2
	The mass canceled from each term; therefore, the rate of acceleration is independent of the object's mass.

b. 1	$2a(x - x_o) = v^2 - v_o^2$
Determine the object's speed at the bottom of the incline. Note: the initial speed at the top $v_o \approx 0$.	$2(3.2 \text{ m/s}^2)(1.0 \text{ m}) = v^2 - (0 \text{ m/s})^2$
	$v^2 = 6.4 \text{ m}^2/\text{s}^2$
	$v = 2.5 \text{ m/s}$

PROBLEM 5. In a device known as an Atwood machine, a massless, unstretchable rope passes over a frictionless peg. One end of the rope is connected to a object $m_1 = 1.0$ kg while the other end is connected to a object $m_2 = 2.0$ kg. The system is released from rest and the 2.0 kg object accelerates downward while the 1.0 kg object accelerates upward. Calculate the a) rate of acceleration and b) tension in the rope.

a. 1	Solution: (Section 4-7)
Draw an accurate diagram locating the forces acting on each object.	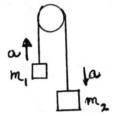  

a. 2	m_2 accelerates downward; therefore, $m_2g > T$.
Use Newton's second law to write a separate equation for the motion of each object.	net $F = m\,a$
	$m_2\,g - T = m_2\,a$
	$(2.0 \text{ kg})(9.8 \text{ m/s}^2) - T = (2.0 \text{ kg})\,a$
	$19.6 \text{ N} - T = (2.0 \text{ kg})\,a$ (equation 2)
	m_1 accelerates upward; therefore, $T > m_1\,g$ and
	$T - m_1\,g = m_1\,a$
	$T - (1.0 \text{ kg})(9.8 \text{ m/s}^2) = (1.0 \text{ kg})\,a$
	$T - 9.8 \text{ N} = (1.0 \text{ kg})\,a$ (equation 1)

a. 3	The above equations have the same two unknowns. To solve for the acceleration, add the two equations together.
Use algebra to solve for the rate of acceleration of the system.	$19.6 \text{ N} - T + T - 9.8 \text{ N} = (2.0 \text{ kg} + 1.0 \text{ kg})\,a$
	$9.8 \text{ N} = (3.0 \text{ kg})\,a$ and $a = 3.3 \text{ m/s}^2$

a. 4 Solve for the acceleration using only the external forces acting on the system.	Tension is an internal force in the system. As can be seen in step a. 3, the tension algebraically cancels when solving for the acceleration. The external force that causes the motion is the weight of m_2 while the weight of m_1 retards the motion. Applying Newton's second law gives net F = m a m_2 g - m_1 g = (m_2 + m_1) a (2.0 kg)(9.8 m/s²) - (1.0 kg)(9.8 m/s²) = (2.0 kg + 1.0 kg) a 19.6 N - 9.8 N = (3.0 kg) a and a = 3.3 m/s²
b. 1 Solve for the tension in the rope.	From equation 1: T - 9.8N = (1.0 kg) a T - 9.8N = (1.0 kg)(3.3 m/s²) T = 13 N

PROBLEM 6. As shown in the diagram, object m_2 = 10.0 kg is connected by a massless, unstretchable string to an object m_1 = 7.0 kg which rests on a 30° incline. The coefficient of friction between the 7.0 kg object and the incline is 0.10. Determine the a) rate of acceleration of the system and b) tension in the rope that connects the boxes. Assume that m_2 accelerates downward and that m_1 accelerates up the incline.

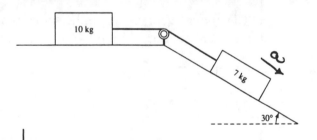

a. 1 Draw a free body diagram locating the forces acting on each object. Note: these forces include tension in the rope, weight, normal force, and frictional force acting on each.	(10.0 kg object) (7.0 kg object)
a. 2 Determine the components of the 7.0 kg object's weight that are perpendicular and parallel to the incline.	$F_\parallel / m_1 g$ = sin 30° $F_\parallel$ = m_1g sin 30° = (7.0 kg)(9.8 m/s²)(0.50) = 34.3 N $F_\perp / m_1 g$ = cos 30° $F_\perp$ = m_1g cos 30° = (0.87)(7.0 kg)(9.8 m/s²) = 59.7 N 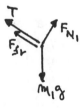

a. 3 Determine the magnitude of the frictional force acting on the 7.0 kg box.	For the box on the 30° incline, the frictional force is given by $F_{fr} = \mu_k F_N$, where F_N is equal in magnitude but opposite in direction to $F_\perp$. $F_{fr} = \mu_k\, m_1 g \cos 30° = (0.10)(59.7\ N) = 5.97\ N$
a. 4 Use Newton's second law to write a separate equation for the motion of each object.	The motion of object 1 is parallel to the incline, i.e., the tension (T) in the string causes the motion while $F_{\parallel}$ and F_{fr} retard the motion. object 1: $\qquad\qquad$ $T - F_{\parallel} - F_{fr} = m_1\, a$ (7.0 kg box) $\qquad\qquad\qquad\qquad$ $T - 34.3\ N - 5.97\ N = (7.0\ kg)\, a$ (equation 1) $\qquad\qquad$ $T - 40.3\ N = (7.0\ kg)\, a$ The motion of the 10.0 kg box, i.e., object 2, is vertically downward. The weight of m_2, i.e., $m_2 g$, causes m_2 to move downward while the tension T in the string retards its motion. object 2: $\qquad\qquad$ $m_2 g - T = m_2\, a$ (10.0 kg box) $\qquad\qquad\qquad$ $(10.0\ kg)(9.8\ m/s^2) - T = (10.0\ kg)\, a$ (equation 2) $\qquad\qquad$ $98.0\ N - T = (10.0\ kg)\, a$
a. 5 Use algebra to solve for the rate of acceleration of the system.	The above equations have the same two unknowns. To solve for the acceleration, add the two equations together. $98.0\ N - T + T - 40.3\ N = (7.0\ kg + 10.0\ kg)\, a$ $57.7\ N = (17.0\ kg)\, a$ and $a = 3.4\ m/s^2$
a. 6 Solve for the acceleration by considering only the external forces acting on the system.	As can be seen in step a. 5, tension does not appear in the final equation used to solve for the acceleration. The forces involved are the weight of object 2 (i.e., $m_2 g$) which causes the motion and $F_{\parallel}$ and F_{fr} which oppose the motion. Apply the second law and solve for the acceleration. net $F = m\, a$ $m_2 g - F_{\parallel} - F_{fr} = (m_1 + m_2)\, a$ $98.0\ N - 34.3\ N - 5.97\ N = (7.0\ kg + 10.0\ kg)\, a$ $57.7\ N = (17.0\ kg)\, a$ $a = 3.40\ m/s^2$

b. 1

Solve for the
tension in the rope.

From equation 1:   98.0 N - T = (10.0 kg) a

98.0 N - T = (10.0 kg)(3.40 m/s^2)

T = 64.0 N

PRACTICE PROBLEMS

PROBLEM 1. A student of mass 50 kg decides to test Newton's laws of motion by standing on a bathroom scale placed on the floor of an elevator. Assume that the scale reads in newtons. The student pushes the down button and the elevator accelerates downward. At one point in the motion she notes that the bathroom scale reads 470 N. Determine the elevator's rate of acceleration.

ANS. 0.40 m/s^2, downward

PROBLEM 2. In a laboratory experiment, a 2.0 kg box is pulled along a horizontal wooden plank by a 10 newton force that acts parallel to the direction of motion. The rate of acceleration of the box is experimentally measured to be 0.50 m/s^2. Determine the coefficient of kinetic friction.

ANS. 0.46

PROBLEM 3.   A truck of mass 1000 kg is connected to a 400 kg log by a weightless, unstretchable rope. The driving force exerted by the ground on the truck is 2000 N. The force of friction exerted by the ground on the log is 600 N.   Assume that the ground is level. Determine the a) rate of acceleration of the system and b) tension in the rope that connects the truck and the log.

ANS. a) a = 1.0 m/s², b) T = 1000 N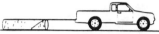

PROBLEM 4. A man decides to push a 2.0 kg box from the bottom to the top of a 30° incline. The coefficient of kinetic friction is 0.20. Determine the minimum amount of force required to to keep the box moving up the incline at a constant speed.

ANS. 13 N

CHAPTER 5

CIRCULAR MOTION: GRAVITATION

KEY TERMS AND PHRASES

uniform circular motion occurs when an object travels in a circle at constant speed.

tangential velocity refers to the direction of the velocity vector of an object as it travels in a curved path. The velocity vector is tangent to the path at each point in the curve.

tangential acceleration refers to the change in speed in time of an object as it travels in a curved path. The tangential acceleration is directed tangent to the path at each point in the curve.

centripetal acceleration is the inward radial acceleration experienced by an object traveling in a circle. The direction of the acceleration vector is perpendicular to the direction of the velocity vector.

period (T) of the circular motion is the time required for an object to complete one revolution.

frequency of rotation (f) is the number of repetitions of the motion per unit time. The frequency is inversely related to the period of the motion.

centripetal force is the term used for the force acting toward the center of the circle. The centripetal force is not some new kind of force. The term merely means that the force is directed toward the circle's center. The force must be applied by some external agent.

Newton's universal law of gravitation is the force of gravitational attraction between any two objects. The law states that the magnitude of the force between two particles is directly proportional to the product of their masses and inversely proportional to the square of the distance between their centers.

Kepler's first law states that the path of each planet about the Sun is an ellipse with the Sun at one of the focal points of the ellipse.

Kepler's second law states that each planet moves such that an imaginary line drawn from the Sun to the planet sweeps out equal areas in equal periods of time.

Kepler's third law states that the ratio of the squares of the periods of any two planets revolving about the Sun is equal to the ratio of the cubes of their average distances from the Sun.

SUMMARY OF MATHEMATICAL FORMULAS

centripetal acceleration	$a_c = v^2/r$	centripetal acceleration (a_c) as related to tangential velocity (v) and the radius of the circle (r)
frequency and period of rotation	$f = 1/T$ $T = 1/f$	Frequency of rotation (f) is inversely related to the period (T) of the motion.
tangential velocity	$v = 2\pi r/T$ $v = 2\pi r f$	tangential velocity (v) as related to the radius of the circle and the period (T) or frequency (f) of the motion
centripetal acceleration	$a_c = 4\pi^2 r/T^2$ $a_c = 4\pi^2 r f^2$	centripetal acceleration (a_c) as related to the radius of the circle and the period (T) or frequency (f) of the motion
centripetal force	$F_c = m v^2/r$	centripetal force (F_c) as related to the object's mass (m), speed (v), and the radius of the circle.
Newton's universal law of gravitation	$F = G m_1 m_2/r^2$	gravitational force between two particles of mass m_1 and m_2 separated by a radial distance (r). G = universal constant = 6.67×10^{-11} N m^2/kg^2.
Kepler's third law	$T^2/r^3 = $ constant or $T_1^2/T_2^2 = r_1^3/r_2^3$	Period of the motion (T) squared of a planet traveling about the Sun divided by the average radius (r) of its orbit cubed is a constant for all planets.

CONCEPT SUMMARY

Kinematics of Uniform Circular Motion

An object traveling in a circle at constant speed is exhibiting **uniform circular motion.** Although the object's speed is not changing, its direction of motion is and therefore its velocity is constantly changing. Note: recall that speed is a scalar quantity having only magnitude while velocity is a vector quantity having both magnitude and direction.

Since its velocity is changing, the object is undergoing accelerated motion. This acceleration is called **centripetal acceleration**. As shown in the diagram, the velocity vector at every point is directed tangent to the circle while the acceleration vector is directed toward the center of the circle. The direction of the acceleration vector is perpendicular to the direction of the velocity vector.

The magnitude of the centripetal acceleration (a_c) is directly proportional to the square of the speed and inversely proportional to the radius of the circle in which the object is traveling.

$$a_c = v^2/r$$

An object moving in uniform circular motion travels a distance equal to the circumference of the circle ($2\pi r$) in a time interval called the **period** (T) of the motion. The period of the motion is the time required for the motion to repeat. The equation for the speed is as follows:

$$v = 2\pi r/T$$

By substituting the above expression for v into the centripetal acceleration equation, it can be shown that the formula for the centripetal acceleration can be written as follows:

$$a_c = (2\pi r/T)^2/r = 4\pi^2 r/T^2$$

Frequency of Rotation

The **frequency of rotation** (f) is the number of repetitions of the motion per unit time. For example, a stereo turntable rotating at 45 rpm has a frequency of rotation of 45 revolutions per minute.

The period of the motion is inversely related to the frequency of rotation ($T = 1/f$ or $f = 1/T$). If the frequency of rotation is 10 revolutions per second, then the period of the motion is 1/10 second or 0.10 seconds.

Since $f = 1/T$, the equation for the speed $v = 2\pi r/T$ can be written as in terms of the frequency as

$$v = 2\pi r f$$

Also, the equation for centripetal acceleration $a_c = 4\pi^2 r/T^2$ can be written as $a_c = 4\pi^2 r f^2$.

Dynamics of Uniform Circular Motion

An object traveling in a circle must have a net force acting on it to keep it in the circle. The magnitude of the object's acceleration can be determined by using Newton's second law, i.e., a = net F/m. For an object traveling in a circle the acceleration is the centripetal acceleration. Since the centripetal acceleration is directed toward the center of the circle, the net force, called the **centripetal force,** must also be directed toward the center of the circle. As stated in the text, "This force is sometimes called a centripetal force. But be aware that 'centripetal force' does not indicate some new kind of force. The term merely means that the force is directed toward the circle's center. The force must be applied by some object."

In solving problems involving uniform circular motion, it is usually necessary to determine the nature of the force which causes the object to travel in the circular path. For example, when a car rounds an unbanked curve, the centripetal force is related to friction. Gravity provides the centripetal force for a satellite in orbit, while a ball on the end of a string is held in a circular path by the tension in the string.

Newton's Universal Law of Gravitation

The weight of an object placed on or very near the Earth's surface can be determined by using the equation   weight = mg   where g is approximately 9.8 m/s^2.

In general, the force (F) of gravitational attraction between any two objects of mass m_1 and mass m_2, their centers separated by a distance r, is given by **Newton's universal law of gravitation**: $F = G \, m_1 m_2/r^2$ where $G = 6.67 \times 10^{-11}$ N m^2/kg^2 (the universal constant).

Gravity Near the Earth's Surface

For an object of mass m on or very near the Earth's surface,

$$mg = G \, m \, m_e/r_e^2 \quad \text{and} \quad g = G \, m_e/r_e^2$$

where m_e is the mass of the Earth and r_e the radius of the Earth. The acceleration due to gravity at some distance from the Earth can be determined by calculating the "effective" value of g. In order to determine the effective value of g, the radius of the Earth is replaced by the value for the radial distance between the object and the center of the Earth.

Kepler's Laws and Newton's Synthesis

Johannes Kepler used data collected by Tycho Brahe to empirically derive three laws for the motion of the planets about the Sun. These laws are now referred to as Kepler's laws of planetary motion and can be summarized as follows:

First law: The path of each planet about the Sun is an ellipse with the sun at one of the focal points of the ellipse.

Second law: Each planet moves so that an imaginary line drawn from the Sun to the planet sweeps out equal areas in equal periods of time.

Third law: The ratio of the square of the period (T^2) of any planet revolving about the sun to the ratio of the cube of the planets average distance distances from the Sun(r^3) is a constant, i.e., T^2/r^3 = constant. Also, the third law can be stated as the ratio of the squares of the periods of any two planets revolving about the Sun is equal to the ratio of the cubes of their average distances from the Sun, i.e., $T_1^2/T_2^2 = r_1^3/r_2^3$

Newton was able to mathematically derive Kepler's laws by using his laws of motion and universal law of gravitation. Thus Kepler's laws, which were derived from empirical data, could be used to support Newton's universal law of gravitation.

Types of Forces in Nature

Physicists now recognize only four different forces in nature. These are the gravitational force, electromagnetic force, strong nuclear force, and weak nuclear force. The electromagnetic force refers to electrical and magnetic forces. The strong nuclear force and weak nuclear force operate at the level of the atomic nucleus.

The ordinary pushes, pulls, and frictional forces that have been considered in previous chapters are referred to as "contact forces" because they involve objects that come into contact with one another, e.g., a boy pulling a wagon. According to modern theory, contact forces are due to the electromagnetic force.

SELECTED TEXT QUESTIONS WITH ANSWERS

QUESTION 2. Will the acceleration of a car be the same if it travels around a sharp curve at 60 km/h as when it travels around a gentle curve at the same speed? Explain.

ANSWER: The acceleration is NOT the same. The centripetal acceleration is directly proportional to the square of the speed and inversely proportional to the radius of the curve. The speed is held constant at 60 km/h but the radius of the sharp curve is less than the gentle curve. Therefore, the SMALLER the radius, the GREATER the centripetal acceleration.

QUESTION 6. Does an apple exert a gravitational force on the Earth? If so, how large a force? Consider an apple a) attached to a tree and b) falling.

ANSWER: According to the universal law of gravitation, the force exerted by the apple on the Earth must be equal in magnitude to the force exerted by the Earth on the apple. According to Newton's third law, the forces must be equal but in opposite directions. The Earth pulls the apple downward while the apple pulls the earth upward.

In part a, the apple is attached to the tree and the apple's weight is balanced by an upward force exerted by the tree branch. However, it still exerts an upward force on the Earth.

In part b, the apple is falling and the net force on the apple equals the gravitational force exerted by the Earth. The apple exerts an equal but opposite force on the Earth. However, based on Newton's second law, the upward acceleration of the Earth is negligible because of its enormous mass.

QUESTION 18. The Earth moves faster in its orbit around the Sun in the winter than in summer. Is it closer to the Sun in summer or in winter? Does this affect the seasons? Explain.

ANSWER: Based on Kepler's second law, the Earth moves slower in the summer when it is farther away from the Sun. The effect on the seasons due to the variation in the distance is minimal compared to the effect of the tilt of the Earth on its axis.

In the summer months the northern hemisphere is tilted toward the Sun. The number of hours of daylight is greater and we receive the more direct rays of the Sun. In the winter, the northern hemisphere is tilted away from the Sun. The Sun's rays are less direct and there are fewer hours of daylight. The combination of number of hours of daylight and the angle of the Sun's rays is the major cause of the variation in the seasons.

PROBLEM SOLVING SKILLS

Problems involving centripetal acceleration relate to Newton's second law of motion. Therefore, the steps followed in Chapter 4 can be applied to this chapter. The following is an outline of the steps to be followed in attempting to solve problems related to uniform circular motion and centripetal acceleration:

1. Draw an accurate diagram locating each of the forces acting on the object in uniform circular motion. Take note of the force(s) that is/are causing the object to travel in a curved path.
2. Determine the magnitude and direction of the net force acting on the object.
3. Complete a data table based on information both given and implied in the statement of the problem.
4. Use Newton's second law and the concept of centripetal acceleration to solve the problem.

For problems involving Newton's universal law of gravitation:

1. Complete a data table based on information both given and implied in the problem.
2. Use the universal law of gravitation and, if necessary, Newton's second law and the concept of centripetal acceleration to solve the problem.

For problems involving Kepler's third law:

1. Complete a data table based on information both given and implied in the problem.
2. Use Kepler's third law to solve the problem.

PROGRAMMED PROBLEMS

PROBLEM 1. A 2.0 kg box is placed on the floor at the edge of a merry-go-round of radius 6.0 m. The coefficient of friction between the box and the floor is 0.30. The merry-go-round accelerates from rest and eventually the box slides off the edge. Determine the speed at which this occurs.

a. 1	Solution: (Section 5-2)
Draw an accurate diagram locating all of the forces acting on the object.	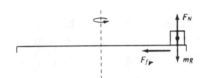

a. 2	$m = 2.0$ kg $\mu_s = 0.30$ $r = 6.0$ m $v = ?$
Complete a data table.	

a. 3	The box is held in position by a frictional force that provides the centripetal acceleration. At the point where the box tends to slide off the circle, the force required to hold the box in position "just" exceeds the maximum possible value of the frictional force. The maximum value of the frictional force can now be determined.
Determine the net force on the object.	$F_{fr} = \mu_s F_N$ but $F_N = mg$
	$F_{fr} = \mu_s mg = (0.30)(2.0$ kg$)(9.8$ m/s$^2)$
	$F_{fr} = 5.9$ N; thus net $F = 5.9$ N

a. 4	net $F = m a$ but net $F = F_{fr}$ and $a = v^2/r$.
Use Newton's second law and the formula for centripetal acceleration to determine the object's velocity.	5.9 N $= (2.0$ kg$) v^2/(6.0$ m$)$
	$v^2 = 17.7$ m^2/s^2 and $v = 4.2$ m/s
	Note: it can be shown that the mass of the box does not affect the maximum velocity.
	$F_{fr} = m v^2/r$ but $F_{fr} = \mu_s F_N = \mu_s mg$
	$\mu_s mg = m v^2/r$
	The mass cancels from both sides of the equation.
	$\mu_s g = v^2/r$
	$v = (\mu_s g r)^{1/2} = [(0.30)(9.8$ m/s$^2)(6.0$ m$)]^{1/2}$ and $v = 4.2$ m/s

PROBLEM 2. A car travels over the crest of a hill at 6.0 m/s. The radius of curvature at the crest is 15 m. a) Determine the force exerted by the car seat on a 30 kg passenger. b) Determine the minimum speed required for the passenger to feel momentarily "weightless."

a. 1 Draw a free body diagram locating the forces acting on the passenger.	Solution: (Section 5-2) 
a. 2 Derive a formula for the net force acting on the person. In which direction is this force acting?	At the crest of the hill the car is traveling in a circle at a constant speed. Since it is traveling in a circle there is a centripetal acceler-ation directed into the center of the circle. This centripetal acceler-acceleration is produced by a net inward force equal to the differ-ence between the passenger's weight, which acts downward, and the upward force exerted by the seat. net F = weight - F_{seat} = mg - F_{seat}
a. 3 Apply Newton's second law and the formula for centripetal acceleration and determine the magnitude of the force exerted by by the seat.	Since the direction of the net force is downward, let the downward direction be positive. net F = m a weight - F_{seat} = m a but a = v^2/r (30 kg)(9.8 m/s²) - F_{seat} = (30 kg)[(6.0 m/s)²/(15 m)] 294 N - F_{seat} = 72 N F_{seat} = 294 N - 72 N = 222 N As the car goes over the crest of the hill, the passenger feels less upward force from the seat.
b. 1 Determine the minimum speed required for the passenger to feel momentarily "weightless."	At this "minimum speed" the seat exerts no upward force on passenger, i.e., F_{seat} = 0. As a result the passenger experiences a feeling of momentary "weightlessness." Since the direction of the net force is downward, let the downward direction be positive. Net F = m a weight - F_{seat} = m a but F_{seat} = 0 and a = v^2/r (30 kg)(9.8 m/s²) - 0 N = (30 kg) v^2/(15 m) v = [(9.8 m/s²)(15 m)]$^{1/2}$ = 12 m/s or approximately 27 mph

PROBLEM 3. In an amusement park ride called the "Round-Up," riders are pressed against a wire-mesh wall that revolves in a vertical circle of radius 5.00 m and a frequency of 15.0 rpm. Determine the force that the wall exerts on a 60.0 kg rider at the a) top of the loop and b) bottom of the loop.

a. 1 Express the frequency of rotation to revolutions per second.	Solution: (Section 5-2) f = (15.0 revolutions/min) (1 min/60 s) = 0.250 revolutions/sec f = 0.250 rps
a. 2 Determine the person's centripetal acceleration.	The centripetal acceleration can be written in terms of the radius of the circle and the frequency of rotation as follows: $a_c = 4\ \pi^2\ r\ f^2 = 4\ \pi^2\ (5.00\ m)(0.250\ rev/s)^2 = 12.3\ m/s^2$
a. 3 Draw an accurate diagram and locate the forces acting on the person at the top and bottom of the loop.	
a. 4 Write a formula for the magnitude and direction of the net force acting on the person at the top of the loop.	Below the minimum velocity, gravity is the only force acting on the rider and the rider tends to fall out of the loop. If the machine rotates with a speed greater than the minimum required to hold the rider in the circle, then the wall, as well as gravity, exerts a downward force. Therefore, net F = F_{wall} + mg
a. 5 Apply Newton's second law to determine the force exerted by the wall.	net F = m a but a = a_c and from step a. 4, net F = F_{wall} + mg Thus F_{wall} + mg = m a_c F_{wall} + (60.0 kg)(9.8 m/s^2) = (60.0 kg)(12.3 m/s^2) F_{wall} + 588 N = 738 N F_{wall} = 150 N

b. 1 Write a formula for the magnitude and direction of the net force acting on the person at the bottom of the loop.	As shown in the diagram, at the bottom of the loop, the wall exerts an upward force while the gravitational force is downward. The net force equals the difference between the upward force exerted by the wall and the downward gravitational force. Therefore, net $F = F_{wall} - mg$
b. 2 Apply Newton's second law to determine the force exerted by the wall.	net $F = m\,a$ but $a = a_c$ and from step b. 1, net $F = F_{wall} - mg$ Thus $F_{wall} - mg = m\,a_c$ $F_{wall} - (60.0 \text{ kg})(9.8 \text{ m/s}^2) = (60.0 \text{ kg})(12.3 \text{ m/s}^2)$ $F_{wall} - 588 \text{ N} = 738 \text{ N}$ $F_{wall} = 1330 \text{ N}$ People frequently hold out their arms during this ride. They readily notice that it is easy to hold their arms out at the top while their arms feel very 'heavy' and hard to hold out at the bottom.

PROBLEM 4. A student whirls a bucket of water in a vertical circle of radius 0.75 m. Determine the minimum a) speed and b) frequency of rotation required for the bucket to completely negotiate the top of the loop without water spilling out of the bucket.

a. 1 Draw an accurate diagram and locate the forces acting on the bucket at the top of the loop.	Solution: (Section 5-2) 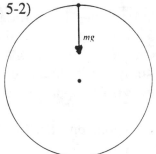
a. 2 Write a formula for the magnitude and direction of the net force acting on the circle.	If the bucket rotates with a speed greater than the minimum required to hold the water in the circle, then the bottom of the bucket exerts a downward force. Below the minimum velocity the water tends to fall out of the bucket. At the point of minimum speed, the bottom of the bucket exerts no force on the water when the bucket + water are at the top of the loop. The only force water at the top ofthe acting on the bucket + water is the force of gravity (mg). Therefore, net $F = mg$

a. 3	net F = ma but a = a_c = v^2/r
Apply Newton's second law and the concept of centripetal acceleration to determine the bucket's minimum speed.	and from the previous step; net F = mg.
	Thus mg = m v^2/r
	The mass of the bucket + water mathematically cancels from both sides of the equation.
	g = v^2/r and upon simplifying
	v^2 = r g = (0.75 m)(9.8 m/s²) = 7.35 m²/s²
	v = 2.7 m/s
b. 1	Since the motion is uniform circular motion,
Determine the period of the motion.	v = 2πr/T and T = 2πr/v
	T = 2π(0.75 m)/(2.7 m/s) and T = 1.7 s
b. 2	The frequency (f) of rotation is inversely proportional to the period of the motion. f = 1/T = 1/1.7 s = 0.57/s
Determine the frequency of rotation.	The frequency is 0.57 revolutions per second (0.57 rps or 0.57/s). It is convenient to express this in revolutions per minute (rpm); therefore, (0.57/s)(60 s/1 min) = 34 rpm.

PROBLEM 5. A 50 kg rider on the "Rotor-Ride" stands along a vertical wall in a circular room 3.0 m in radius. The room begins to rotate and at a certain point the floor drops down several meters, leaving the rider pressed against the wall. If the coefficient of static friction between the rider and the wall is 0.40, determine the minimum a) speed and b) frequency of rotation required to prevent the rider from slipping down the wall.

a. 1	Solution: (Section 5-2)
What forces must balance if the rider does not slip down the wall?	Since the rider does not slip down the wall, the upwardly directed frictional force must be equal to but opposite the rider's weight.
	F_{fr} = m g
a. 2	The net force on the rider is due to the normal force the wall exerts on the rider. This normal force is directed toward the center of circular motion and provides the centripetal acceleration experienced by the rider.
What force provides the acceleration?	net F = F_N = m a_c = m v^2/r

a. 3 Draw an accurate diagram and locate the forces acting on the rider.	
a. 4 Write an equation relating the normal force to the frictional force.	$F_{fr} = \mu_s\, F_N$
a. 5 Use Newton's second law and the concept of centripetal acceleration to solve for the minimum velocity of the "Rotor-Ride."	net $F = F_N = m\, a_c = m\, v^2/r$ $F_{fr} = \mu_s\, F_N$ where $F_{fr} = mg$ $F_{fr} = \mu_s\, F_N = \mu_s\, m\, v^2/r$ $mg = \mu_s\, m\, v^2/r$ Note: the rider's mass algebraically cancels. $g = \mu_s\, v^2/r$ and rearranging gives $v^2 = r\, g/\mu_s = (3.0\text{ m})(9.8\text{ m/s}^2)/(0.40) = 74\text{ m}^2/\text{s}^2$ $v = 8.6$ m/s
b. 1 Determine the period of the motion.	Since the motion involves uniform circular motion $v = 2\pi r/T$ and $T = 2\pi r/v = 2\pi(3.0\text{ m})/(8.6\text{ m/s}) = 2.2$ s
b. 2 Determine the frequency of rotation.	$f = 1/(2.2\text{ s}) = 0.46/\text{s}$ or 0.46 rps Converting to revolutions per min (rpm), $f = (0.46\text{ rev/s})(60\text{ s/min}) = 28$ rpm

PROBLEM 6. On December 24, 1968, the Apollo 8 command module became the first manned vehicle to go into orbit above the surface of the Moon. Assuming that the orbit was approximately circular and the vehicle was 110 km (69 miles) above the lunar surface, determine the a) orbital velocity of the spacecraft and b) period of the motion.
Note: $m_{Moon} = 7.40 \times 10^{22}$ kg, $r_{Moon} = 1.74 \times 10^6$ m, and 110 km $= 1.10 \times 10^5$ m.

a. 1 Draw a diagram of the spacecraft's orbit about the Moon.	Solution: (Sections 5-7 and 5-8) 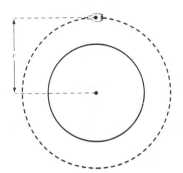
a. 2 Determine the radius of the orbit as measured from the center of the Moon.	$r = r_{Moon}$ + altitude above surface $= 1.74 \times 10^6$ m $+ 1.10 \times 10^5$ m $r = 1.85 \times 10^6$ m
a. 3 What force holds the spacecraft in its circular orbit about the Moon?	Lunar gravity is the only force acting on the spacecraft and this force causes it to travel in uniform circular motion. $F = G\ m\ M/r^2$
a. 4 To solve for the orbital velocity apply Newton's second law and the concept of centripetal acceleration.	net $F = F_{grav} = m\ a_c$ $G\ m\ M/r^2 = m\ v^2/r$ The mass of the orbiting object (m) appears on both sides of the equation and cancels algebraically. $v^2 = G\ M/r$ $= [(6.67 \times 10^{-11}\ N\ m^2/kg^2)(7.4 \times 10^{22}\ kg)]/(1.85 \times 10^6\ m)$ $v^2 = 2.7 \times 10^6\ m^2/s^2$ $v = 1.6 \times 10^3$ m/s; approximately 3600 mph
b. 1 Determine the period of the motion.	$v = 2\pi r/T$ 1.6×10^3 m/s $= 2\ \pi\ (1.85 \times 10^6\ m)/T$ $T = 7.12 \times 10^3$ s or 1.97 hours

PROBLEM 7. The mean distance from the Earth to the Sun is 1.496×10^8 km and the period of its motion about the Sun is 1.00 year. The period of Jupiter's motion about the Sun is observed to be 11.86 years. Determine the mean distance from the Sun to Jupiter.

a. 1 Complete a data table.	Solution: (Section 5-8) $T_1 = 11.86$ y $\qquad r_1 = ?$ $T_2 = 1.00$ y $\qquad r_2 = 1.496 \times 10^8$ km
a. 2 Use Kepler's third law to solve for the distance.	$T_1^2/T_2^2 = r_1^3/r_2^3$ $(11.86 \text{ y})^2/(1.00 \text{ y})^2 = r_1^3/(1.496 \times 10^8 \text{ km})^3$ $r_1^3 = (11.86)^2(1.496 \times 10^8 \text{ km})^3$ $r_1^3 = 4.70 \times 10^{26} \text{ km}^3$ $r_1 = 7.8 \times 10^8$ km or approximately 480 million miles

PRACTICE PROBLEMS

PROBLEM 1. An object travels at a constant speed in a circle of radius 9.0 m and completes one revolution in 3.0 s. Determine the object's a) speed and b) centripetal acceleration.

ANS. a) 19 m/s,   b) 40 m/s^2

PROBLEM 2. During a laboratory experiment involving circular motion, a 0.150 kg object whirls in a horizontal circle of radius 0.055 m. The frequency of rotation is determined to be 600 revolutions per minute. Determine the magnitude of the a) centripetal acceleration and b) net force acting on the object.

ANS. a) 220 m/s^2, b) 33 N

PROBLEM 3. A 30 kg child swings at the end of a 5.0 m long rope. The child is traveling at 3.0 m/s at the lowest point of the motion. Determine the tension in the rope at this point.

ANS. 350 N

PROBLEM 4. The Viking I spacecraft went into orbit about the planet Mars in June 1976. Assume that the spacecraft's orbit was approximately circular at 1000 km above the planet's surface. Determine the a) orbital velocity of the spacecraft and b) period of the spacecraft's motion. The mass of Mars is 6.6×10^{23} kg and the radius of Mars is 3400 km.

ANS. a) 3.2×10^3 m/s or 7100 mph, b) 8.6×10^3 s or 2.4 h

CHAPTER 6

WORK AND ENERGY

KEY TERMS AND PHRASES

work is done on an object when the energy of the object changes. The work done equals the product of the net force (F), the displacement (d), and the cosine of the angle between the force vector and the displacement vector.

joule (J) is the unit associated with work and energy. 1 J = 1 N m = 1 kg m/s^2.

kinetic energy (KE) is energy due to an object's motion. Translational kinetic energy of an object depends on the object's mass and the square of its speed.

work-energy theorem states that the net work done on an object is equal to its change in kinetic energy.

gravitational potential energy (PE) is energy stored by an object due to its position. Near the surface of the Earth, the gravitational potential energy relative to some reference point, e.g., the ground, is given by the product of the object's weight and the height above the reference level.

elastic potential energy is energy that results from an object being stretched or compressed (e.g., a spring), or twisted as in case of a wire. The change in a spring's elastic potential energy as it is stretched (or compressed) is related to the force (F) required to stretch (or compress) the spring and the distance that the spring is stretched (or compressed).

conservative force does the same amount of work independent of the path taken between two points. An example of a conservative force is gravity. The work done equals the change in potential energy ($W = \Delta PE = mg \, \Delta y$) and depends only on the initial and final positions above the ground and not on the path taken.

nonconservative force results in different amounts of work being expended in moving an object. The net work required depends on the path taken moving the object between points. For example, the work done in sliding a box of books against friction from one end of a room to the other depends on the path taken.

mechanical energy refers to an object's kinetic energy and potential energy.

law of conservation of energy states that energy is neither created nor destroyed. Energy can be transformed from one kind to another, but the total amount remains constant.

Power is the rate at which work is done and is measured in watts where 1 watt = 1 J/s.

SUMMARY OF MATHEMATICAL FORMULAS

work	$W = F\,d\,\cos\theta$	The work done depends on the net force, displacement, and the angle between the force vector and the displacement vector.
kinetic energy	$KE = \tfrac{1}{2}\,m\,v^2$	Kinetic energy is energy of motion. Kinetic energy depends on an object's mass and the square of its speed.
work-energy theorem	$W = \tfrac{1}{2}\,m\,v_f^2 - \tfrac{1}{2}\,m\,v_i^2$	The work-energy theorem states that the net work done on an object is equal to its change in kinetic energy.
gravitational potential energy	$PE = m\,g\,y$	For an object near the surface of the Earth, the gravitational potential energy depends on its weight and its height above a reference level.
spring or elastic potential energy	$PE = \tfrac{1}{2}\,k\,x^2$	Elastic potential energy depends on the force constant of the spring (k) and the displacement from equilibrium position (x).
power	$P = W/t$ or $P = F\,\bar{v}\,\cos\theta$	Power is the rate at which work is done, or for an object traveling at constant speed, the power expended is related to the net force and the average speed.

CONCEPT SUMMARY

Work and Energy

The **work** (W) done by a force (F) that is constant in both magnitude and direction is given by the following equation:

$W = F\,d\,\cos\theta$

where d is the displacement of the object and θ is the angle between the force vector and the displacement vector.

Work is a scalar quantity. Since work is the product of a force and displacement, the unit of work is newton meter. The newton meter has been named the **joule** (J), and 1 J = 1 N m.

If the force acting on an object varies in magnitude and/or direction during the object's displacement, graphical analysis can be used to determine the work done. F cos θ is plotted on the y axis and the distance through which the object moves is plotted on the x axis. The work done is represented by the area under the curve. The sum of the complete and partial blocks under the curve is determined and the work done equals the product of the work represented by one block and the total number of blocks.

A moving object is said to have **kinetic energy** (KE). The translational kinetic energy of an object depends on the object's mass and the square of its speed. The formula for kinetic energy is

$$KE = \tfrac{1}{2} m v^2$$

When a force does work on an object, the energy of the object will be changed by an amount equal to the amount of work done. The work done is positive if the energy of the object increases, for example, a car accelerating from rest. The work done is negative if the energy of the object decreases; for example, the kinetic energy of a sliding object on a rough surface is gradually lost as the object comes to a halt. In this case the kinetic energy is dissipated into the form of heat energy. The **work-energy theorem** states that the net work done on an object is equal to its change in kinetic energy, i.e., $W = \Delta KE = \tfrac{1}{2} m v_2^2 - \tfrac{1}{2} m v_1^2$.

Gravitational potential energy (PE) is energy stored by an object due to its position. The formula for the gravitational potential energy of an object relative to some reference point, e.g., the ground, is given by

$$PE = m g y \quad \text{where } y = \text{the difference in height and the reference level}$$

If an object of mass m is raised from an initial height y_1 and raised to a height y_2 above its original position, the change in gravitational potential energy is given by

$$\Delta PE = m g y_2 - m g y_1$$

Elastic potential energy is energy that results from an object being stretched or compressed (e.g., a spring), or twisted as in case of a wire. The force (F) required to stretch or compress a spring is given by $F = k x$ where k is the force constant measured in newtons/meter (N/m) and x is the stretch or compression. The change in a spring's elastic potential energy as it is stretched or compressed is given by

$$\Delta PE = \tfrac{1}{2} k x_2^2 - \tfrac{1}{2} k x_1^2 \quad \text{where} \quad x_1 = \text{the initial displacement from the equilibrium}$$
$$\text{and } x_2 = \text{the final displacement from equilibrium}$$

Conservative and Nonconservative Forces

The work done by a **conservative force** depends only on the initial and final position of the object acted upon. An example of a conservative force is gravity. The work done equals the change in potential energy ($W = \Delta PE = mg\ \Delta y$) and depends only on the initial and final positions above the ground and NOT on the path taken.

Friction is a **nonconservative force** and the work done in moving an object against a nonconservative force depends on the path. For example, the work done in sliding a box of books against friction from one end of a room to the other depends on the path taken.

Law of Conservation of Energy

The **law of conservation of energy** states that energy is neither created nor destroyed. Energy can be transformed from one kind to another, but the total amount remains constant. For mechanical systems involving conservative forces, the total **mechanical energy** equals the sum of the kinetic and potential energies of the objects that make up the system.

$E = KE + PE$

An example of the law of conservation of energy involving a mechanical system is a ball thrown vertically upward. Neglecting air resistance, kinetic energy is gradually transformed into potential energy and then back into kinetic energy. At each point of the motion the sum of the kinetic energy and potential energy remains constant.

Power

Power is defined as the rate at which work is done. The average power can be determined by applying the following formula:

power = work/time = energy transformed/time.

$P = W/t$

The unit of power is the **watt**, where 1 watt = 1 joule/sec. The watt is related to the English unit of horsepower: 746 watts = 1 horsepower. The power output of a constant force (F) applied to an object is given by

$P = F\ \overline{v}$ where $\overline{v} = d/t$ is the average speed of the object during the time interval being considered

SELECTED TEXT QUESTIONS WITH ANSWERS

QUESTION 2. Can a centripetal force ever do work? Explain.

ANSWER: For an object in traveling in CIRCULAR motion, the centripetal force NEVER does work. The centripetal force is directed toward the center of the circle while the displacement vector is tangent to the circle. Therefore, the centripetal force vector is always perpendicular to the displacement vector ($\theta = 90°$). However, $\cos 90° = 0$; therefore, $W = F\ d \cos 90° = 0$.

An alternative answer can be given from the work-energy theorem. The centripetal force changes the direction of motion of an object traveling in a circle without changing the object's speed. If the object's speed does not change, then its kinetic energy does not change and no work was done on the object. Note: if the motion is elliptical instead of circular, then θ is not equal to 90° and work is done on the object.

QUESTION 8. Can kinetic energy ever be negative? Explain.

ANSWER: The CHANGE in kinetic energy can be negative but an object's kinetic energy can NEVER be negative. Kinetic energy is a scalar quantity and has no direction assigned to the quantity.

If an moving object is slowing down, its kinetic energy is decreasing and the change in kinetic energy is negative. The minimum value for the kinetic energy is zero (object at rest).

QUESTION 12. Water balloons are tossed from the roof of a building, all with the same speed but with different launch angles. Which one has the highest speed on impact?

ANSWER: When the water balloon is initially released it has both potential and kinetic energy. Assuming that air resistance is negligible and that the surface of the ground below is the zero of potential energy, the balloon's energy is all kinetic energy when it is about to strike the ground.

$$PE_{building} + KE_{building} = KE_{ground}$$

Each balloon's kinetic energy as it strikes the ground depends only on the total initial energy and not the angle at which it was thrown. The balloon's speed is a scalar quantity and at the ground (v_{ground}) is related to the kinetic energy KE_{ground} by the equation $KE = \frac{1}{2} m v^2$. Therefore, the balloon's speed as it is about to strike the ground is independent of the launch angle.

If air resistance is NOT negligible, then the speed does depend on the launch angle. A balloon thrown upward at some angle to the horizontal will encounter more air resistance because it travels a greater distance through the air than a balloon thrown vertically downward. Because of this, if air resistance is a factor, a balloon thrown vertically downward will have the greatest speed.

QUESTION 16. What happens to the gravitational potential energy when water at the top of a waterfall falls to the pool below?

ANSWER: If air resistance is negligible, then the gravitational potential energy and kinetic energy of the water at the top of the waterfall is converted into kinetic energy at the pool below. When the falling water strikes the pool below then it is converted into thermal energy. Therefore, the water in the pool below should be at a higher temperature than the water at the top of the waterfall.

The British physicist, James Joule, measured the temperature of the water at both the top and bottom of a high waterfall and found that the temperature at both locations was about the same. He theorized that evaporative cooling of the falling water as it mixed with the air led to essentially no change in temperature.

PROBLEM SOLVING SKILLS

For problems involving the work-energy theorem:

1. Draw an accurate diagram locating all of the forces, both conservative and nonconservative, acting on the object.
2. Determine the magnitude and direction of the net force acting on the object and then determine the net work done on the object.
3. Apply the work energy theorem and solve the problem.

For problems involving the law of conservation of energy:

1. Draw an accurate diagram locating all of the forces, both conservative and nonconservative, acting on the object.
2. Apply the law of conservation of energy and solve the problem.

For problems involving power:

1. Draw an accurate diagram locating all of the forces, both conservative and nonconservative, acting on the object.
2. Apply the formulas for power and solve the problem.

PROGRAMMED PROBLEMS

a. 1 Draw an accurate diagram locating all of the forces acting on the box.	Solution: (Section 6-3) 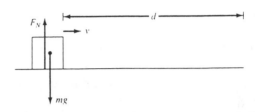
a. 2 Calculate the work done. Note: the angle between the force and the displacement vector is 0°.	$W = F\,d\cos\theta = (10.0\text{ N})(3.0\text{ m})\cos 0° = 30$ J Note: the weight of the box is balanced by the normal force exerted by the floor.
a. 3 Use the work-energy theorem to solve for the final velocity.	$W = \Delta KE = \frac{1}{2}\,m\,v'^2 - \frac{1}{2}\,m\,v^2$ $30\text{ J} = \frac{1}{2}(5.0\text{ kg})\,v'^2 - \frac{1}{2}(5.0\text{ kg})(0\text{ m/s})^2$ $v'^2 = 2(30\text{ J})/(5.0\text{ kg}) = 12\text{ m}^2/\text{s}^2$ $v' = 3.5$ m/s

a. 1 Draw an accurate diagram locating the forces on the object.	Solution: (Sections 6-1 and 6-3) 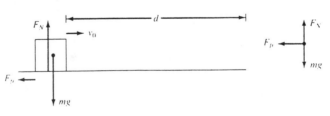

a. 2 Determine the net force acting on the object.	Based on the diagram, we can see that the object's weight is balanced out by the normal force exerted by the ice. The work done on the object is due to the force of friction acting on the object, where $F_{fr} = \mu_k F_N$ and $F_N = mg$. $F_{fr} = (0.10)(1.0 \text{ kg})(9.8 \text{ m/s}^2) = 0.98 \text{ N}$
a. 3 Apply the work-energy theorem and solve for the distance.	$W = \Delta KE$ $F_{fr} \, d \cos \theta = \frac{1}{2} m \, v'^2 - \frac{1}{2} m \, v^2$ $(0.98 \text{ N}) \, d \, (\cos 180°) = \frac{1}{2}(1.0 \text{ kg})(0 \text{ m/s})^2 - \frac{1}{2}(1.0 \text{ kg})(5.0 \text{ m/s})^2$ $- (0.98 \text{ N}) \, d = 0 \text{ J} - 12.5 \text{ J}$ $d = 13 \text{ m}$ Note: the value of θ is 180° because the frictional force is directed opposite from the direction of motion.

PROBLEM 3. a) Determine the kinetic energy in joules of a 2000 kg car that is traveling at 50.0 miles per hour. b) The same car is lifted vertically upward and then dropped from rest. Determine the height from which it is released if its velocity just before it strikes the ground is 50.0 miles per hour.

a. 1 Convert the car's speed from miles per hour to to meters per second.	Solution: (Sections 6-3 and 6-7) $v = (50.0 \text{ mi/1 h})(1609 \text{ m/1 mi})(1 \text{ h/3600 s})$ $v = 22.3 \text{ m/s}$
a. 2 Determine the car's kinetic energy in joules.	$KE = \frac{1}{2} m \, v^2 = \frac{1}{2} (2000 \text{ kg})(22.3 \text{ m/s})^2$ $KE = 4.97 \times 10^5 \text{ J}$
b. 1 Use the law of conservation of energy to solve for the car's initial height above the ground.	The potential energy at the point where the car is released (point 1) is completely converted into kinetic energy just before it strikes the ground (point 2). $KE_1 + PE_1 = KE_2 + PE_2$ $0 \text{ J} + m \, g \, h_1 = \frac{1}{2} m \, v_2^2 + 0 \text{ J}$ The mass of the car cancels algebraically; simplifying gives $h_1 = v_2^2/2g = (22.3 \text{ m/s})^2/(2)(9.80 \text{ m/s}^2)$ $h_1 = 25.3 \text{ m}$ or 83 feet

PROBLEM 4. A 10.0 kg box is pushed from the top of an incline and travels down the incline with an initial speed of 5.00 m/s. The incline is 4.00 m long and the angle of the incline is 37.0°. The coefficient of friction between the box and the incline is 0.400. Determine the a) loss of potential energy and b) work done by friction. c) Use the work-energy principle to determine the velocity of the box at the bottom of the incline.

a. 1 Complete a data table.	Solution: (Sections 6-1, 6-4, and 6-9) h_1 = (4.00 m)sin 37° = 2.40 h_2 = 0 m v_1 = 5.00 m/s v_2 = ? F_{fr} = μ mg cos θ = (0.400)(10.0 kg)(9.8 m/s²)cos 37.0° = 31.4 N
a. 2 Determine the loss of potential energy.	ΔPE = mgh_2 - mgh_1 = $mg(h_2 - h_1)$ = (10.0 kg)(9.8 m/s²)(0 m - 2.40 m) ΔPE = - 235 J
b. 1 Determine the work done by friction.	The frictional force and the direction of motion are in opposite directions; therefore, θ = 180°. W = F d cos θ = (31.4 N)(4.00 m)(cos 180°) = - 125 J The mechanical energy lost by the system equals the work done by friction.
c. 1 Determine the total mechanical energy at the top of the incline.	The total energy at the top of the slide is the sum of the PE and KE at that point. total energy = (10.0 kg)(9.8 m/s²)(2.40 m) + ½(10 kg)(5.0 m/s)² total energy = 235 J + 125 J = 360 J
c. 2 Use the law of conservation of energy to solve for the child's speed at the bottom of the incline.	The total kinetic + potential energy at the top of the final section is greater than that at the bottom by an amount equal to the mechanical energy lost due to friction. PE_1 + KE_1 + energy lost(friction) = PE_2 + KE_2 (360 J) + (- 125 J) = (10.0 kg)(9.8 m/s²)(0 m) + ½ (10 kg) v_2^2

6-9

$v_2{}^2 = 45.0 \text{ m}^2/\text{s}^2$

$v_2 = 6.7 \text{ m/s}$

PROBLEM 5. As shown in the diagram, a 2.0 kg wooden block is on a level board and held against a spring of force constant 100 N/m which has been compressed 0.10 m. The block is released and propelled horizontally across the board. The coefficient of friction between the block and the board is 0.20. Determine the a) velocity of the block just as it leaves the spring and b) distance that the block travels after it leaves the spring. Assume that friction between the block and board is negligible until the point where the block leaves the spring.

a. 1	Solution: (Sections 6-7 and 6-8)
Use the law of conservation of energy to determine the block's speed.	$\Delta PE_{spring} = \Delta KE_{block}$
	$\frac{1}{2} k x^2 = \frac{1}{2} m v^2$ and rearranging gives
	$v = (k/m)^{\frac{1}{2}} x = [(100 \text{ N/m})/(2.0 \text{ kg})]^{\frac{1}{2}} (0.10 \text{ m})$
	$v = 0.71 \text{ m/s}$

b. 1	work = change in kinetic energy
Use the work-energy theorem to determine the distance the block slides.	$F\, d \cos \theta = \frac{1}{2} m v'^2 - \frac{1}{2} m v^2$
	The force acting on the block is due to friction.
	$F = F_{fr} = \mu_k F_N$, where $F_N = mg$.
	The frictional force is directed opposite from the direction of motion; therefore, $\cos \theta = \cos 180° = -1$.
	The block comes to rest; therefore, the final velocity of the block is zero ($v' = 0$).
	$\mu_k mg\, d \cos 180° = \frac{1}{2} m v'^2 - \frac{1}{2} m v^2$
	The mass cancels; therefore,
	$(0.20)(9.8 \text{ m/s}^2)\, d\, (-1) = 0 - \frac{1}{2} (0.71 \text{ m/s})^2$
	$d = 0.13 \text{ m}$

a. 1	Solution: (Section 6-10)
Calculate the energy content in 1.0 lb of fat. Note: 1.0 lb = 454 g.	energy content = (9.0 kcal/1.0 gram) x (454 grams/1.0 lb) energy content = 4090 kcal
a. 2 Solve using the equation power = work//time	power = work/time where work = energy content of the fat Rearranging the equation gives time = work/power = energy content/power time = (4090 kcal)/(600 kcal/h) = 6.8 hours

a. 1	Solution: (Section 6-10)
Draw an accurate diagram locating all of the forces acting on the box.	
a. 2 Complete a data table.	m = 100 kg μ_k = 0.20 v = 2.0 m/s t = 10 s
a. 3 Apply Newton's second law of motion and determine the force (F) exerted by the man.	net F = m a but a = 0 (speed is constant) F - F_{fr} = m (0 m/s^2) but F_{fr} = μ_k F_N F - μ_k F_N = 0 and F_N = mg F - μ_k m g = 0 F - (0.20)(100 kg)(9.8 m/s^2) = 0 F = 200 N

a. 4

Determine the average power output of the man.

$P = W/t = F\ d/t \cos 0°$

The speed is constant; therefore, $d/t = v$

$P = F\ v \cos 0° = (200\ \text{N})(2.0\ \text{m/s})$

$P = 400\ \text{W}$

PRACTICE PROBLEMS

PROBLEM 1. A man exerts a horizontal force of 10.0 N and pushes a 5.00 kg box a distance of 3.00 m across a level floor. The coefficient of friction between the box and the floor is 0.100. Calculate the a) magnitude of the net work done on the box, and b) velocity of the box at the 3.00 m mark.

ANS. a) 15.3 J,   b) 2.47 m/s

PROBLEM 2. A 1.0 kg object is thrown vertically upward with an initial velocity of 19.6 m/s from the top of a building 29.4 m high. On the way down, the object just misses the edge of the building and strikes the ground. Use the law of conservation of energy to determine the a) maximum height above the ground reached by the object and b) object's speed just before it strikes the ground.

ANS. a) 49 m, b) 31 m/s

PROBLEM 3. As shown in the diagram below, a 10.0 kg mass slides from rest down a frictionless inclined plane which makes an angle of 37° with the horizontal. After traveling 5.0 meters it makes contact with a spring that is arranged along the incline. The force constant of the spring is 100 N/m. Determine the distance that the spring is compressed before the mass comes momentarily to a halt.

ANS. 3.1 m

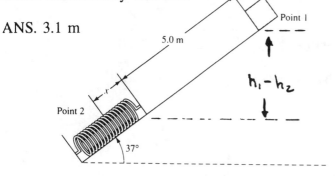

PROBLEM 4. Determine the average power in watts and horsepower employed by an engine which lifts a 100 kg object vertically upward to a height of 10.0 m in 15.0 s.

ANS. 650 W = 0.88 hp

CHAPTER 7

LINEAR MOMENTUM

KEY TERMS AND PHRASES

Linear momentum (p) is defined as the product of an object's mass (m) and its velocity (**v**).

law of conservation of momentum states that in any collision between two or more objects, the vector sum of the momenta before impact equals the vector sum of the momenta after impact. The total momentum remains constant, i.e., the momentum is conserved.

impulse is defined as the product of the force (**F**) acting on an object and the time (Δt) during which the force acts.

perfectly elastic collisions occur when the kinetic energy as well as the linear momentum is conserved.

completely inelastic collisions occur when objects stick together after impact. Most of the kinetic energy is converted into heat. This type of problem can be solved by applying the law of conservation of momentum.

center of gravity (cg) of an object is the point where the entire weight of the object can be considered to be concentrated. At the center of gravity the entire weight of the object can be balanced by a single vertical force equal to the object's weight.

center of mass (CM) is the point where all of the mass of an object can be considered to be concentrated. For almost all objects, the center of mass is at the same point in the object as the center of gravity.

translational motion is where, at any instant, the motion of all points of a moving object have the same velocity and direction of motion. This type of motion, as well as the concept of center of mass, has been implied in solving problems in previous chapters.

SUMMARY OF MATHEMATICAL FORMULAS

momentum	$\mathbf{p} = m\,\mathbf{v}$	Linear momentum ($\mathbf{p}$) is defined as the product of an object's mass (m) and its velocity ($\mathbf{v}$).
Newton's second law	$\bar{\mathbf{F}} = \Delta p/\Delta t$	The average force on an object equals the rate of change of the object's momentum.
conservation of momentum	$m_1\,\mathbf{v}_1 + m_2\,\mathbf{v}_2 = m_1\,\mathbf{v}_1' + m_2\,\mathbf{v}_2'$	In a collision between two or more objects the total momentum is conserved.
impulse	$\mathbf{F}\,\Delta t$	Impulse equals the product of the force and the time during which the force acts.
impulse and momentum	$\mathbf{F}\,\Delta t = \Delta \mathbf{p} = m\,\mathbf{v}' - m\,\mathbf{v}$	An impulse equals the change in an object's momentum.
coordinate positions of the center of mass	$x_{CM} = (\Sigma m_i\,x_i)/(\Sigma m_i)$ $y_{CM} = (\Sigma m_i\,y_i)/(\Sigma m_i)$	These equations are used to determine the horizontal and/or vertical position of the center of mass.
translational velocity of the center of mass	$M\,\mathbf{v}_{cm} = m_1\,\mathbf{v}_1 + m_2\,\mathbf{v}_2 + m_3\,\mathbf{v}_3$	This equation is used to determine the translational velocity of the center of mass.

CONCEPT SUMMARY

Linear Momentum and Force

Linear momentum ($\mathbf{p}$) is defined as the product of an object's mass (m) and its velocity ($\mathbf{v}$). Momentum is a vector quantity and the direction of the momentum vector is the same as the velocity vector. The formula for linear momentum is

$\mathbf{p} = m\,\mathbf{v}$ The unit of linear momentum is kg m/s. This is NOT the same as the unit of force, which is $1\ N = 1\ kg\ m/s^2$.

Newton's second law may be written in terms of the average force required to change an object's momentum as follows:

$\bar{\mathbf{F}} = \Delta p/\Delta t$ where $\Delta \mathbf{p}$ is the change in momentum and Δt is the time interval during which the change in momentum occurs

Conservation of Linear Momentum

In any collision between two or more objects, the vector sum of the momenta before impact equals the vector sum of the momenta after impact. The total momentum remains constant, i.e., the momentum is conserved. This principle is known as the **law of conservation of momentum**. For a collision between two objects, the law of conservation of momentum can be written as follows:

$$m_1 \, v_1 + m_2 \, v_2 = m_1 \, v_1' + m_2 \, v_2'$$

where v_1 and v_2 are the velocities of the objects before impact and v_1' and v_2' are the velocities of the objects after impact.

Impulse

An **impulse** is defined as the product of the force (F) acting on an object and the time (Δt) during which the force acts.

impulse $= F \, \Delta t$

The unit of impulse is newton seconds (N s) where 1 N s = 1 kg m/s. Impulse is a vector quantity and the direction of the impulse vector is the same as that of the force vector.

An impulse causes a change in an object's momentum. The mathematical formula that connects impulse and change in momentum is as follows:

$$F \, \Delta t = \Delta p = m \, v' - m \, v$$

In a collision between two or more objects the force each object exerts on the other is usually very large compared to any other force acting and the time interval during which the interaction occurs is usually very short. Usually, both the magnitude of the force and the time interval remain unknown; however, the impulse can be determined if it is possible to determine the change in momentum.

Energy and Momentum Conservation in Collisions

In collisions between two (or more) objects, the total momentum is always conserved. However, some of the kinetic energy may be converted into other forms of energy, usually thermal energy. Problems involving two types of collisions can be solved by using the methods of this chapter: perfectly elastic collisions and completely inelastic collisions.

In **perfectly elastic collisions,** the kinetic energy as well as the linear momentum is conserved. The sum of the kinetic energies of the objects before the collision is equal to the sum of the kinetic energies of the objects after the collision. No heat energy results from the collision. Certain perfectly elastic collision problems can be solved by applying both the law of conservation of energy and the law of conservation of momentum.

In **completely inelastic collisions**, the objects stick together after impact. Most of the kinetic energy is converted into heat. This type of problem can be solved by applying the law of conservation of momentum.

Center of Gravity and Center of Mass

The **center of gravity** (cg) of an object is the point where the entire weight of the object can be considered to be concentrated. At that point the entire weight of the object can be balanced by a single vertical force equal to the object's weight. For example, you can test for the center of gravity of a textbook by attempting to balance the book with one finger. The center of gravity of a textbook should be very close to the center of the book. For a baseball bat, the balance point and therefore the center of gravity is displaced from the center toward the fat end of the bat.

The **center of mass,** which is abbreviated CM, is the point where all of the mass of an object can be considered to be concentrated. For almost all objects, the center of mass is at the same point in the object as the center of gravity. An extreme example to show the difference is a thin, uniform rod that extends from the surface of the Earth vertically upward. The center of mass of the rod would be at its center. However, since the value of g decreases with altitude and the rod is extremely long, the lower half of the rod would weigh more than the upper half. In this case the center of gravity would fall below the center of mass.

The following formula can be used to determine the distance from one end to the center of mass of a massless rod along which a number of point masses have been attached.

$$x_{CM} = (m_1 \, x_1 + m_2 \, x_2 + ... + m_n \, x_n)/(m_1 + m_2 + ... + m_n)$$

Center of Mass and Translational Motion

The term **translational motion** is used in situations where the motion of all points of a moving object have at any instant the same velocity and direction of motion. This type of motion, as well as the concept of center of mass, has been implied in solving problems in previous chapters. For example, the free diagrams used in chapter 3 assumed that the mass of an object could be considered to be concentrated at a point with its weight acting at that point.

The concept of center of mass is also useful when discussing the motion of a group of particles. For example, as stated in the text, "The total linear momentum of a system of particles is equal to the product of the total mass M and the velocity of the center of mass of the system. Or, the linear momentum of an extended body is the product of the body's mass and the velocity of its CM". Therefore

$$M \, v_{cm} = m_1 \, v_1 + m_2 \, v_2 + m_3 \, v_3 + ... + m_n \, v_n$$

Also, Newton's second law can be written

$$F_{net} = M \, a_{CM}$$

And as stated in the text, "The sum of all of the forces acting on the system is equal to the total mass of the system times the acceleration of the center of mass".

SELECTED TEXT QUESTIONS WITH ANSWERS

QUESTION 4. It is said that in ancient times a rich man with a bag of gold coins was frozen to death stranded on the surface of a frozen lake. Because the ice was frictionless, he could not push himself to shore. What could he have done to save himself had he not been so miserly?

ANSWER: Assuming that the man is at rest on the ice, the initial momentum of the system is zero. If he should throw the bag of coins, he would recoil in the opposite direction. His mass is greater than that of the coins and therefore his recoil speed would be much smaller than the speed of the bag of coins. The final momentum of the coins is equal but opposite the man's momentum and the total final momentum equals zero. Thus, the law of conservation of momentum is upheld. Since the ice is frictionless, his recoil velocity would remain constant until he reached the bank. At that point he could walk around to the opposite bank and retrieve the bag of coins.

An alternate possibility would be to keep the coins in his pocket and throw his shoe or boot. The result would be the same and he would save himself and his gold.

QUESTION 11. A light body and a heavy body have the same kinetic energy. Which has the greater momentum?

ANSWER: Assume that the light body has a mass of 1.0 kg and the mass of the heavy body is 4.0 kg. If the speed of the light body is 4.0 m/s then its KE = ½ (1.0 kg)(4.0 m/s)^2 = 8.0 J. The speed of the heavy body can now be determined because it has an equal amount of KE:

8.0 J = ½ (4.0 kg) v^2 and v = 2.0 m/s.

Momentum is the product of the mass and the velocity. The momentum of the light body is (1.0 kg)(4.0 m/s) = 4.0 kg m/s and the momentum of the heavy body is (4.0 kg)(2.0 m/s) = 8.0 kg m/s. Thus, while both objects have the same KE, the heavy object has twice as much momentum.

QUESTION 12. Is it possible for an object to have momentum without having kinetic energy? Can it have kinetic energy but no momentum? Explain.

ANSWER: If an object has momentum, it must be moving. If it is moving, it must have kinetic energy. Thus, an object cannot have momentum without having kinetic energy, and also an object cannot have kinetic energy without having momentum.

However, it is possible for an object to have energy without momentum. For example, an object held motionless two meters above the ground has gravitational potential energy but no momentum.

PROBLEM SOLVING SKILLS

For problems involving impulse-change of momentum:

1. Draw an accurate diagram locating all of the forces acting on the system.
2. If a net external force acts on the object(s), then the momentum of the system will change. Determine the magnitude and direction of the net force.
3. Apply the impulse-momentum equation taking note that force and velocity are vectors and that direction of the vector plays an important part in the solution.

For problems involving no external force acting on the system:

1. Use the law of conservation of momentum to solve the problem. Take note that momentum is a vector quantity and must be considered in the solution.

For problems involving graphical integration:

1. Determine the sum of the partial and complete blocks that lie under the curve.
2. Determine the magnitude of the impulse represented by one block.
3. Determine the magnitude of the total impulse by multiplying the impulse represented by one block by the total number of blocks found in step 2.
4. Use the impulse-momentum equation and solve the problem.

For problems involving perfectly elastic and completely inelastic collisions:

1. Determine which type of collision is described in the problem.
2. If the collision is completely inelastic and the objects stick together, use the law of conservation of momentum to solve the problem.
3. If the collision is perfectly elastic use, both conservation of momentum and conservation of mechanical energy. Each law produces an algebraic equation with two unknowns. The final velocity of each object can be determined by applying standard algebraic techniques.

PROGRAMMED PROBLEMS

> **PROBLEM 1.** A 0.200 kg baseball is traveling at 40.0 m/s. After being hit by a bat, the ball's velocity is 50.0 m/s in the opposite direction. Compute the a) change in the ball's momentum and b) average force exerted by the bat if the ball and bat were in contact for 0.00200 s. Hint: assume that the ball's initial direction of motion is the positive direction.

a. 1 Determine the change in the baseball's momentum.	Solution: (Section 7-3) Velocity is a vector quantity. Therefore, if the initial velocity is +40.0 m/s, the final velocity is a -50.0 m/s. $\Delta \mathbf{p} = m\,\mathbf{v}' - m\,\mathbf{v}$ $\quad = (0.200 \text{ kg})(-50.0 \text{ m/s}) - (0.200 \text{ kg})(+40.0 \text{ m/s})$ $\Delta \mathbf{p} = -18.0$ kg m/s
b. 1 Use the impulse-momentum equation to determine the net force acting on the ball.	$\mathbf{F}\,\Delta t = m\,\mathbf{v}' - m\,\mathbf{v}$ $\mathbf{F}\,(0.00200 \text{ s}) = -18.0$ kg m/s $\mathbf{F} = -9000$ N Note: the negative sign indicates that the direction of the force is in the direction arbitrarily defined as the negative direction, i.e., in the same direction that the ball is moving after impact.

> **PROBLEM 2.** A 1000 kg car traveling at 22.0 m/s (about 50 mph) strikes a concrete bridge support and comes to a complete halt in 0.500 seconds. a) Determine the magnitude of the average force acting on the car. b) Suppose the bridge support had been surrounded by a barrier that contains a material which is gradually crushed during impact so that the stopping time was increased to 3.00 s. What would have been the magnitude of the average force?

a. 1 Apply the impulse-momentum equation and solve for the average force.	Solution: (Section 7-3) $\overline{\mathbf{F}}\,\Delta t = \Delta \mathbf{p} = m\,\mathbf{v}' - m\,\mathbf{v}$ $\overline{\mathbf{F}}(0.500 \text{ s}) = (1000 \text{ kg})(0 \text{ m/s}) - (1000 \text{ kg})(22.0 \text{ m/s})$ $\overline{\mathbf{F}} = -44{,}000$ N The negative sign indicates that the force opposes the car's motion.
b. 1 Apply the impulse-momentum equation and solve for the average force.	The stopping time (Δt) is increased by a factor of $(3.00 \text{ s})/(0.500 \text{ s}) = 6.00$. The average force is reduced by a factor of 6.00 to -7300 N. This could easily be the difference between life and death during this type of accident.

PROBLEM 3. In the ballistic pendulum experiment, a bullet of mass 0.0600 kg is fired horizontally into a wooden block of mass 0.200 kg. The wooden block is suspended from the ceiling by a long string as shown in the diagram. The collision is completely inelastic and after impact the bullet + pendulum move together until the center of mass of the system rises 0.120 m above its initial position. Calculate the a) velocity of the bullet + wooden block just after impact and b) the velocity of the bullet just before impact.

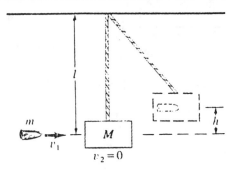

a. 1	Solution: (Section 7-6)
The collision is completely elastic. The objects stick together after impact. Use the law of conservation of energy to solve for the velocity just after impact.	$KE_{after\ impact} = PE_{at\ top\ of\ swing}$ $\frac{1}{2}(m_b + m_w)v'^2 = (m_b + m_w)\ g\ h$ the mass cancels out, and rearranging gives $v' = (2\ g\ h)^{\frac{1}{2}} = [2\ (9.8\ m/s^2)(0.120\ m)]^{\frac{1}{2}} = 1.53\ m/s$
b. 1 Use the law of conservation of momentum to determine the velocity of the bullet just before impact.	The initial momentum equals the final momentum. The bullet and wooden block stick together after impact; therefore, $v_b'= v_w'= v' = 1.53\ m/s$ $m_b\ v_b + m_w\ v_w = (m_b + m_w)\ v'$ $(0.0600\ kg)\ v_b + (0.200\ kg)(0\ m/s) =$ $\qquad\qquad (0.0600\ kg + 0.200\ kg)(1.53\ m/s)$ $v_b = (0.260\ kg)(1.53\ m/s)/0.0600\ kg)$ $v_b = 6.63\ m/s$

PROBLEM 4. A 1.0 kg object traveling at 1.0 m/s collides head-on with a 2.0 kg object initially at rest. Determine the a) velocity of the objects after the impact if the collision is completely inelastic and the objects stick together after impact and b) fraction of the original kinetic energy lost during the collision.

a. 1	Solution: (Sections 7-2 and 7-6)
The objects stick together after impact. Solve for the final velocity using the law of conservation of momentum.	$m_1 v_1 + m_2 v_2 = m_1 v_1' + m_2 v_2'$ but $v_1' = v_2' = v'$ (objects move together) $(1.0 \text{ kg})(1.0 \text{ m/s}) + (2.0 \text{ kg})(0 \text{ m/s}) = (1.0 \text{ kg} + 2.0 \text{ kg}) v'$ $v' = \frac{1}{3} \text{ m/s}$
b. 1 Determine the initial kinetic energy of each object, the total initial kinetic energy, and the total final kinetic energy.	$KE_1 = \frac{1}{2} m_1 v_1^2 = \frac{1}{2} (1.0 \text{ kg})(1.0 \text{ m/s})^2 = 0.50 \text{ J}$ $KE_2 = \frac{1}{2} m_2 v_2^2 = \frac{1}{2} (2.0 \text{ kg})(0 \text{ m/s})^2 = 0 \text{ J}$ total initial KE = 0.50 J + 0 J = 0.50 J final KE = $\frac{1}{2}(m_1 + m_2) v'^2 = \frac{1}{2} (3.0 \text{ kg})(\frac{1}{3} \text{ m/s})^2$ final KE = 1/6 J = 0.17 J
b. 2 Determine the % of the original KE that remains.	% of KE (remaining) = (total final KE)/(total initial KE) x 100 % % of KE (remaining) = (0.17 J)/(0.50 J) x 100 % = 34 %
b. 3 Determine the fraction of kinetic energy converted into heat energy.	% of KE lost = 100 % - 34 % = 66 % The lost KE was converted into heat energy.

PROBLEM 5. A 2.0 kg object traveling at 1.0 m/s collides head-on with a 1.0 kg object initially at rest. Determine the velocity of each object after the impact if the collision is perfectly elastic.

a. 1	Solution: (Sections 7-4 and 7-5)
Write an equation for the collision using the law of conservation of momentum.	$m_1 v_1 + m_2 v_2 = m_1 v_1' + m_2 v_2'$ $(2.0 \text{ kg})(1.0 \text{ m/s}) + (1.0 \text{ kg})(0 \text{ m/s}) = (2.0 \text{ kg}) v_1' + (1.0 \text{ kg}) v_2'$ $2 \text{ m/s} = 2 v_1' + 1 v_2'$ (equation 1)

a. 2	The collision is perfectly elastic; therefore, kinetic energy is conserved.
Write an equation for the collision using the law of conservation of energy.	$\tfrac{1}{2} m_1 v_1^2 + \tfrac{1}{2} m_2 v_2^2 = \tfrac{1}{2} m_1 v'^2 + \tfrac{1}{2} m_2 v'^2$
	Since $\tfrac{1}{2}$ appears in each term, the equation may be simplified by dividing each term by $\tfrac{1}{2}$. Substituting values gives
	$(2.0 \text{ kg})(1.0 \text{ m/s})^2 + (1.0 \text{ kg})(0 \text{ m/s})^2 = (2.0 \text{ kg}) v_1'^2 + (1.0 \text{ kg}) v_2'^2$
	Upon simplifying $\quad 2 \text{ m}^2/\text{s}^2 = 2 v_1'^2 + 1 v_2'^2 \qquad$ (equation 2)

---

a. 3	from equation 1 $\qquad v_2' = 2 - 2 v_1'$
Application of the conservation laws has resulted in two algebraic equations with the same two unknowns. Solve for the final velocity of each object.	Substituting for v_2' in equation 2
	$2 = 2 v_1'^2 + (2 - 2 v_1')^2$
	$2 = 2 v_1'^2 + 4 - 8 v_1' + 4 v_2'^2$
	$0 = 6 v_1'^2 - 8 v_1' + 2 \quad$ and simplifying gives
	$0 = 3 v_1'^2 - 4 v_1' + 1 \qquad$ and factoring gives
	$0 = (3 v_1' - 1)(v_1' - 1)$
	Either $\;0 = 3 v_1' - 1 \;$ or $\; v_1' - 1 = 0$
	and either $\; v_1' = \tfrac{1}{3} \text{ m/s} \;$ or $\; v_1' = 1 \text{ m/s}$
	According to Newton's third law, after impact the velocity of object 1 must change. Therefore, the solution, $v_1' = 1$, while a valid solution, is uninteresting because it means that object 1 missed object 2, so object 2 is still at rest. The interesting answer is $v_1' = \tfrac{1}{3} \text{ m/s}$. Substitute the value of v_1' into equation 1 and determine v_2'.
	$v_2' = 2 - 2 v_1' = 2 - 2 (\tfrac{1}{3})$
	$v_2' = 4/3 \text{ m/s} = 1.3 \text{ m/s}$

PROBLEM 6. The mass of the Sun is 2.0×10^{30} kg while the mass of the Earth is 6.0×10^{24} kg. The center-to-center distance between the Earth and the Sun is 1.50×10^{11} m. Determine the distance from the center of the Sun to the center of mass of the Earth-Sun system.

a. 1

The center of mass is to be measured from the center of the Sun. Therefore, $x_{Sun} = 0$ m. Complete a data table listing the mass and position of each object.

Solution: (Section 7-8)

$m_{Sun} = 2.0 \times 10^{30}$ kg, $x_{Sun} = 0$ m

$m_{Earth} = 6.0 \times 10^{24}$ kg, $x_{Earth} = 1.50 \times 10^{11}$ m

---

a. 2

Use the formula for determining the position of the center of mass of objects arranged along a horizontal line.

$x_{CM} = (m_{Sun} x_{Sun} + m_{Earth} x_{Earth})/(m_{Sun} + m_{Earth})$

$m_{Sun} x_{Sun} = (2.0 \times 10^{30}$ kg$)(0.0$ m$) = 0$ kg m

$m_{Earth} x_{Earth} = (6.0 \times 10^{24}$ kg$)(1.50 \times 10^{11}$ m$) = 9.0 \times 10^{35}$ kg m

$m_{Sun} + m_{Earth} = 2.0 \times 10^{30}$ kg $+ 6.0 \times 10^{24}$ kg $\approx 2.0 \times 10^{30}$ kg

$x_{CM} = (0$ kg m $+ 9.0 \times 10^{35}$ kg m$)/(2.0 \times 10^{30}$ kg$)$

$x_{CM} = 4.5 \times 10^{5}$ m ≈ 280 miles

Note: The Earth-Sun system revolves around the center of mass. The text gives the radius of the Sun as 6.96×10^{8} m (approximately 433,000 miles). Therefore, the center of mass of the Earth-Sun system is very close (approximately 280 miles) from the center of the Sun. For all practical purposes, the Earth revolves about the center of the Sun.

PRACTICE PROBLEMS

PROBLEM 1. An object at rest explodes into three pieces, each of which travels parallel to the ground after the explosion. The first piece has a mass of 3.0 kg and travels at 4.0 m/s at an angle of 30° north of east. The second piece has a mass of 4.0 kg and travels at 3.0 m/s at an angle of 30° south of east. Determine the velocity, both magnitude and direction, of the third piece just after the explosion. The third piece has a mass of 5.0 kg.

ANS. 4.2 m/s due west

PROBLEM 2. An object of mass 3.0 kg and traveling at 2.0 m/s collides head-on with a 2.0 kg object traveling in the opposite direction at 3.0 m/s. Determine the velocity of each object after impact if the collision is perfectly elastic. Assume that the 3.0 kg object is traveling in the positive direction.

ANS. Final velocity of the 3.0 kg object is - 2.0 m/s, while the final velocity of the 2.0 kg object is + 3.0 m/s.

PROBLEM 3. a) Determine the velocity after impact of the objects in problem 2 if the collision in problem 2 is completely inelastic and the objects stick together after impact. b) Determine the percentage of kinetic energy lost as a result of the impact.

ANS. a) zero,   b) 100 %

PROBLEM 4. A bullet of mass 0.0020 kg and traveling at 300 m/s strikes the center of a 3.0 kg block of wood that is sitting on a fence post. The block of wood is 2.0 m above the ground. a) Determine the velocity of the bullet + block of wood just after impact if the bullet remains imbedded in the wood. b) Determine the horizontal distance from the base of the post to the point where the block of wood first hits the ground.

ANS. a) 0.20 m/s,   b) 0.13 m

CHAPTER 8

ROTATIONAL MOTION

KEY TERMS AND PHRASES

radian is a dimensionless quantity which is a unit of angular measurement. 1 radian = 57.3° while 1 revolution = 2 π radians = 360°.

angular displacement (θ) is the angle through which a point on a rotating object moves during a time interval. The angular displacement is measured in radians, degrees, or revolutions.

angular velocity (ω) is measured in radians per second. The average angular velocity is measured by dividing the angular displacement by the time required to travel through the displacement.

angular acceleration (α) is measured in radians per second per second (rad/s²). The average angular acceleration is defined as the rate of change of the angular velocity in time.

torque (τ) is the measure of the effectiveness of a force in producing rotation of an object about an axis.

moment of inertia or rotational inertia (I) is the measure of the tendency of an object to resist any change in its state of rotation.

radius of gyration is the distance from the axis of rotation to a point where all of the object's mass could be concentrated. It is convenient to determine the moment of inertia of irregularly shaped objects by using the object's radius of gyration.

rotational kinetic energy is energy due to an object's rotational motion. The rotational kinetic energy is related to the object's moment of inertia and the square of its angular velocity.

angular momentum (L) is a quantity that is found from the product of an object's moment of inertia (I) and angular velocity (ω).

law of conservation of angular momentum states that in the absence of a net torque acting on an object, the object's angular momentum must remain constant in both magnitude and direction, i.e., I ω = constant.

SUMMARY OF MATHEMATICAL FORMULAS

angular displacement	$\theta = \ell/r$ or $\ell = r\,\theta$	Angular displacement is directly proportional to the length of arc (ℓ) and inversely related to the radius (r).
angular velocity	$\omega = v/r$ or $v = r\,\omega$	Angular velocity (ω) is directly proportional to the tangential velocity (v) and inversely related to the radius (r).
angular acceleration	$\alpha = a_T/r$ or $a_T = r\,\alpha$.	Angular acceleration (α) is directly proportional to the tangential acceleration (a_T) and inversely proportional to the radius (r).
centripetal acceleration	$a_c = r\,\omega^2$	Centripetal acceleration equals the product of the radius and the square of the angular velocity.
equations for uniformly accelerated rotational motion	$\omega = \omega_o + \alpha\,t$	angular speed as related to initial angular speed, angular acceleration, and time
	$\theta = \omega_o\,t + \frac{1}{2}\,\alpha\,t^2$	angular displacement as related to initial angular velocity, angular acceleration, and time
	$\omega^2 - \omega_o^2 = 2\,\alpha\,\theta$	angular velocity as related to initial angular velocity, angular acceleration and angular displacement
	$\bar{\omega} = (\omega + \omega_o)/2$	average angular velocity as related to the initial angular velocity and the final angular velocity
	$\theta = \bar{\omega}\,t$	angular displacement as related to the average angular velocity and time
torque	$\tau = \ell_{\perp}\,F = \ell\,F\sin\theta$	torque as related to the lever arm distance and the magnitude of the applied force
torque	$\tau = I\,\alpha$	torque as related to the moment of inertia and the angular acceleration

moment of inertia	$I = M R^2$	thin ring of radius R
Note: M represents the object's mass, R represents the object's radius, and L represents the object's length.	$I = \frac{1}{2} M R^2$	solid cylinder of radius R
	$I = 2/5\ M R^2$	solid sphere of radius R
	$I = 1/12\ M L^2$	long rod of length L (axis through center and perpendicular to rod)
	$I = \frac{1}{3} M L^2$	long rod of length L (axis through end and perpendicular to rod)
moment of inertia	$I = M k^2$	moment of inertia written in terms of the object's radius of gyration (k)
rotational kinetic energy	$KE = \frac{1}{2} I \omega^2$	Rotational kinetic energy is related to an object's moment of inertia and the square of its angular velocity.
angular momentum	$L = I \omega$	Angular momentum equals the product of an object's moment of inertia and angular velocity.
law of conservation of angular momentum	$I \omega = \text{constant}$	In the absence of a net torque acting on an object, the object's angular momentum must remain constant in both magnitude and direction.

CONCEPT SUMMARY

Circular Motion in Terms of Angular Quantities

Angular displacement (θ) is measured in degrees, revolutions or radians. In this chapter, the radian will be used most often in solving problems. 1 **radian** is the angle subtended at the center of a circle by a length of arc (ℓ) equal to the radius of the circle (r), i.e, $\theta = \ell/r$. Since ℓ and r are both units of length, the ratio of ℓ to r results in a quantity which is dimensionless. Although the radian is a dimensionless quantity, it is most useful when dealing with angular measurement. A useful conversion factor that should be noted is 1 revolution = 2 π radians = 360° and 1 radian = 57.3°

Angular velocity (ω) is measured in radians per second. The average angular velocity is measured by dividing the angular displacement by the time required to travel through the

displacement, $\overline{\omega} = \theta/t$. If the time interval is small, $\Delta t \to 0$, then $\omega = \Delta\theta/\Delta t$ defines the instantaneous angular velocity.

Angular acceleration (α) is measured in radians per second per second (rad/s^2). The average angular acceleration is defined as the rate of change of the angular velocity in time. If the time interval is small, $\Delta t \rightarrow 0$ and $\alpha = \Delta \omega / \Delta t$ defines the instantaneous angular acceleration.

As an object travels in a circle, the motion of a point a distance r from the center of the circle can be described in terms of tangential quantities. Since $\theta = \ell/r$, the displacement of the object from its initial position can be found by the formula $\ell = r \theta$. The **tangential velocity** is given by the formula $v = r \omega$ while the **tangential acceleration** $a_T = r \alpha$.

The total linear acceleration of a particle traveling in a circle is $a = a_T + a_c$. The centripetal acceleration $a_c = v^2/r$, but $v = r \omega$, thus $a_c = \omega^2 r$.

The relationship between the frequency of rotation (f) and the angular velocity (ω) is given by the equation: $\omega = 2 \pi f$. If the frequency of rotation is 1.0 revolution per second, then the angular frequency $\omega = 2 \pi (1.0 \text{ rev/s}) = 6.28$ radians per second.

Kinematics Equations for Uniformly Accelerated Rotational Motion

The equations used for uniformly accelerated rotational motion are analogous to the equations used for uniformly accelerated linear motion. θ is analogous to $x - x_o$, ω to v, and α to a.

$v = v_o + at$	$\omega = \omega_o + \alpha t$
$x - x_o = v_o t + \frac{1}{2} a t^2$	$\theta = \omega_o t + \frac{1}{2} \alpha t^2$
$v^2 - v_o^2 = 2 a (x - x_o)$	$\omega^2 - \omega_o^2 = 2 \alpha \theta$
$\bar{v} = (v + v_o)/2$	$\bar{\omega} = (\omega + \omega_o)/2$
$x - x_o = \bar{v} t$	$\theta = \bar{\omega} t$

In solving problems using the above formulas for rotational motion, it is important to be consistent in the use of units. It is suggested that the student always convert from revolutions to radians. Also, time is to be expressed in seconds.

Torque

Torque (τ) is the measure of the effectiveness of a force in producing rotation of an object about an axis. It is measured by the product of the force and the perpendicular distance from the axis of rotation to the line along which the force acts. This perpendicular distance is often referred to as the **lever arm distance** ($\ell_\perp$).

$$\tau = \ell_\perp F = \ell F \sin \theta$$

The SI unit of torque is the meters newtons (m N).

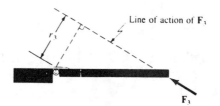

For example, from everyday experience the student realizes that the most effective manner in which to open a door is to push on the side of the door opposite from the hinge point (axis of rotation) and to apply the force perpendicular to the plane of the door. The student is applying an unbalanced torque to the door which causes it to rotate.

Rotational Dynamics; Torque and Rotational Inertia

When an unbalanced torque acts on an object, it tends to change an object's state of rotation, i.e., it produces an angular acceleration or deceleration. However, the magnitude of the angular acceleration or deceleration depends on the object's **moment of inertia** (I) as well as the magnitude of the torque (τ).

$$\tau = I \alpha \quad \text{and} \quad \alpha = \tau/I$$

Moment of Inertia or Rotational Inertia

Just as objects tend to resist any change in translational motion (straight line motion), an object tends to resist any change in its rotational motion. The tendency to resist any change in translational motion is referred to as inertia and is measured by measuring an object's mass in kg. **Moment of inertia** or **rotational inertia** is the measure of the tendency of an object to resist any change in its state of rotation.

The moment of inertia is determined by calculating the sum of the moments of inertia of the particles that make up the object. It is determined not only by the mass of the object but also by the distribution of the mass about the axis of rotation. In general, it is necessary to use integral calculus to determine an object's moment of inertia.

The object's moment of inertia is determined by applying the formula $I = \Sigma\, m\, r^2$, where I is the symbol for moment of inertia in kg m^2, Σ is the Greek letter sigma and means "the sum of," m is the mass of a particle of the object and r is the distance from the axis of rotation to a particular particle.

The following is a list of the moments of inertia of certain solids of uniform composition. Except for the last object, the axis of rotation in each case is through the center of the object. It is worthwhile to commit these formulas to memory.

Thin Ring of Radius R $\qquad\qquad\qquad I = M R^2 \qquad$ Note: M is the object's total mass and
$\qquad\qquad\qquad\qquad\qquad\qquad\qquad\qquad\qquad\qquad\qquad\quad$ R is the object's radius.

Solid Cylinder of Radius R $\qquad\qquad I = \frac{1}{2} M R^2$

Solid Sphere of Radius R $I = 2/5\ M\ R^2$

Long Rod of Length L $I = 1/12\ M\ L^2$
(axis through center and
perpendicular to rod)

Long Rod of Length L $I = \frac{1}{3}\ M\ L^2$
(axis through end and
perpendicular to rod)

The moment of inertia of an object may also be written in terms of the object's **radius of gyration** (k) as $I = M\ k^2$. The radius of gyration is the distance from the axis of rotation to a point where all of the object's mass could be concentrated. The moment of inertia has the same magnitude as that determined by using the more general formula $I = \Sigma\ mr^2$. It is convenient to determine the moment of inertia of irregularly shaped objects by using the object's radius of gyration.

Rotational Kinetic Energy

A rotating object has the ability to do work and therefore has energy. This energy is in the form of **rotational kinetic energy** and is given by the formula

$$KE = \tfrac{1}{2}\ I\ \omega^2$$

The total kinetic energy of an object that has both translational as well as rotational kinetic energy, for example, a wheel on a moving car, can be expressed as follows:

$$KE = \tfrac{1}{2}\ M\ v_{CM}^2 + \tfrac{1}{2}\ I_{CM}\ \omega^2$$

v_{CM} is the linear velocity of the center of mass and I_{CM} is the object's moment of inertia about an axis through the object's center of mass.

Angular Momentum

Angular momentum (L) is a quantity that is found from the product of an object's moment of inertia and angular velocity.

$$L = I\ \omega$$

The units of angular momentum are $kg\ m^2/s$ and angular momentum is a vector quantity. The direction of the vector is along the axis of rotation and is found by the right-hand rule. The right-hand rule states that when the fingers of the right hand curl in the direction in which the object is rotating, the thumb of the right hand points in the direction of the angular momentum vector.

Conservation of Angular Momentum

A torque tends to change an object's angular momentum and the relationship between torque and angular momentum is described by the following equation:

$\tau = \Delta L / \Delta t$

The **law of conservation of angular momentum** states that in the absence of a net torque acting on an object, the object's angular momentum must remain constant in both magnitude and direction, i.e. $I\omega$ = constant   or $I_{initial}\ \omega_{initial} = I_{final}\ \omega_{final}$.

This principle is quite useful in explaining a number of phenomena. For example, the change in angular velocity of a spinning ice skater can be explained by using the law of conservation of angular momentum. As the skater brings her arms and legs close to her body, her moment of inertia decreases and as a result her angular velocity increases.

SELECTED TEXT QUESTIONS WITH ANSWERS

QUESTION 5. Can a small force exert a greater torque than a large force? Explain.

ANSWER: Yes, a small force can exert a greater torque than a large force. Torque equals the product of the force and the lever arm distance. A force of 5.0 N with a lever arm distance of 3.0 m produces a torque of 15 m N. A 10 N force with a lever arm distance of 1.0 m produces a torque of 10.0 m N. Therefore, if the lever arm distance is large enough, the torque exerted by the small force will be greater than that exerted by a large force with a small lever arm distance.

QUESTION 16. A sphere and a cylinder have the same radius and the same mass. They start from rest at the top of an incline. Which reaches the bottom first? Which has the greater speed at the bottom? Which has the greater total kinetic energy at the bottom? Which has the greater rotational KE?

ANSWER: Based on the law of conservation of energy, the total kinetic energy at the bottom of the incline equals the potential energy at the top. Therefore, the total KE of each object is the same at the bottom of the incline.

However, the law of conservation of energy can be used to show that each object has a different speed at the bottom.

$PE_{top} = [KE_{translational} + KE_{rotational}]_{bottom}$

$mgh = \frac{1}{2} mv^2 + \frac{1}{2} I \omega^2$

The rotational KE depends on the object's moment of inertia. The larger the moment of inertia, the smaller the object's translational velocity (v) at the bottom of the incline. The cylinder has the larger moment of inertia of the two objects ($I = \frac{1}{2} mr^2$) and it will arrive after the sphere.

For example, $I_{cylinder} = \frac{1}{2} m r^2$. Substituting this value gives

$mgh = \frac{1}{2} m v^2 + \frac{1}{2} (\frac{1}{2} m r^2) \omega^2$ but $v^2 = r^2 \omega^2$

$mgh = \frac{1}{2} m v^2 + \frac{1}{4} m v^2$

$mgh = \frac{3}{4} m v^2$ The mass cancels and solving for v gives

$v = (4/3 \ g \ h)^{\frac{1}{2}} = 1.15 \ (g \ h)^{\frac{1}{2}}$

For the sphere, $I_{sphere} = 2/5 \ m r^2$, and substituting gives

$mgh = \frac{1}{2} m v^2 + \frac{1}{2} (2/5 \ m r^2) \omega^2$ but $v^2 = r^2 \omega^2$

$mgh = \frac{1}{2} m v^2 + 1/5 \ m v^2$

$mgh = 7/10 \ m v^2$ The mass cancels and solving for v gives

$v = (10/7 \ g \ h)^{\frac{1}{2}} = 1.20 \ (g \ h)^{\frac{1}{2}}$

Because the translational speed (v) of the sphere is greater than that of the cylinder, the translational kinetic energy $(KE = \frac{1}{2} m v^2)$ of the sphere will be greater than that of the cylinder. However, since the total KE is the same for both, if the sphere arrives with the greater translational kinetic energy, it must have a smaller rotational KE as compared to the cylinder.

QUESTION 21. In what direction is the Earth's angular velocity for its daily rotation on its axis?

ANSWER: The right-hand rule is used to answer this question. The fingers of the right hand point toward the east, which is the direction of Earth's rotation. The thumb of the right hand points toward the north. Therefore, the angular velocity vector points toward the north.

PROBLEM SOLVING SKILLS

For problems involving angular kinematics:

1. Complete a data table using information both given and implied in the problem.
2. Memorize the kinematics formulas for uniformly accelerated angular motion. Using the data from the completed data table, determine which formula or combination of formulas can be used to solve the problem.

For problems involving torque:

1. Draw a diagram locating the axis of rotation of the object.
2. Determine the magnitude of the force and the lever arm distance.

3. Solve for the magnitude of the torque. Use the right-hand rule to determine the direction of rotation and also the direction of the torque vector.

For problems involving torque and rotational dynamics:

1. Determine the moment of inertia of the object(s).
2. Use $\tau = \ell_{\perp} F$ and $\tau = I \alpha$ to solve for the object's angular acceleration.
3. If necessary, use the kinematics equations to solve for the angular velocity and linear velocity of the object(s).

If the problem involves a system of objects:

1. Draw a free body diagram and locate the forces acting on each object.
2. Use Newton's second law as well as $\tau = \ell_{\perp} F$ and $\tau = I \alpha$ to write an equation for each object.
3. Use the kinematics relationships as well as standard algebraic methods to solve for the rate of acceleration of the system.
4. If necessary, use the kinematics equations to solve the problem.

For problems involving the law of conservation of angular momentum:

1. If a net external torque acts on the system, the law of conservation of angular momentum does not apply and cannot be used.
2. Convert the angular velocity from revolutions per minute to radians per second.
3. Determine the initial and final moment of inertia of the system.
4. Use the law of conservation of angular momentum to solve the problem.
5. Calculate the change in rotational kinetic energy and explain why the kinetic energy changes while the angular momentum remains constant.

PRACTICE PROBLEMS

PROBLEM 1. A wheel is rotating at 30.0 rpm. The wheel then accelerates uniformly to 50.0 rpm in 10.0 seconds. Determine the a) rate of angular acceleration in rad/s² and b) angular displacement of the wheel during the 10.0 s.

---

a. 1	Solution: (Sections 8-1 and 8-2)
Complete a data table based on information given in the problem.	Since the angular acceleration and displacement are to be expressed in radians, it is necessary to convert the initial and final angular velocities from revolutions per minute to radians per second.
	ω_o = (30.0 rev/min)(2π rad/rev)(1.00 min/60.0 s) = 3.14 rad/s
	ω = (50.0 rev/min)(2π rad/rev)(1.00 min/60.0 s) = 5.23 rad/s
	t = 10.0 s
a. 2	ω = α t + ω_o
Solve for the angular acceleration.	5.23 rad/s = α (10.0 s) + 3.14 rad/s
	α = 0.209 rad/s²
b. 1	The angular acceleration is now known and can be added to the data table. Substitute the values into the appropriate equation and solve for θ.
Determine the angular displacement θ.	
	$\theta = \omega_o t + \frac{1}{2} \alpha t^2$
	= (3.14 rad/s)(10.0 s) + ½ (0.209 rad/s²)(10.0 s)²
	θ = 41.9 rad

PROBLEM 2. A wheel 0.500 m in radius is rotating at 200 rad/s. The wheel decelerates uniformly from 200 rad/s to 50.0 rad/s in 50.0 revolutions. Determine the a) rate of angular deceleration, b) time required for the wheel to decelerate to 50.0 radians per second, c) tangential acceleration of a point on the rim as the wheel is decelerating, and d) radial acceleration of a point on the rim before the deceleration begins.

---

a. 1	Solution: (Sections 8-1 and 8-2)
Complete a data table based the information given in the problem.	Since the angular velocities are given in radians per second, it is necessary to convert the angular displacement to radians.
	θ = (50.0 rev)(2π rad/1 rev) = 314 rad ω_o = 200 rad/s
	α = ? ω = 50.0 rad/s t = ?

a. 2 Solve for the angular deceleration.	$2\,\alpha\,\theta = \omega^2 - \omega_o{}^2$ $2\,\alpha\,(314\text{ rad}) = (50.0\text{ rad/s})^2 - (200\text{ rad/s})^2$ $\alpha = -\,59.7\text{ rad/s}^2$ The negative sign indicates that the wheel is decelerating.
b. 1 Determine the time required for the wheel to decelerate to 50.0 rad/s.	The angular acceleration is now known and can be added to the data table. Substitute the values into the appropriate equation and solve for the time. $\omega = \omega_o + \alpha\,t$ $50.0\text{ rad/s} = 200\text{ rad/s} + (-\,59.7\text{ rad/s}^2)\,t$ $t = 2.51\text{ s}$
c. 1 Determine the tangential acceleration.	$a_T = r\,\alpha = (0.500\text{ m})(-\,59.7\text{ rad/s}^2)$ $a_T = -\,29.9\text{ m/s}^2$ The negative sign indicates that the wheel is decelerating.
d. 1 Determine the magnitude and direction of the centripetal acceleration a point on the rim.	The centripetal acceleration is directed radially inward toward the center of the circle and is related to the angular velocity and radius of the circle as follows: $a_c = r\,\omega_o{}^2 = (0.500\text{ m})(200\text{ rad/s})^2$ $a_c = 2.00 \times 10^4\text{ m/s}^2$

PROBLEM 3. A 3.0 m long rod of mass 2.0 kg is acted upon by three forces each of magnitude 10.0 newtons as shown in the diagram. a) Determine the torque produced about the left end of the rod if each force acts individually. b) Determine the angular acceleration which results from the torque produced by F_2.

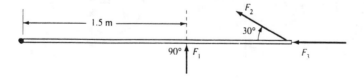

a. 1 Determine the lever arm distance of each force. Note: the axis of rotation is at the left end of the rod.	Solution: (Sections 8-4 and 8-5) $\ell_{\perp 1} = \ell_1 \sin 90° = (1.5\text{ m})(\sin 90°) = 1.5\text{ m}$ $\ell_{\perp 2} = \ell_2 \sin 30° = (3.0\text{ m})(\sin 30°) = 1.5\text{ m}$ $\ell_{\perp 3} = \ell_3 \sin 0° = (3.0\text{ m})(\sin 0°) = 0\text{ m}$

8-11

a. 2 Determine the torque produced by each force, $\tau = \ell_\perp F$.	$\tau_1 = (1.5 \text{ m})(10.0 \text{ N}) = 15 \text{ m N}$ $\tau_2 = (1.5 \text{ m})(10.0 \text{ N}) = 15 \text{ m N}$ $\tau_3 = (0 \text{ m})(10.0 \text{ N}) = 0 \text{ m N}$ F_3 does not cause the rod to rotate; thus it does not produce a torque.
b. 1 Determine the moment of inertia of the rod.	The moment of inertia of a rod hinged at one end is $I = \frac{1}{3} M L^2$. $I = \frac{1}{3} (2.0 \text{ kg})(3.0 \text{ m})^2 = 6.0 \text{ kg m}^2$
b. 2 Determine the angular acceleration produced by F_2.	$\tau = I \alpha$ $15 \text{ m N} = (6.0 \text{ kg m}^2) \alpha$ $\alpha = 2.5 \text{ rad/s}^2$ Note: the radian is a dimensionless quantity which does not appear in either the units of torque or moment of inertia. It is necessary to remember to include radians in your answer.

PROBLEM 4. One end of a string is attached to a 1.0 kg object while the other end is wrapped around a solid cylinder. The 1.0 kg object is released from rest and accelerates downward while the solid cylinder rotates about a point located at its center. The mass of the solid cylinder is 2.0 kg and its radius is 0.030 m. Use rotational dynamics to determine the a) rate of acceleration of the object and the tension in the string. Determine the object's speed when it is 0.50 m below its initial position using the b) kinematics equations and c) law of conservation of energy.

a. 1 Draw a free body diagram and locate the forces acting on each object.	Solution: (Sections 8-5, 8-6, and 8-7) 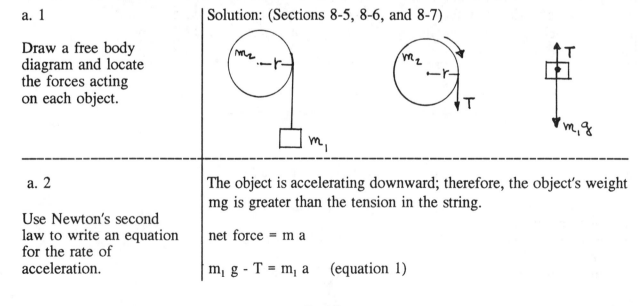
a. 2 Use Newton's second law to write an equation for the rate of acceleration.	The object is accelerating downward; therefore, the object's weight mg is greater than the tension in the string. net force = m a $m_1 g - T = m_1 a$ (equation 1)

a. 3	$\tau = r\, F \sin 90°$ but $F = T$ where T = tension in the string
Torque produced by tension in the string causes the cylinder to rotate. Use the torque equations and rotational dynamics to derive an equation for the tension in the string in terms of the cylinder's rate of acceleration.	But $\tau = I\,\alpha$ where $I = \frac{1}{2} m_2\, r^2$ Therefore; $r\,T \sin 90° = \frac{1}{2} m_2\, r^2\, \alpha$ and $T = \frac{1}{2} m_2\, r\, \alpha$ but $r\,\alpha = a$ Therefore; $T = \frac{1}{2} m_2\, a$ (equation 2)
a. 4	From step a. 3, $T = \frac{1}{2} m_2\, a$.
Solve for the rate of acceleration.	Substituting $T = \frac{1}{2} m_2\, a$ into equation 1 gives $m_1\, g - \frac{1}{2} m_2\, a = m_1\, a$ $(1.0\text{ kg})(9.8\text{ m/s}^2) = (1.0\text{ kg})\, a + \frac{1}{2}(2.0\text{ kg})\, a$ $9.8\text{ N} = (2.0\text{ kg})\, a$ $a = (9.8\text{ N})/(2.0\text{ kg}) = 4.9\text{ m/s}^2$
a. 5	$T = \frac{1}{2} m_2\, a = \frac{1}{2}(2.0\text{ kg})(4.9\text{ m/s}^2) = 4.9\text{ N}$
Solve for the tension.	
b. 1	$2\, a\, (x - x_o) = v^2 - v_o^2$
Use the kinematics equations to solve for the velocity of the cylinder after it travels 0.50 m.	$2\,(4.9\text{ m/s}^2)(0.50\text{ m}) = v^2 - (0\text{ m/s})^2$ $v^2 = 4.9\text{ m}^2/\text{s}^2$ $v = 2.2\text{ m/s}$
c. 1	$KE_{initial} + PE_{initial} = KE_{final} + PE_{final}$
Solve for the velocity using the law of conservation of energy. Note: the initial KE = 0 and assume that the final PE = 0. Also, for a solid cylinder $I = \frac{1}{2} m\, r^2$.	$0 + m_1\, g\, y = \frac{1}{2} m_1\, v^2 + \frac{1}{2} I_2\, \omega^2 + 0$ But $I = \frac{1}{2} m_2\, r^2$ and $v = r\,\omega$ therefore, $\frac{1}{2} I\, \omega^2 = \frac{1}{2}\left[\frac{1}{2} m_2\, r^2\right]\omega^2 = \frac{1}{4} m_2\, v^2$ $m_1\, g\, y = \frac{1}{2} m_1\, v^2 + \frac{1}{4} m_2\, v^2$

$(1.0 \text{ kg})(9.8 \text{ m/s}^2)(0.50 \text{ m}) = \frac{1}{2} (1.0 \text{ kg}) \text{ v}^2 + \frac{1}{4} (2.0 \text{ kg}) \text{ v}^2$

$4.9 \text{ J} = (0.50 \text{ kg}) \text{ v}^2 + (0.50 \text{ kg}) \text{ v}^2$

$4.9 \text{ J} = (1.0 \text{ kg}) \text{ v}^2$

$\text{v}^2 = 4.9 \text{ m}^2/\text{s}^2$ and $\text{v} = 2.2 \text{ m/s}$

PROBLEM 5. One end of a string is attached to a student's finger while the other end is wrapped around a solid cylinder. The cylinder is released from rest while the student holds her finger steady. The string unwinds as the cylinder rotates and accelerates downward. The mass of the cylinder is 0.20 kg and its radius is 0.030 m. Use rotational dynamics to determine the a) rate of acceleration of the object and the tension in the string. Determine the cylinder's speed when it is 0.50 m below its initial position using the b) kinematics equations and c) law of conservation of energy.

a. 1 Draw a free body diagram and locate the forces acting on the cylinder.	Solution: (Sections 8-5, 8-6, and 8-7) 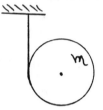 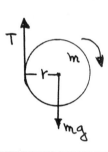
a. 2 Torque produced by tension in the string causes the cylinder to rotate. Use the torque equations and rotational dynamics to derive an equation for the tension in the string in terms of the cylinder's. rate of acceleration.	$\tau = \text{r F sin } 90°$ but $\text{F} = \text{T}$ where T = tension in the string Also $\tau = \text{I } \alpha$ where $\text{I} = \frac{1}{2} \text{ m r}^2$ Therefore $\text{r T sin } 90° = \frac{1}{2} \text{ m r}^2 \alpha$ and $\text{T} = \frac{1}{2} \text{ m r } \alpha$ but $\text{r } \alpha = \text{a}$ Therefore $\text{T} = \frac{1}{2} \text{ m a}$ (equation 1)
a. 3 Use Newton's second law to write an equation for the rate of acceleration.	The cylinder is accelerating downward. therefore, mg is greater than the tension in the string. net force = m a $\text{m g - T} = \text{m a}$ (equation 2)
a. 4 Solve for the rate of acceleration.	From step a. 2, $\text{T} = \frac{1}{2} \text{ m a}$. Substituting $\text{T} = \frac{1}{2} \text{ m a}$ into equation 2 gives $\text{m g - } \frac{1}{2} \text{ m a} = \text{m a}$

	$m\,g = m\,a + \tfrac{1}{2}\,m\,a$
	$m\,g = 3\,m\,a/2$ The mass cancels and
	$a = \tfrac{2}{3}\,g = \tfrac{2}{3}(9.8\ \text{m/s}^2) = 6.5\ \text{m/s}^2$
a. 5	$T = \tfrac{1}{2}\,m\,a = \tfrac{1}{2}\,(0.20\ \text{kg})(6.5\ \text{m/s}^2)$
Solve for the tension.	$T = 0.65\ \text{N}$
b. 1	$2\,a\,(x - x_0) = v^2 - v_0^2$
Use the kinematics equations to solve for the velocity of the cylinder after it travels 0.50 m.	$2\,(6.5\ \text{m/s}^2)(0.50\ \text{m}) = v^2 - (0\ \text{m/s})^2$
	$v^2 = 6.5\ \text{m}^2/\text{s}^2$
	$v = 2.6\ \text{m/s}$
c. 1	$KE_{initial} + PE_{initial} = KE_{final} + PE_{final}$
Solve for the velocity using the law of conservation of energy. Note: the initial KE = 0 and assume that the final PE = 0. Also, for a solid cylinder $I = \tfrac{1}{2}\,m\,r^2$.	$0 + m\,g\,y = \tfrac{1}{2}\,m\,v^2 + \tfrac{1}{2}\,I\,\omega^2 + 0$
	but $I = \tfrac{1}{2}\,m\,r^2$ and $v = r\,\omega$ Therefore,
	$\tfrac{1}{2}\,I\,\omega^2 = \tfrac{1}{2}\,(\tfrac{1}{2}\,m\,r^2)\,\omega^2 = \tfrac{1}{4}\,m\,v^2$
	$mgy = \tfrac{1}{2}\,m\,v^2 + \tfrac{1}{4}\,m\,v^2$
	$m\,g\,y = \tfrac{3}{4}\,m\,v^2$ m cancels, rearranging
	$v^2 = 4/3\,g\,y = 4/3\,(9.8\ \text{m/s}^2)(0.50\ \text{m})$
	$v^2 = 6.5\ \text{m}^2/\text{s}^2$
	$v = 2.6\ \text{m/s}$

PROBLEM 6. A small ball of mass m and radius r rolls from rest and without slipping along the loop apparatus shown in the diagram. Determine the minimum height H from which the ball can be released and still negotiate the loop. Express your answer in terms of the radius of the loop (R). Note: assume that the radius of the loop is much greater than the radius of the sphere.

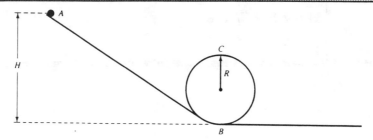

a. 1	Solution: (Section 8-7)
Determine the ball's rate of acceleration at the top of the track, i.e., point C.	At the top of the loop the ball is just "touching" the track but the track exerts no force on the ball. The net force on the ball is the gravitational force. The gravitational force produces a centripetal acceleration and a_c = g.
a. 2	a_c = g = v^2/R Solving for v:
Derive a formula for the object's speed at point C.	v^2 = gR
	v = $(Rg)^{1/2}$

a. 3

Use the law of conservation of energy to solve for H.

energy at point A = energy at point C

Since the ball is initially at rest, KE_A = 0.

The ball rolls without slipping; therefore,

KE_C = $KE_{translation}$ + $KE_{rotation}$

KE_C = ½ m v^2 + ½ I ω^2

PE_A + KE_A = PE_C + KE_C

m g H + 0 = m g h_C + ½ m v^2 + ½ I ω^2

but h_C = 2R,   I = 2/5 mr^2 and $r^2 \omega^2$ = v^2

m g H = m g (2R) + ½ m v^2 + ½(2/5 m r^2) ω^2

The mass appears in each term. It cancels algebraically from each side of the equation.

g H = 2 g R + ½ v^2 + 1/5 v^2 = 2 g R + 7/10 v^2

but, as shown in step a. 2, v^2 = R g

g H = 2 g R + 7/10 R g

g appears in each term and cancels from each side of the equation.

H = 2 R + 7/10 R

H = 2.7 R

PROBLEM 7. A student stands on a rotating platform that has frictionless bearings. He has a 2.00 kg object in each hand, held 1.00 m from the axis of rotation of the system. The system is initially rotating at 10.0 rpm. Determine the a) initial angular velocity of the system in radians per second, b) angular velocity of the system in radians per second if the objects are brought to a distance of 0.200 m from the axis of rotation, and c) change in the rotational kinetic energy of the system as the objects are pulled closer to the center of rotation. d) What causes the increase in the rotational kinetic energy? Assume that the moment of inertia of the platform + student remains constant at 1.00 kg m².

a. 1 Determine the initial angular velocity in radians per second.	Solution: (Section 8-8) 1 rev = 2 π radians and 1 minute = 60.0 sec ω = (10.0 rev/min)(2 π rad/rev)(1 min/60.0 s) = 1.05 rad/s
b. 1 Determine the initial and final moment of inertia of the system.	$I = \Sigma mr^2$ and includes the moment of inertia of each object as well as the moment of inertia of the student + platform. $I = m_1 r_1^2 + m_2 r_2^2 + I_{student + platform}$ I_i = (2.00 kg)(1.00 m)² + (2.00 kg)(1.00 m)² + 1.00 kg m² I_i = 5.00 kg m² I_f = (2.00 kg)(0.200 m)² + (2.00 kg)(0.200 m)² + 1.00 kg m² I_f = 1.16 kg m²
b. 2 Determine the final angular velocity of the system.	Since no external torque acts on the system, apply the law of conservation of angular momentum, I ω = constant. $I_i \omega_i = I_f \omega_f$ (5.00 kg m²)(1.05 rad/s) = (1.16 kg m²) ω_f ω_f = 4.53 rad/s
c. 1 Determine the change in KE of the system as the objects are brought closer to the axis of rotation.	$\Delta KE = KE_f - KE_i$ $\Delta KE = \frac{1}{2} I_f \omega_f^2 - \frac{1}{2} I_i \omega_i^2$ $\Delta KE = \frac{1}{2}$ (1.16 kg m²)(4.53 rad/s)² - $\frac{1}{2}$ (5.00 kg m²)(1.05 rad/s)² ΔKE = 11.9 J - 2.76 J = 9.14 J
d. 1 What causes the increase in the rotational kinetic energy?	As the student pulls the objects in toward the center of rotation, he is doing work on the system. The energy he expends in doing this work appears in the form of increased kinetic energy.

PRACTICE PROBLEMS

PROBLEM 1. A wheel, revolving at 30.0 rpm, accelerates uniformly to 45.0 rpm in 10.0 s. Determine the a) angular acceleration in rad/s^2 and b) angle in radians through which the wheel turns in 10.0 s.

ANS. a) 0.157 rad/s^2,   b) 39.3 radians

PROBLEM 2. A school playground contains a uniform, horizontal disc of mass 20 kg and radius 1.0 m. The axis of rotation of the disc is located at its center. A child applies a 20 N force tangent to the rim for 5.0 seconds. Determine the a) angular acceleration and b) change in the angular velocity of the disc.

ANS. a) 2.0 rad/s^2,   b) 10 rad/s

PROBLEM 3. A uniform sphere of mass 5.0 kg and radius 0.10 m is revolving at 10 radians per second about an axis through the center of the sphere. The sphere decelerates uniformly to a halt in 5.0 s. Determine the a) rate of angular deceleration, b) moment of inertia of the sphere, and c) magnitude of the torque causing the deceleration.

ANS. a) - 2.0 rad/s^2,   b) 2.0 x 10^{-2} kg m^2,   c) - 0.04 m N

PROBLEM 4. A student stands on a rotating platform that has frictionless bearings. He has a 2.00 kg object in each hand, held 1.00 m from the axis of rotation of the system. The system is initially rotating at 10.0 rpm. Determine the a) angular velocity of the system in radians per second if the objects are brought to the center of rotation and b) change in the rotational kinetic energy of the system as the objects are pulled to the center of rotation. Assume that the moment of inertia of the platform + student remains constant at 1.00 kg m^2.

ANS. a) 5.25 rad/s,   b) 11.0 J

CHAPTER 9

BODIES IN EQUILIBRIUM; ELASTICITY AND FRACTURE

KEY TERMS AND PHRASES

equilibrium occurs when an object is not accelerated and is not rotating. In order for an object to be in equilibrium the following two conditions for equilibrium must be satisfied.

first condition of equilibrium is satisfied when the vector sum of the forces acting on an object equals zero, $\Sigma \mathbf{F} = 0$.

second condition of equilibrium is satisfied when the sum of all torques acting on an object about any axis perpendicular to the plane of the forces equals zero, i.e., $\Sigma \tau = 0$.

static equilibrium occurs when an object at equilibrium is at rest.

dynamic equilibrium occurs when an object at equilibrium is moving in a straight line at a constant speed.

concurrent forces are aligned such that the line of action of each force acting on the object passes through a common point. If the forces are concurrent, only the first condition of equilibrium ($\Sigma \mathbf{F} = 0$) is needed to solve the problem.

nonconcurrent forces are aligned such that the line of action of each force does not pass through a common point. The result is a net torque which tends to cause the object to rotate. If the forces are nonconcurrent, and yet the object is in equilibrium, then both conditions of equilibrium are required to solve the problem, i.e., $\Sigma \mathbf{F} = 0$ and $\Sigma \tau = 0$.

machines are devices used to change the magnitude and/or direction of a force. In the process some practical advantage is achieved. Examples of simple machines include pulleys, levers, and incline planes.

ideal mechanical advantage (IMA) is the advantage if the work done by friction equals zero.

actual mechanical advantage (AMA) is the advantage when the work done by friction is included.

efficiency of a machine is defined as the amount of work output from the machine divided by the amount of work input. The efficiency also equals the ratio of the actual mechanical advantage to the ideal mechanical advantage.

stable equilibrium occurs when a small displacement from its undisturbed position results in a torque that tends to restore the object to its original position. An object remains in stable equilibrium as long as the object's center of gravity remains within its base of support.

unstable equilibrium occurs when any displacement from the object's undisturbed position results in a torque that tends to cause the object to move farther away from its original position. An example is a pencil balanced vertically on its point.

neutral equilibrium occurs if the object remains in stable equilibrium independent of the object's orientation. An example is a spherical ball placed on a flat surface. No matter how the ball is displaced the cg remains over the support point.

Hooke's law states that the deformation of an object is proportional to the magnitude of the applied force and a proportionality constant called the force constant.

stress is the ratio of the force causing an object's distortion to the area on which the force acts.

strain is the ratio of the object's change in length to the object's original length.

fracture occurs when the stress on an object exceeds the elastic limit of the object. The force exerted per unit area at the fracture point is called the ultimate strength of the material.

SUMMARY OF MATHEMATICAL FORMULAS

first condition of equilibrium	$\Sigma F = 0$	the vector sum of the forces acting on an object equals zero, $\Sigma \mathbf{F} = 0$
second condition of equilibrium	$\Sigma \tau = 0$	the vector sum of all torques acting about any axis perpendicular to the plane of the forces equals zero, i.e., $\Sigma \tau = 0$
ideal mechanical advantage (IMA)	$IMA = d_i/d_o$	The "ideal" mechanical advantage (IMA) is the advantage if the work done by friction equals zero. IMA is the ratio of the distance (d_i) moved by the input force to the distance (d_o) moved by the output force.
actual mechanical advantage (AMA)	$AMA = F_o/F_i$	The actual mechanical advantage (AMA) is the advantage when the work done by friction is included. AMA is the ratio of the output force (F_o) to the input force (F_i).

efficiency	$e = W_o/W_i = (F_o\,d_o)/(F_i\,d_i)$ $e = (F_o/F_i)/(d_i/d_o)$ or $e = AMA/IMA$	The efficiency (e) of a machine is defined as the amount of work output from the machine divided by the amount of work input. The efficiency can also be written as the ratio of the actual mechanical advantage (AMA) to the ideal mechanical advantage (IMA).
Hooke's law	$\Delta F = k\,\Delta L$	The deformation (ΔL) of an elastic object is related to the magnitude of the applied force (ΔF) and the force constant (k).
tensile and compressive stress	$\Delta F/A = E\,(\Delta L/L_o)$	Tensile (or compressive) stress ($\Delta F/A$) is the force per unit area on which the force acts. Strain is the ratio of the change in length to the original length; strain = $\Delta L/L_o$. The stress is related to the strain by elastic modulus (E).
shear stress	$\Delta F/A = G\,(\Delta L/L_o)$	Shear stress ($\Delta F/A$) occurs when equal but opposite forces are applied tangentially across the opposite faces of an object. Shear stress is related to the strain ($\Delta L/L_o$) by the shear modulus (G).
volume stress	$\Delta P = -\,B\,(\Delta V/V_o)$	A compressive stress (ΔP) that acts over the entire surface of an object will cause a decrease in the object's volume (ΔV). V_o is the original volume and B is a proportionality constant called the bulk modulus. The negative sign indicates that the volume of the object decreases with increasing pressure.

CONCEPT SUMMARY

A body is in equilibrium if it is at rest (**static equilibrium**) or moving in a straight line at a constant speed (**dynamic equilibrium**).

First Condition of Equilibrium

In order for an object to be in equilibrium, the vector sum of the forces acting on it must be zero, $\Sigma\mathbf{F} = 0$. To solve problems involving equilibrium, the force vectors are resolved into x, y, and z components. The equation $\Sigma\mathbf{F} = 0$ is replaced by $\Sigma\mathbf{F}_x = 0$, $\Sigma\mathbf{F}_y = 0$, and $\Sigma\mathbf{F}_z = 0$.

If the line of action of each force acting on the object passes through a common point, the forces are termed **concurrent** and $\Sigma \mathbf{F} = 0$ is the only condition needed to solve the problem. Below is an example of forces acting on an object which are concurrent.

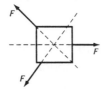

Second Condition of Equilibrium

If the forces are **nonconcurrent**, that is, the line of action of each force does not pass through a common point, then a net torque will cause a change in the state of rotation of the object. As can be seen in the diagram below, the object would tend to rotate in a clockwise direction.

The **second condition of equilibrium** states that the vector sum of all torques acting about any axis perpendicular to the plane of the forces must be zero, i.e., $\Sigma \tau = 0$.

A torque that causes a counterclockwise rotation is arbitrarily defined as positive while a clockwise torque is defined as negative. Thus the sum of clockwise and counterclockwise torques must add to zero. In other words, the sum of the clockwise torques about any axis equals the sum of the counterclockwise torques about the axis.

Since no rotation occurs, any point may be selected to be the location of the reference axis. In order to simplify the solution to most problems, the axis of rotation is usually taken to be a point through which the line of action of an unknown force passes. In this situation, the lever arm distance from the rotation point to the line of action of the unknown force is zero. Therefore, the torque produced by the unknown force is zero and at least one unknown has been eliminated from the torque equation. For nonconcurrent force problems, both conditions of equilibrium must usually be applied to solve the problem.

Simple Machines: Levers and Pulleys

Machines are devices used to change the magnitude and/or direction of a force. In the process some practical advantage is achieved. Examples of simple machines include pulleys, levers, and incline planes.

The principle of work applies to machines. The work input is equal to the sum of the useful work output plus the work done by friction. When discussing machines it is convenient to refer to the mechanical advantage. The **"ideal" mechanical advantage** (IMA) is the advantage if the work done by friction equals zero. The IMA can be determined by using the following formula:

IMA = d_i/d_o where d_i is the distance moved by the input force and d_o is the distance moved by the output force.

The **actual mechanical advantage** (AMA) is the advantage when the work done by friction is included. The formula for AMA is

AMA = F_o/F_i where F_o is the output force acting on the object and F_i is the input force exerted by the object causing the motion.

The **efficiency** (e) of a machine is defined as the amount of work output from the machine divided by the amount of work input.

efficiency = e = work output/work input

efficiency = e = W_o/W_i = $(F_o\ d_o)/(F_i\ d_i)$ = $(F_o/F_i)/(d_i/d_o)$ = AMA/IMA

Stable, Unstable, and Neutral Equilibrium

An object whose center of gravity is below its support point is said to be in **stable equilibrium**. Any displacement from its undisturbed position results in a torque that tends to restore it to its original position. An example of this is a ball on the end of a string. When displaced from its undisturbed position, a component of its weight ($F_{\parallel}$) tends to return the ball to its original orientation.

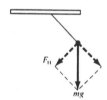

If the center of gravity is above the support point, the question of stable equilibrium depends on the location of the position of the center of gravity (cg) relative to the support point. As stated in the text, "In general, a body whose cg is above its base of support will be stable if a vertical line projected downward from its cg falls within the base of support." To clarify this idea try the following experiment. Stand with the heels of your shoes against a wall. If you slowly bend over to touch your toes with your fingers without bending your knees, you will find your toes gradually supporting more and more of your weight. Past a certain point you will fall forward. This is because past a certain point the cg of the upper portion of your body falls outside your shoes, which form your base of support. Once your center of gravity is outside your shoes, an unbalanced torque is produced which causes you to topple over.

If you move away from the wall and bend over, you will not fall forward. This is because your hips move backward as the upper portion of your body moves forward. Thus your cg

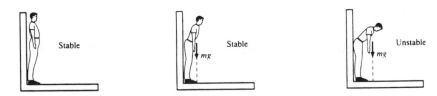

remains within your base of support and you will be able to touch your toes.

An example of an object in neutral equilibrium is a spherical ball placed on a flat surface. No matter how the ball is displaced the cg remains over the support point.

Elasticity: Stress, Strain, and Fracture

Robert Hooke (1635-1703) first stated the relationship that connects the deformation of a body to the magnitude of the applied force.

$\Delta F = k \, \Delta L$ where ΔF is the applied force, ΔL is the deformation of the object, and k is the force constant in N/m

Hooke's law is found to hold for most materials, e.g., a wire, rubber band, or a spring, up to a point which is called the elastic limit. Beyond this limit the object will not return to its original shape when the force is removed. An object deformed beyond the elastic limit will eventually reach its breaking point and fracture.

Tension, Compression, and Shear

Hooke's law can be stated in terms of **tensile stress** and **strain** and a proportionality constant called the **elastic modulus** (E), Young's modulus, or the tensile modulus.

Stress is defined as the force per unit area on which the force acts, stress = $\Delta F/A$. Strain is defined as the ratio of the change in length to the original length, strain = $\Delta L/L_o$ and elastic modulus = stress/strain.

$$E = (\Delta F/A)/(\Delta L/L_o) \quad \text{and} \quad \Delta F/A = E \, (\Delta L/L_o)$$

Compressive Stress

Compressive stress is the compression of a body by an outside force that acts inwardly on the object. The equation for tensile stress also applies to compressive stress, and the values of E are usually the same for both tensile stress and compressive stress.

Shear Stress

Shear stress occurs when equal but opposite forces are applied tangentially across the opposite faces of an object.

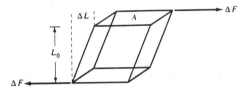

The shear modulus (G) equals the ratio of the shear stress to the shear strain and is given by the formula:

$$G = (\Delta F/A)/(\Delta L/L_o) \quad \text{and} \quad \Delta F/A = G\,(\Delta L/L_o)$$

The values of E and G will be given as needed in the programmed and practice problems; however, Table 9-1 in the text lists the values for a number of materials.

Fracture

If the stress on an object is too great, the object fractures or breaks. The force exerted per unit area at the fracture point is called the ultimate strength of the material.

Bulk Modulus

A compressive stress that acts over the entire surface of an object will cause a decrease in the object's volume. If the force per unit area is uniform, the relationship that connects the volume stress and the change in volume is given by

$$B = -\,\Delta P/(\Delta V/V_o) \quad \text{and} \quad \Delta P = -\,B\,(\Delta V/V_o)$$

where ΔP is the volume stress, which for a liquid is the change in pressure. ΔV is the change in volume. The negative sign appears because the volume of the object decreases with increasing pressure. V_o is the original volume and B is a proportionality constant called the bulk modulus.

ANSWERS TO SELECTED TEXT QUESTIONS

QUESTION 5. A bear sling, Fig. 9-41, is used in some national parks for placing backpackers' food out of the reach of bears. Explain why the force needed to pull the backpack up increases as the backpack gets higher and higher. Is it possible to pull the rope hard enough so that it doesn't sag at all?

Based on the diagram, the backpack is placed in the middle of the rope. Let us assume that the angle made between each section of the rope and the horizontal is the same.

step 1. Apply the first condition of equilibrium to show that $T_1 = T_2$.

There is no motion in the horizontal direction; therefore $\Sigma F_x = 0$ $T_{1x} - T_{2x} = 0$

$T_1 \cos \theta - T_2 \cos \theta = 0$ but the $\cos \theta$ appears in each term. Therefore, $T_1 = T_2$.

step 2. There is no motion in the vertical direction; therefore, $\Sigma F_y = 0$.

$T_{1y} + T_{2y} - mg = 0$ where mg is the weight of the backpack

$T_1 \sin \theta + T_2 \sin \theta - mg = 0$ but $T_1 = T_2 = T$.   Therefore simplifying,

$2 T \sin \theta = mg$ and $T = (mg)/(2 \sin \theta)$

As the backpack gets higher and higher, the value of θ becomes smaller. As θ becomes smaller, $\sin \theta$ becomes smaller and T becomes larger. As a result the force required to pull the backpack up increases as it gets higher above the ground. As θ approaches $0°$, T approaches infinity. Even if the camper were able to pull with enough force to remove the sag, the rope would snap before the no sag point was reached.

QUESTION 13. Why do you tend to lean backward when carrying a heavy load in your arms?

ANSWER: A heavy load in your arms tends to move the center of gravity of your body + load outward from the center of your body. If a line from the center of gravity to the floor passes within the base of your feet, you will be in equilibrium. If the line passes beyond your toes, you will tend to tip forward and fall on your face. By leaning backward you are shifting the center of mass of your body + load so that it remains within the base of your feet.

QUESTION 14. Place yourself facing the edge of an open door. Position your feet astride the door with your nose and abdomen touching the door's edge. Try to rise on your tiptoes. Why can't this be done?

In order to rise onto your tiptoes, the center of mass of your body must be shifted forward until it is over your toes. Since your nose and abdomen are already touching the door's edge, it is impossible to shift your weight forward. You feel that your feet are nailed to the floor.

Note: if you try to do this you will find that you may be able to momentarily shift your weight but you immediately tend to fall back onto your heels.

PROBLEM SOLVING SKILLS

For problems where the forces are concurrent and the object is in static equilibrium:

1. Draw an accurate diagram locating the forces acting on the object or system of objects.
2. Draw a free body diagram locating the forces acting on the object(s) in question.
3. Resolve each force vector into x and y components.
4. Apply the first condition of equilibrium ($\Sigma F_x = 0$, $\Sigma F_y = 0$) and solve the problem.

For problems where the forces are nonconcurrent and the object or system of objects is in static equilibrium:

1. Repeat steps 1, 2, and 3 listed above.
2. Write an equation(s) using the first condition of equilibrium.
3. Select a convenient point for the axis of rotation. A convenient point is located at a position where an unknown force acts.
4. Use the tip of pencil method (see problem 2) to determine the direction of the rotation (CW or CCW) produced by the force about the axis of rotation.
5. Write an equation using the second condition of equilibrium.
6. Use the equations written using the two conditions of equilibrium to solve the problem.

For problems involving mechanical advantage and efficiency:

1. Draw a diagram locating the forces acting on the object.
2. Determine the input force, output force, input distance, and output distance.
3. Use the appropriate formula to determine the IMA, AMA, and efficiency.

For problems involving stress and strain:

1. Draw an accurate diagram locating the forces acting on the object.
2. Determine whether the problem involves tensile stress, compressive stress, or shear stress.
3. Apply the appropriate formula and solve the problem.

PROGRAMMED PROBLEMS

PROBLEM 1. A 15.0 kg child holds on to a rope attached to a tree. The child's brother applies a horizontal force and pulls the child back a certain distance before releasing the child. Calculate the magnitude of the force exerted by the brother when the rope makes an angle of 37.0° with the vertical.

a. 1 Draw an accurate diagram locating the forces acting on the swing.	Solution: (Section 9-2)
a. 2 Apply the first condition of equilibrium to the vertical components of force and solve for T.	$\Sigma F_y = 0$ $\quad T_y - m g = 0$ $T \sin 53.0° - (15.0 \text{ kg})(9.8 \text{ m/s}^2) = 0$ $(0.799) T - 147 \text{ N} = 0$ $T = (147 \text{ N})/(0.799) = 184 \text{ N}$
a. 3 Apply the first condition of equilibrium to the horizontal components of force and solve for F.	$\Sigma F_x = 0$ $\quad F - T_x = 0$ $F - T \cos 53.0° = 0$ $F - (184 \text{ N})(0.602) = 0$ $F = 111 \text{ N}$

PROBLEM 2. As a classroom demonstration, the instructor asks Andy and Bob to lift a table on which a third student, Chuck, is sitting. The table is 2.0 m long and weighs 150 N. The table is uniform and it is held so that it remains horizontal even though Chuck is sitting on it. Chuck weighs 700 N and is sitting 0.60 m from Andy. Determine the force exerted by Andy and Bob.

a. 1

Draw an accurate free body diagram locating the forces acting on the table.

Solution: (Section 9-2 and 9-3)

Let F_A be the force that Andy exerts and F_B be the force exerted by Bob.

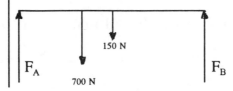

a. 2

Apply the first condition of equilibrium, $\Sigma F = 0$.

$\Sigma F_x = 0$ No horizontal forces are present in this problem.

$\Sigma F_y = 0$ $F_A + F_B - 700\ N - 150\ N = 0$

$F_A + F_B = 850\ N$ (equation 1)

---

a. 3

Apply the second condition of equilibrium, $\Delta \tau = 0$. Choose the center of the table as the axis of rotation.

In order to determine whether the torque produced by a force about the rotation point is clockwise (CW) or counterclockwise (CCW), let us use the tip of pencil method. Place the point of a pencil at the axis of rotation and hold it fixed. The rest of the pencil is parallel to the table on which the forces act. Applying the force to the pencil at the point in the diagram where the force is located will cause the pencil to rotate either CW or CCW.

For example, place the point of the pencil at the center of the line representing the table. Pushing on the pencil at the point where Andy's force is located and in the direction of Andy's force causes the pencil to rotate CW. Thus, Andy's force causes a clockwise torque about the center of the board. Now arrange the pencil so that Bob's force acts. Note that this force produces a counterclockwise torque. Repeating this process we find that Chuck's weight produces a CCW torque. The weight of the board creates no torque. The weight is at the rotation point. The lever arm distance is zero; therefore, the torque is zero.

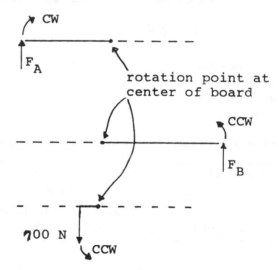

Since no rotation occurs,

$\Sigma \tau_{CCW} - \Sigma \tau_{CW} = 0$

$F_B\ (1.0\ m) + (700\ N)(0.40\ m) - F_A\ (1.0\ m) = 0$

Simplifying: $F_B - F_A = -280\ N$ (equation 2)

a. 4	Adding the two equations:
There are now two equations and the same two unknowns. Solve for each unknown.	$F_A + F_B + F_B - F_A = 850 \text{ N} - 280 \text{ N}$
	$2 F_B = 570 \text{ N}$
	$F_B = 285 \text{ N}$
	Substituting $F_B = 285 \text{ N}$ into equation 1,
	$F_A + 285 \text{ N} = 850 \text{ N}$, then
	$F_A = 565 \text{ N}$

---

a. 5

The choice of an axis of rotation is arbitrary. Select the left end and determine the force exerted by Andy and Bob.

It is convenient to select a point for the axis of rotation where an unknown force acts, e.g., the left end. Drawing the free body diagram and using the point of pencil method, we determine that the 700 N and 150 N forces create a CW torque, F_B produces a CCW torque, while F_A produces no torque since its lever arm distance equals zero.

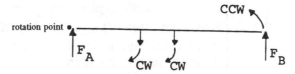

Applying the second condition of equilibrium:

$\Sigma \tau_{CCW} - \Sigma \tau_{CW} = 0$

$F_B(2.0 \text{ m}) - (700 \text{ N})(0.60 \text{ m}) - (150 \text{ N})(1.0 \text{ m}) = 0$

$F_B = 285 \text{ N}$

To determine F_A substitute the value of F_B into equation. 1.

$F_A + 285 \text{ N} = 850 \text{ N}$

$F_A = 565 \text{ N}$

PROBLEM 3. A 70 kg stunt man intends to walk to the end of uniform 10.0 meter long plank which extends horizontally over the edge of a roof. If the mass of the plank is 150 kg, determine the maximum amount of the plank that can overhang the roof and have the system still be in static equilibrium. Note: at the point that the board would begin to tip only a point at the edge of the roof makes contact with the board.

a. 1 Draw an accurate free body diagram locating the forces acting on the plank.	Solution: (Section 9-3) Let F represent the upward force that the edge of the roof exerts on the plank and the man. 
a. 2 Apply the first condition of equilibrium, $\Sigma F = 0$.	$\Sigma F_x = 0$ No horizontal forces are present in this problem. $\Sigma F_y = 0$ F - (150 kg)(9.8 m/s^2) - (70 kg)(9.8 m/s^2) = 0 F = 2160 N
a. 3 Choose the left end of the plank as the rotation point. Use the tip of pencil method to determine the direction (CW or CCW) the torque produced by each force.	According to theory, since no rotation occurs, any point may be selected to be the location of the reference point. Therefore, let the left end of the plank be chosen as the rotation point. Using the point of pencil method, place the point of the pencil at the left end of the line representing the plank. Pushing on the pencil at the point where F is located and in the direction of F causes the pencil to rotate CCW. Thus F causes a counterclockwise torque about the about the left end of the plank. Now repeat for the stunt man's weight and the weight of of the board. Arrange the pencil so that the stunt man's weight and the weight of the board act. Note that each force produces a clockwise torque (CW) about the left end of the board.
a. 4 Use the second condition of equilibrium to solve for length of overhang (x).	$\Sigma \tau_{CCW} - \Sigma \tau_{CW} = 0$ (2160 N)(10.0 m - x) sin 90° - (1470 N)(5.00 m) sin 90° - (686 N)(10.0 m) sin 90° = 0 21,600 Nm - (2160 N) x - 7350 Nm - 6860 Nm = 0 7390 Nm - (2160 N) x = 0 x = (7390 Nm)/2160 N) = 3.42 m
a. 5 Choose the edge of the roof as the reference point. Use the tip of pencil method to determine the direction of the torque produced by each force.	Using the point of pencil method, place the point of the pencil at the point where the plank touches the edge of the roof, i.e., at point F in the diagram. Pushing on the pencil at the point where F is located does not produce rotation about this point. Therefore, force F does not produce a torque about the rotation point because the lever arm distance from F to the reference point is zero. The weight of the plank causes a CCW rotation while the weight of the man causes a CW rotation about the reference point.

a. 6

Use the second condition of equilibrium to solve for length of the overhang (x).

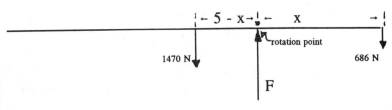

1470 N

rotation point

686 N

F

$$\Sigma \tau_{CCW} - \Sigma \tau_{CW} = 0$$

$$F (0 \text{ m}) \sin 90° + (1470 \text{ N})(5.00 \text{ m} - x) \sin 90°$$
$$- (686 \text{ N})(x) \sin 90° = 0$$

$$7350 \text{ Nm} - (1470 \text{ N}) x - (686 \text{ N}) x = 0$$

$$7390 \text{ Nm} - (2160 \text{ N}) x = 0$$

$$x = (7390 \text{ Nm})/2160 \text{ N} = 3.42 \text{ m}$$

Although some points may be more convenient to use, it should be noted that the selection of the position of the reference point is arbitrary.

PROBLEM 4. A uniform pole 6.0 meters long weighs 800 N and is attached at one end to a vertical wall. A load of 400 N hangs from the other end of the pole. A horizontal guy wire attached to the outer end of the pole holds the pole at a 37° angle with the horizontal. Determine the a) tension in the wire, and b) horizontal and vertical components of the force exerted by the wall on the pole.

---

a. 1

Draw an accurate diagram locating each of the forces acting on the pole.

Solution: (Section 9-3)

The pole is uniform; therefore, the weight is concentrated at the center.

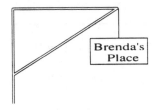

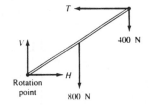

T

400 N

V

H

Rotation point

800 N

---

a. 2

Apply the first condition of equilibrium.

$$\Sigma F_x = 0 \quad H - T = 0$$

$$\Sigma F_y = 0 \quad V - 800 \text{ N} - 400 \text{ N} = 0$$

$$V = 1200 \text{ N}$$

a. 3

Apply the second condition of equilibrium, $\Sigma\tau = 0$, and solve for the tension in the wire. Note: choose the point where the pole is attached to the wall as the rotation point.

The convenient point for the axis of rotation would be a point where an unknown force is located. Therefore, either end of the pole would be a convenient point. Let us choose the end of the pole that meets the wall as the rotation point. At this point the wall exerts a force which is represented by the components, H and V.

Use the point of pencil method to determine the direction of each torque. From the diagram it can be seen that T produces a CCW torque while the 800 N weight of the pole and the 400 N load each produce a CW torque. Since V and H act at the rotation point, the lever arm distance is zero and they produce no torque.

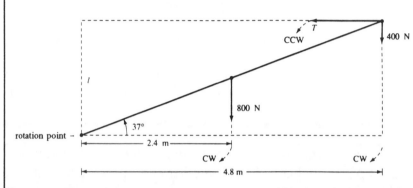

$\Sigma\tau_{ccw} - \Sigma\tau_{cw} = 0$

T(6.00m)(sin 37°)(sin 90°) - (800 N)(3.00 m)(cos 37°)(sin 90°)
                                    - (400 N)(6.00 m)(cos 37°)(sin 90°) = 0

(3.60 m) T - 1920 m N - 1920 m N = 0

T = 1070 N

b. 1

The value of V was obtained in step a. 2. T is now known; solve for H.

From step a. 2   H - T = 0

                      H - 1070 N = 0

                      H = 1070 N

PROBLEM 5. The machine shown in the diagram consists of two wheels that move together and are mounted on the same axle. The radii of the wheels are 0.10 m and 0.20 m, respectively. A 300 N force must be applied at F in order to raise 400 N at W. Determine the a) IMA, b) AMA, and c) efficiency of the machine.

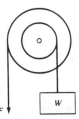

F   W

a. 1 Determine the ideal mechanical advantage.	Solution: (Section 9-4) In one complete turn, a length of string equal to the circumference of each wheel is wound or unwound. Apply the formula for IMA. $IMA = d_i/d_o$ where $d_i = 2 \pi R$ and $d_o = 2 \pi r$ $= 2 \pi R/2 \pi r = (0.20 \text{ m})/(0.10 \text{ m})$ $IMA = 2.0$
b. 1 Determine the actual mechanical advantage.	$F_o = W = 400 \text{ N}$ and $F_i = F = 300 \text{ N}$ $AMA = F_o/F_i = (400 \text{ N})/(300 \text{ N})$ $AMA = 1.3$
c. 1 Determine the efficiency(e).	$e = AMA/IMA \times 100\%$ $e = (1.3)/(2.0) \times 100\% = 65\%$

PROBLEM 6. A metal wire 0.50 m long and 1.00×10^2 meters in diameter stretches 3.2×10^4 m when a load of 10.0 kg is attached at the end. Determine the a) stress, b) strain, and c) elastic modulus for the material of the wire.

a. 1 Determine the tensile stress.	Solution: (Section 9-6) stress $= \Delta F/A$ and $A = \pi r^2$ where $r = dia/2$ $r = (1.00 \times 10^{-2} \text{ m})/2 = 0.50 \times 10^{-2} \text{ m}$ stress $= [(10.0 \text{ kg})(9.8 \text{ m/s}^2)]/[\pi(0.50 \times 10^{-2} \text{ m})^2]$ stress $= 1.25 \times 10^6 \text{ N/m}^2$
b. 1 Determine the strain.	strain $= \Delta L/L_o$ $= (3.20 \times 10^{-4} \text{ m})/(0.50 \text{ m})$ strain $= 6.40 \times 10^{-4}$

c. 1	E = stress/strain
Determine the elastic modulus.	$= (1.25 \times 10^6 \text{ N/m}^2)/(6.40 \times 10^{-4})$
	$E = 1.95 \times 10^9 \text{ N/m}^2$

PROBLEM 7. $1.000 \times 10^{-3} \text{ m}^3$ (1.000 liter) of water is placed in a flexible container which is attached to lead weights and allowed to sink in the ocean. Determine the a) change in volume of the water at a point where the water pressure is $3.0 \times 10^6 \text{ N/m}^2$ and b) % change in the volume of the water. c) Based on your answer to part b, is the phrase "water is incompressible" approximately correct? The bulk modulus of water is $2.0 \times 10^9 \text{ N/m}^2$.

a. 1	Solution: (Section 9-6)
Apply the bulk modulus formula and determine the change in volume of the water. Note: assume that all of the air in the container was removed before it was lowered.	$B = - \Delta P/(\Delta V/V_o)$ and $\Delta V = - (\Delta P \, V_o)/B$
	$\Delta V = - [(3.0 \times 10^6 \text{ N/m}^2)(1.0 \times 10^{-3} \text{ m}^3)]/(2.0 \times 10^9 \text{ N/m}^2)$
	$\Delta V = - 1.5 \times 10^{-6} \text{ m}^3$ or $- 1.5 \text{ cm}^3$
	The change in volume is negative because the water is compressed as the pressure increases.
b. 1	% change = $(- 1.5 \text{ cm}^3)/(1000 \text{ cm}^3) \times 100\%$
Determine the % change in the volume.	% change = $- 0.15\%$
c. 1	To a good approximation the phrase is correct.
Is the phrase "water is incompressible" correct?	

PRACTICE PROBLEMS

PROBLEM 1.   Determine the weight W required to keep the system shown below in static equilibrium.

ANS. 52 N

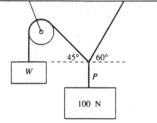

PROBLEM 2. A uniform ladder 5.0 m long and weighing 200 N is placed against a very smooth wall so that the lower end of the ladder is 3.0 m from the wall. If the coefficient of friction between the ladder and the ground is 0.50, determine the maximum distance that a 700 N man could climb along the length of the ladder before the ladder slips.

ANS. 3.5 m

PROBLEM 3. A solid cylindrical steel column, E = 200 x 10^9 N/m², is 5.0 m long and 0.20 m in diameter. What will be its decrease in length when carrying a load of 100,000 kg?

ANS. 7.8 x 10^{-4} m or 0.78 mm

PROBLEM 4. An incline plane 10.0 m long is inclined at 30° above the horizontal. A 10 kg block is pushed from rest from the bottom to the top of the incline. Calculate the a) IMA, b) AMA if the coefficient of kinetic friction is 0.20, and c) efficiency of the machine.

ANS. a) 2.0, b) 1.5, c) 75%

CHAPTER 10

FLUIDS

KEY TERMS AND PHRASES

density (ρ) is the quantity of mass (m) per unit volume (V).

weight density is the ratio of the weight of a substance to its volume.

specific gravity (SG) is the ratio of the density of the substance to the density of pure water.

pressure is the ratio of the force acting perpendicular to a surface to the surface area (A) on which the force acts.

gauge pressure measures the difference in pressure between an unknown pressure and atmospheric pressure.

absolute pressure is the sum of the gauge pressure and the atmospheric pressure.

Pascal's principle states that pressure applied to a confined fluid is transmitted throughout the fluid and acts in all directions.

Archimedes' principle states that a body wholly or partly immersed in a fluid is buoyed up by a force equal to the weight of the fluid it displaces.

buoyant force is the force caused by the displaced fluid. The buoyant force is considered to be acting upward through the center of gravity of the displaced fluid.

streamline flow assumes that as each particle in the fluid passes a certain point it follows the same path as the particles that preceded it. There is no loss of energy due to internal friction in the fluid.

turbulent flow is the irregular movement of particles in a fluid and results in loss of energy due to internal friction in the fluid. Turbulent flow tends to increase as the velocity of a fluid increases.

flow rate is the mass (or volume) of fluid that passes a point per unit time.

Bernoulli's equation states that the quantity $P + \frac{1}{2} \rho v^2 + \rho g h$ is the same at every point in the moving fluid. This equation is derived through the use of the work-energy principle and ignores the effect of internal friction in a moving fluid. It should be noted that P, $\frac{1}{2} \rho v^2$, and $\rho g h$ all have dimensions of pressure.

SUMMARY OF MATHEMATICAL FORMULAS

density	$\rho = m/V$	ratio of object's mass (m) to its volume (V)
specific gravity	$SG = \rho_{object} / \rho_{water}$	ratio of the density of the substance to the density of pure water
pressure	$P = F/A$	ratio of the force acting perpendicular to a surface to the surface area (A) on which the force acts
fluid pressure	$P = \rho g h$	gauge pressure as a function of depth (h)
Archimedes' principle	$F_b = m_f g = \rho_f g V_f$	buoyant force acting on an object placed in a fluid
equation of continuity	$\rho A v = constant$	Mass of fluid moving through any point is constant.
volume flow rate	$R = v A$	volume of fluid that passes a point per unit time
Bernoulli's equation	$P + \frac{1}{2} \rho v^2 + \rho g h = constant$	equation for the pressure at any point in a moving fluid

CONCEPT SUMMARY

Density, Weight Density, and Specific Gravity

The **density** (ρ) of a substance is defined as the quantity of mass (m) per unit volume (V) and is given by

$\rho = m/V$

For solids and liquids, the density is usually expressed in grams per cm^3 (g/cm^3) or kilograms per cubic meter (kg/m^3). The density of gases is usually expressed in grams per liter (g/l).

The **weight density** of a substance is the ratio of the weight of a substance to its volume. weight density = weight/volume = $m g/V$ = $\rho V g/V$ = ρg. Weight density is commonly used in problems involving English units. The weight density of distilled water at 39°F (4°C) is 62.4 lb/ft^3.

The **specific gravity** (SG) of a substance is the ratio of the density of the substance to the density of another substance which is taken as a standard. The density of pure water at 4°C is usually taken as the standard and this has been defined to be exactly 1.000 g/cm^3 or 1.000 x 10^3 kg/m^3.

Specific gravity is a dimensionless quantity. This is because it is the ratio of the density of one substance to the density of another. The specific gravity of an object is the same in any system of measurement.

SG = density of substance/density of water

For example, the density of aluminum is 2.7 x 10^3 kg/m^3; therefore, the SG of aluminum is

SG = 2.7 x 10^3 kg/m^3/1.0 x 10^3 kg/m^3 = 2.7

Pressure

Pressure (P) is defined as the ratio of the force acting perpendicular to a surface to the surface area (A) over which the force acts.

P = F/A

The SI unit of pressure is N/m^2 or pascal (Pa) while the English unit is lb/ft^2. A number of other units are used, e.g., lb/in^2 (or psi) and atmospheres. One atmosphere is the average pressure exerted by the earth's atmosphere at sea level.

1.00 atm = 14.7 lb/in^2 = 1.01 x 10^5 N/m^2 = 101.3 kPa

Pressure in Fluids

The pressure (P) produced by a column of fluid of height h and density ρ is given by

P = ρ g h

The density of liquids and solids is considered to be constant. In reality, the density of a liquid will increase slightly with increasing depth. The variation in density is usually negligible and can be ignored.

Gauge Pressure and Absolute Pressure

Ordinary pressure gauges measure the difference in pressure between an unknown pressure and atmospheric pressure. The pressure measured is called the **gauge pressure** and the unknown pressure is referred to as the **absolute pressure.**

$P_{absolute} = P_{gauge} + P_{atmosphere}$

Thus, if a tire gauge registers 193 kPa, the absolute pressure would be approximately 193 kPa + 101 kPa = 294 kPa.

Pascal's Principle

Pascal's principle states that pressure applied to a confined fluid is transmitted throughout the fluid and acts in all directions. The principle means that if the pressure on any part of a confined fluid is changed, then the pressure on every other part of the fluid must be changed by the same amount. This principle is basic to all hydraulic systems.

Buoyancy and Archimedes' Principle

Archimedes' Principle states that a body wholly or partly immersed in a fluid is buoyed up by a force equal to the weight of the fluid it displaces. An object lowered into a fluid "appears" to lose weight. The force that causes this apparent loss of weight is referred to as the buoyant force. The **buoyant force** is considered to be acting upward through the center of gravity of the displaced fluid.

Fluids in Motion

The equations that follow are applied when a moving fluid exhibits **streamline flow**. Streamline flow assumes that as each particle in the fluid passes a certain point it follows the same path as the particles that preceded it. There is no loss of energy due to internal friction in the fluid.

In reality, particles in a fluid exhibit **turbulent flow**, which is the irregular movement of particles in a fluid and results in loss of energy due to internal friction in the fluid. Turbulent flow tends to increase as the velocity of a fluid increases.

Flow Rate

The **flow rate** is the mass of fluid that passes a point per unit time (m/t). The formula for flow rate is

$$m/t = \rho \, A \, v$$

where ρ is the density of the fluid, A is the cross-sectional area of the tube of fluid at the particular point in question, and v is the velocity of the fluid at the point in question.

Since fluid cannot accumulate at any point, the flow rate is constant. This is expressed as the **equation of continuity.**

$$\rho \, A \, v = \text{constant}$$

In streamline flow, the fluid is considered to be incompressible and the density is the same throughout. The equation of continuity can then be written in terms of the volume rate of flow (R) which is constant throughout the fluid: R = Av = constant

Bernoulli's Equation

A fluid in streamline flow follows **Bernoulli's equation**:

$$P + \tfrac{1}{2} \rho v^2 + \rho g h = \text{constant}$$

P, ρ, and h are the absolute pressure, density, and height at a particular point in a fluid. The equation states that the quantity $P + \tfrac{1}{2} \rho v^2 + \rho g h$ is the same at every point in the stream. This equation is derived through the use of the work-energy principle and ignores the effect of internal friction in a moving fluid. It should be noted that P, $\tfrac{1}{2} \rho v^2$, and ρgh all have dimensions of pressure.

SELECTED TEXT QUESTIONS WITH ANSWERS

QUESTION 6. A small amount of water is boiled in a one-gallon gasoline can. The can is removed from the heat and the lid put on. Shortly thereafter, the can collapses. Explain.

ANSWER: When the lid is placed on the can, the internal pressure exerted by the hot gases, water vapor, and air inside the can balances the external atmospheric pressure acting on the can. As the gases inside the can cool, the pressure they exert on the side walls of the can decreases while the external atmospheric pressure remains constant. Eventually, the difference in pressure becomes large enough that the can collapses.

QUESTION 13. Does the buoyant force on a diving bell deep beneath the ocean have precisely the same value as when the bell is just beneath the surface? Explain.

ANSWER: If the density of the seawater remained constant, then the answer would be yes. According to Archimedes' principle, the buoyant force is equal to the weight of the fluid displaced. The weight of the fluid displaced equals the product of the density of the fluid, the volume of the fluid displaced, and the acceleration due to gravity. However, the pressure is so great deep beneath the ocean that the density of seawater is greater than at the surface. Therefore, the buoyant force is greater at great depths than it is at the surface. It should be noted that the assumption is made that the volume of the diving bell has not changed due to the increased pressure of the seawater.

QUESTION 18. Roofs of houses are sometimes pushed off during a tornado or hurricane. Explain, using Bernoulli's equation.

ANSWER: Air moving over the top of a roof causes a decrease in the downward pressure (P_2) on the top of the roof. Inside the house the air is moving very little and therefore the upward pressure (P_1) exerted by the air inside the house is greater than the downward pressure. If the pressure difference ($P_1 - P_2$) is great enough, the roof can be torn off the house. If 2 is a point above roof and 1 is a point inside the roof, then

$$P_1 + \tfrac{1}{2}\, \rho\, v_1^2 + \rho\, g\, h_1 = P_2 + \tfrac{1}{2}\, \rho\, v_2^2 + \rho\, g\, h_2$$

but $h_1 \approx h_2$ and $v_1 = 0$. Then $P_1 - P_2 = \tfrac{1}{2}\, \rho\, v_2^2$.

PROBLEM SOLVING SKILLS

For problems involving density and specific gravity:

1. If necessary, use information given in the problem to calculate the object's mass and volume.
2. Determine the object's density ($\rho = m/V$).
3. Determine the object's specific gravity by dividing the object's density by the density of water.

For problems involving Pascal's principle:

1. Identify which quantities are related to the input and which relate to the output and complete a data table based on information given and implied in the problem.
2. Use the concept of pressure in a fluid and Pascal's principle to solve the problem.

For problems where Archimedes' principle is used to calculate the buoyant force on an object completely immersed in a fluid:

1. If the mass and density of the object are known, then the volume of the object can be determined.
2. The volume of the object equals the volume of the fluid the object displaces.
3. Use Archimedes' principle to determine the buoyant force.

For problems where Archimedes' principle is used to calculate the buoyant force on an object that floats in the fluid:

1. For a floating object the buoyant force equals the object's weight.
2. The weight of the fluid displaced equals the buoyant force.

For problems involving the rate flow equation and the equation of continuity:

1. Complete a data table listing the cross-sectional area of the closed pipe at each point, as well as the velocity and density of the fluid at the each point in question.
2. Use the rate flow equation and equation of continuity to solve the problem.

For problems involving Bernoulli's equation:

1. Complete a data table listing the pressure, velocity of the fluid, and height of the fluid above a reference point for each point in question. If the problem involves a fluid moving through a closed pipe, then determine the cross-sectional area of the closed pipe at each point in question.
2. If necessary, use the equation of continuity to determine the velocity of the fluid at a particular point.
3. Use Bernoulli's equation to solve the problem.

PROGRAMMED PROBLEMS

PROBLEM 1. A solid cylinder be of material 0.100 m in radius and 0.0500 m high has a mass of 5.0 kg. Determine the a) density and b) specific gravity of the object.

a. 1	Solution: (Section 10-1)
Determine the volume of the cylinder.	$V = \pi r^2 h$ $V = \pi (0.100 \text{ m})^2 (0.0500 \text{ m}) = 1.570 \times 10^{-3} \text{ m}^3$
a. 2 Apply the formula for density.	$\rho = m/V$ $\rho = 5.00 \text{ kg}/1.57 \times 10^{-3} \text{ m}^3 = 3180 \text{ kg/m}^3$
b. 1 Apply the formula for specific gravity.	SG = density of object/density of water $SG = (3180 \text{ kg/m}^3)/(1000 \text{ kg/m}^{3)} = 3.18$

PROBLEM 2. As shown in the diagram, a 30.0 kg piston holds compressed gas in a tank of volume 100 m³. The radius of the piston is 0.050 m. Determine the a) absolute pressure and b) gauge pressure of the gas in the tank.

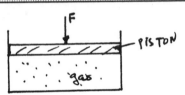

a. 1	Solution: (Section 10-3)
Determine the absolute pressure.	The absolute pressure is the sum of the atmospheric pressure and the pressure exerted by the piston. $P_{abs} = P_{atm} + P_{piston}$ $= 1.01 \times 10^5 \text{ N/m}^2 + [(30.0 \text{ kg})(9.80 \text{ m/s}^2)]/[(3.14)(0.050 \text{ m})^2]$ $= 1.01 \times 10^5 \text{ N/m}^2 + 3.75 \times 10^4 \text{ N/m}^2$ $P_{abs} = 1.38 \times 10^5 \text{ N/m}^2$
b. 1 Determine the gauge pressure.	The gauge pressure on the tank reads the difference between the absolute pressure and the atmospheric pressure. Thus, the gauge pressure equals the pressure applied by the piston. $P_{gauge} = P_{piston} = 3.75 \times 10^4 \text{ N/m}^2$

PROBLEM 3. In the hydraulic system shown in the diagram, the 200 kg cylinder has a cross-sectional area of 100 cm². The cylinder on the right has a cross-sectional area of 10.0 cm². a) Determine the weight (F) required to hold the system in equilibrium. b) If the left-hand cylinder is pushed down 5.0 cm, determine through what distance F will move. c) Determine the mechanical advantage of this system, assuming 100% efficiency.

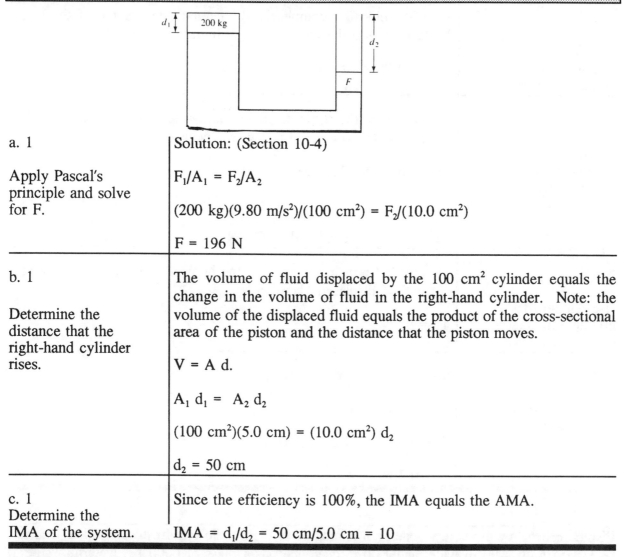

a. 1	Solution: (Section 10-4)
Apply Pascal's principle and solve for F.	$F_1/A_1 = F_2/A_2$
	$(200 \text{ kg})(9.80 \text{ m/s}^2)/(100 \text{ cm}^2) = F_2/(10.0 \text{ cm}^2)$
	F = 196 N
b. 1	The volume of fluid displaced by the 100 cm² cylinder equals the change in the volume of fluid in the right-hand cylinder. Note: the volume of the displaced fluid equals the product of the cross-sectional area of the piston and the distance that the piston moves.
Determine the distance that the right-hand cylinder rises.	
	V = A d.
	$A_1 d_1 = A_2 d_2$
	$(100 \text{ cm}^2)(5.0 \text{ cm}) = (10.0 \text{ cm}^2) d_2$
	$d_2 = 50 \text{ cm}$
c. 1 Determine the IMA of the system.	Since the efficiency is 100%, the IMA equals the AMA.
	IMA = d_1/d_2 = 50 cm/5.0 cm = 10

PROBLEM 4. An aluminum object has a mass of 27.0 kg and a density of 2.70 x 10³ kg/m³. The object is attached to a string and immersed in a tank of water. Determine the a) volume of the object and b) tension in the string when it is completely immersed.

a. 1	Solution: (Sections 10-1 and 10-6)
Determine the volume of the object.	$\rho = m/V$ and
	$V = m/\rho = 27.0 \text{ kg}/2.70 \times 10^3 \text{ kg/m}^3 = 1.00 \times 10^{-2} \text{ m}^3$

b. 1 Draw an accurate diagram and locate all of the forces acting on the object.	T is the tension in the string, mg is the object's weight, and the buoyant force is F_B.
b. 2 Determine the volume of water displaced by the object.	Since the object is completely submerged, the volume of the water displaced equals the volume of the aluminum object. $V_{water} = 1.00 \times 10^{-2} \text{ m}^3$
b. 3 Determine the weight of the water displaced by the submerged object.	The weight of the water displaced can be determined since both the density and the volume of the water displaced are known. weight of water displaced $= \rho_w \, V \, g$ $\qquad = (1.00 \times 10^3 \text{ kg/m}^3)(1.00 \times 10^{-2} \text{ m}^3)(9.80 \text{ m/s}^2)$ weight of water displaced $= 98.0$ N
b. 4 Use Archimedes' principle to determine the buoyant force.	Archimedes' principle states that the buoyant force equals the weight of the fluid displaced; therefore, $F_B = 98$ N
b. 5 Apply Newton's second law and solve for the tension in the string.	net F = m a $\quad$ but $\quad$ a = 0 m/s^2 $T + F_B - mg = 0$ $T + 98.0 \text{ N} - (27.0 \text{ kg})(9.80 \text{ m/s}^2) = 0$ $T = 167$ N

PROBLEM 5. A solid-core wooden raft is moored in the middle of a freshwater lake. The raft is 3.00 m long, 2.00 m wide, and 1.00 m deep and has a mass of 6000 kg. Determine the a) depth of the raft submerged below the water line and b) depth that is submerged if 10 adults stand on top of the raft. Assume that the average adult has a mass of 70.0 kg.

a. 1 Calculate the volume of the raft and the volume of the water displaced by the raft.	Solution: (Section 10-6) The raft is a rectangular solid; therefore, the volume is V = length x width x depth = 3.00 m x 2.00 m x 1.00 m V_{raft} = 6.0 m^3 The volume of the water displaced by the raft equals the fraction of the raft below the water surface and is given by V_{water} = 3.00 m x 2.00m x h′
a. 2 Apply Archimedes' principle.	According to Archimedes' principle, the buoyant force equals the weight of the water displaced. However, since the raft floats on the water, the buoyant force of the water equals the raft's weight. F_B = weight of water displaced = weight of the raft m_{water} g = m_{raft} g but g cancels from each side of the equation and m = ρV $\rho_{water} V_{water}$ = $\rho_{wood} V_{wood}$ (1000 kg/m^3)(3.0 m x 2.0 m x h′) = (600 kg/m^3)(6.0 m^3) h′ = 0.60 m Of the 1.0 m depth of the raft, 0.60 m is below the water line while 0.40 m is above the water line.
b. 1 Apply Archimedes' principle.	The raft now floats lower in water and the bouyant force equals the weight of the raft + the weight of the adults. F_B = weight of water displaced = weight of the raft + adults m_{water} g = m_{raft} g + m_{adults} g and again g cancels $\rho_{water} V_{water}$ = $\rho_{wood} V_{wood}$ + (10 adults)(70 kg/adult) (1000 kg/m^3)(3.0 m x 2.0 m x h′) = (600 kg/m^3)(6.0 m^3) + 700 kg (1000 kg/m^3)(3.0 m x 2.0 m x h′) = 3600 kg + 700 kg h′ = 0.72 m Of the 1.0 m depth of the raft, 0.72 m is below the water line while 0.28 m is above the water line.

During the early morning hours of August 24, 1992, the winds of Hurricane Andrew visited the home of the author of this study guide. It is estimated that wind gusts in excess of 150 mph (67 m/s) passed over the roof of the author's house. Calculate the a) difference in pressure between the air in the attic and the air passing over the roof and b) upward force exerted on the roof if area of the roof is 275 m². Assume that the air inside the attic was not moving and that the thickness of the roof was negligible. Note: assume that the density of air is 1.29 kg/m³.

a. 1

Use Bernoulli's equation to determine the pressure.

Let P_1 represent the pressure inside the attic and $v_1 = 0$. Also, since the thickness of the roof is negligible $h_1 = h_2$.

Solution: (Section 10-8)

$$P_1 + \tfrac{1}{2} \rho \, v_1^2 + \rho \, g \, h_1 = P_2 + \tfrac{1}{2} \rho \, v_2^2 + \rho \, g \, h_2$$

but $h_1 = h_2$ while $v_1 = 0$. Therefore, the equation becomes

$$P_1 = P_2 + \tfrac{1}{2} \rho \, v_2^2$$

$$P_1 - P_2 = \tfrac{1}{2} \rho \, v_2^2 = \tfrac{1}{2} (1.29 \text{ kg/m}^3)(67 \text{ m/s})^2$$

$$P_1 - P_2 = 2900 \text{ N/m}^2$$

b. 1

Calculate the upward force (lift) exerted by the moving air on the roof.

pressure = force/area and force = pressure x area

$$F = (2900 \text{ N/m}^2)(275 \text{ m}^2) = 8.0 \times 10^5 \text{ N}$$

Note: this upward force is approximately equal to 90 tons of lift. According to the Dade County building code all roof trusses must be attached to the tie beam of the building by metal straps. The tile on the author's roof needed be replaced but the roof did stay on the house.

A horizontal pipe has a diameter of 0.15 m at point 1 and 0.050 m at point 2. The velocity of water at point 1 is 0.80 m/s and the pressure is 1.01 x 10⁵ N/m². Determine the a) volume flow rate, b) velocity of the water at point 2, and c) pressure at point 2.

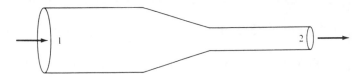

a. 1

Calculate the volume flow rate.

Solution: (Sections 10-7 and 10-8)

$$R = A_1 \, v_1 = [\pi(0.15 \text{ m}/2)^2](0.80 \text{ m/s}) = 0.014 \text{ m}^3/\text{s}$$

b. 1	Water cannot accumulate at any point in the hose, the rate of volume flow (R) must be the same throughout.
Apply the rate flow equation and determine the velocity of the water at point 2.	$R = A_1 v_1 = A_2 v_2$ where $A = \pi(\text{diameter}/2)^2$
	$[\pi(0.15 \text{ m}/2)^2](0.80 \text{ m/s}) = [\pi(0.050 \text{ m}/2)^2] v_2$
	$v_2 = 7.2 \text{ m/s}$

c. 1	$P_1 + \frac{1}{2} \rho v_1^2 + \rho g h_1 = P_2 + \frac{1}{2} \rho v_2^2 + \rho g h_2$
Use Bernoulli's equation to determine the pressure. Note: the pipe is horizontal; therefore, $h_1 = h_2$.	but $h_1 = h_2$, therefore,
	$P_1 + \frac{1}{2} \rho v_1^2 = P_2 + \frac{1}{2} \rho v_2^2$
	$1.01 \times 10^5 \text{ N/m}^2 + \frac{1}{2}(1000 \text{ kg/m}^3)(0.80 \text{ m/s})^2 =$ $P_2 + \frac{1}{2}(1000 \text{ kg/m}^3)(7.2 \text{ m/s})^2$
	$1.01 \times 10^5 \text{ N/m}^2 + 320 \text{ N/m}^2 = P_2 + 2.59 \times 10^4 \text{ N/m}^2$
	$P_2 = 7.54 \times 10^4 \text{ N/m}^2$

PROBLEM 8. The flow rate in the pipe shown below is 0.10 m³/s. At point 1 the diameter of the pipe is 0.15 m and the pressure is 1.2 x 10⁵ N/m². At point 2, which is 3.0 m higher than point 1, the diameter is 0.20 m. Determine the pressure at point 2.

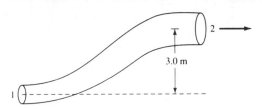

3.0 m

a. 1	Solution: (Sections 10-7 and 10-8)
Complete a data table using information both given and implied in the problem.	Assume that point 1 is at $h_1 = 0$ and that $h_2 = 3.0$ m.
	$P_1 = 1.2 \times 10^5 \text{ N/m}^2 \qquad \rho = 1.0 \times 10^3 \text{ kg/m}^3$
	$v_1 = ? \qquad\qquad\qquad P_2 = ?$
	$h_1 = 0 \text{ m} \qquad\qquad\quad h_2 = 3.0 \text{ m}$
	$g = 9.8 \text{ m/s}^2 \qquad\qquad v_2 = ?$
	$R = 0.10 \text{ m}^3/\text{s}$

a. 2 Use the rate flow equation to solve for the velocity at each point.	The volume rate flow remains constant throughout the pipe. Determine v_1 and v_2 using the rate flow equation. $R = A_1 v_1$ where $A = \pi(\text{diameter}/2)^2$ $0.10 \text{ m}^3/\text{s} = (3.14)(0.15 \text{ m}/2)^2 v_1$ $v_1 = 5.7 \text{ m/s}$ $R = A_2 v_2$ $0.10 \text{ m}^3/\text{s} = (3.14)(0.20 \text{ m}/2)^2 v_2$ $v_2 = 3.2 \text{ m/s}$
a. 3 Use Bernoulli's equation and solve for the pressure at point 2.	$h_1 = 0$, then $\rho g h_1 = 0$ $1.2 \times 10^5 \text{ N/m}^2 + \frac{1}{2}(1.0 \times 10^3 \text{ kg/m}^3)(5.7 \text{ m/s})^2 + 0 =$ $P_2 + \frac{1}{2}(1.0 \times 10^3 \text{ kg/m}^3)(3.2 \text{ m/s})^2 + (1.0 \times 10^3 \text{ kg/m}^3)(9.8 \text{ m/s}^2)(3.0 \text{ m})$ $1.2 \times 10^5 \text{ N/m}^2 + 1.6 \times 10^4 \text{ N/m}^2 = P_2 + 5.1 \times 10^3 \text{ N/m}^2 + 2.9 \times 10^4 \text{ N/m}^2$ $P_2 = 1.02 \times 10^5 \text{ N/m}^2$

PRACTICE PROBLEMS

PROBLEM 1. A container weighs 25.0 grams when empty, 50.0 grams when completely filled with water, and 48.0 grams when completely filled with an unknown liquid. Determine the a) specific gravity and b) density of the unknown liquid.

ANS. a) 0.92, b) 0.92 g/cm^3

PROBLEM 2. A piece of metal attached to a spring scale weighs 9.8 N in air and 4.0 N when completely immersed in water. Determine the density of the metal.

ANS. 1.7 x 10^3 kg/m^3

PROBLEM 3. The specific gravity of a solid piece of wood is determined to be 0.70. What fraction of the volume of the wood will be submerged if it floats in a liquid of specific gravity 1.2?

ANS. 0.58

PROBLEM 4. A garden hose of internal diameter 0.030 m is connected to a nozzle which has an internal diameter of 0.0050 m in diameter. The hose lies horizontally on the ground. The water in the hose travels at 1.00 m/s and the pressure of the water is 1.0×10^5 N/m². Determine the a) speed at which the water leaves the nozzle and b) gauge pressure of the water as it leaves the nozzle.

ANS. a) 36 m/s, b) 3.5×10^5 N/m²

CHAPTER 11

VIBRATIONS AND WAVES

KEY TERMS AND PHRASES

simple harmonic motion (SHM) refers to periodic vibrations or oscillations that exhibit two characteristics: 1) the force acting on the object and the magnitude of the object's acceleration are always directly proportional to the displacement of the object from its equilibrium position, and 2) both the force vector and the acceleration vector are directed opposite to the displacement vector and therefore in toward the object's equilibrium position.

amplitude (A) of an object undergoing SHM refers to the maximum displacement of the object from the equilibrium position.

period (T) of the motion is the time required for the motion to repeat.

frequency (f) refers to the number of complete repetitions of the motion that occur each second. The frequency is inversely related to the period.

simple pendulum is assumed to have its entire mass concentrated at the end of its length. The simple pendulum undergoes SHM if the maximum angle that it is displaced from equilibrium is small (approximately 15° or less).

damping is the loss of mechanical energy as the amplitude of motion in a simple harmonic oscillator gradually decreases.

forced vibrations have the same frequency as the external force and not necessarily equal to the natural frequency of the object.

resonance occurs if the external forced vibrations have the same frequency as one of the natural frequencies of the object. The natural frequency (or frequencies) at which resonance occurs is called the **resonant frequency**.

transverse wave is a wave in which the particles of the medium move at right angles to the direction of motion of the wave.

crest is the highest point of that portion of a transverse wave above the equilibrium position.

trough is the lowest point of that portion of a transverse wave below the equilibrium position.

longitudinal wave is a wave in which the particles of the medium move back and forth, parallel to the direction of motion of the wave.

compressions or **condensations** are regions in a longitudinal wave where the density of particles of the medium is greater than when the medium is at equilibrium. It is convenient to compare a region of compression with the crest of a transverse wave.

expansions or **rarefactions** are regions in a longitudinal wave where the density of particles of the medium is less than when the medium is at equilibrium. It is convenient to compare an expansion with the trough of a transverse wave.

wavelength (λ) is the distance between any two repeating points on a periodic wave, e.g., the distance between adjacent crests or adjacent compressions.

intensity (I) of a wave is defined as the power transmitted across a unit area (A) perpendicular to the direction of energy flow.

interference results when two waves pass through the same region of space at the same time.

principle of superposition states that when two waves pass through a medium at the same time, the resultant displacement of the medium at any particular moment of time equals the algebraic sum of the displacements of the component waves at that point.

destructive interference occurs if the amplitude of the resultant of two interfering waves is smaller than the displacement of either wave.

constructive interference occurs if the amplitude of the resultant of two interfering waves is larger than the displacement of either wave.

diffraction refers to the ability of waves to bend around obstacles. The amount of diffraction depends on the wavelength of the waves and the size of the obstacle.

standing waves are produced by the superposition of two periodic waves having identical frequencies and amplitudes which are traveling in opposite directions.

nodal points are fixed positions along the entire length of a standing wave where the displacement is always zero. The nodal points are found at half wavelength ($\frac{1}{2}\lambda$) intervals along the length of the medium.

antinodal points are points of maximum amplitude located halfway between adjacent nodal points. The displacement of the medium at the point fluctuates between a crest and a trough. The amplitude of the wave at the antinodal points equals the sum of the amplitudes of the two component waves which produce the standing wave pattern.

first harmonic or **fundamental frequency** or **first mode of vibration** refers to the lowest possible frequency that can produce a standing wave. Overtones refer to higher frequencies which also produce standing waves.

SUMMARY OF MATHEMATICAL FORMULAS

period of an object in SHM at the end of a spring	$T = 2\pi (m/k)^{1/2}$	The period (T) is related to the object's mass (m) as well as the force constant (k) of the spring.
velocity of an object in SHM at the end of a spring	$v = v_o (1 - x^2/A^2)^{1/2}$ or $v = [(k/m)(A^2 - x^2)]^{1/2}$	v is the velocity of the object at displacement x from equilibrium and v_o is the maximum velocity of the object. A is the amplitude of the motion.
total energy of an object in SHM at the end of a spring	$E = \frac{1}{2} m v^2 + \frac{1}{2} k x^2$	The total energy(E) equals the sum of the kinetic energy and the potential energy.
period of a simple pendulum	$T = 2\pi (L/g)^{1/2}$	The period of a simple pendulum depends on its length and the acceleration due to gravity.
wave speed	$v = f \lambda$	The speed (v) of a periodic wave equals the product of the frequency (f) and the wavelength (λ).
displacement of a particle in SHM	$x = A \cos \omega t$ $\omega = 2\pi f = 2\pi/T$	The displacement (x) of a particle in a transverse wave in SHM as a function of time t. Note: at t = 0, x = A.
velocity of a particle in SHM	$v = -\omega A \sin \omega t$	The velocity (v) of a particle in a transverse wave in SHM as a function of time t. Note: at t = 0, v = 0.
acceleration of a particle in SHM	$a = -\omega^2 A \cos \omega t$	The acceleration (a) of a particle in a transverse wave in SHM as a function of time t. Note: at t = 0, the acceleration is a maximum but the direction is opposite that of the displacement.
velocity of a transverse wave on a string	$v = [F_T/(m/L)]^{1/2}$	The velocity of a transverse wave on a string is related to the tension (F_T) and the mass per unit length (m/L) of the string.
velocity of a longitudinal wave	solid rod $v = (E/\rho)^{1/2}$ liquid $v = (B/\rho)^{1/2}$ or gas	The velocity of a longitudinal wave through a solid depends on the elastic modulus (E) and the density of the medium (ρ). For a wave traveling through a liquid or gas, the velocity depends on the **bulk modulus** (B) and the density (ρ).

energy (E) transmitted by a wave	$E = 2 \pi^2 \rho A v t f^2 x_o^2$	The energy (E) transmitted by a wave depends on the: density of the medium (ρ), cross-sectional area (A) through which the wave travels, distance (vt) the wave travels in time t, frequency (f), and amplitude x_o.
power of the wave	$\bar{P} = 2 \pi^2 \rho A v f^2 x_o^2$	Power is the average rate of energy transfer.
intensity (I) of a wave	$I = 2 \pi^2 v \rho f^2 x_o^2$	The intensity (I) of a wave is the power transmitted across a unit area (A) perpendicular to the direction of energy.
The frequencies of harmonics of standing waves on a string.	$f_n = v/\lambda_n = v/(2L/n)$	The frequency of the particular harmonic depends of the velocity (v) of the wave, the length of the string, and the number of the harmonic (n = 1,2,3, etc.).

CONCEPT SUMMARY

Simple Harmonic Motion

Simple harmonic motion (SHM) refers to periodic vibrations or oscillations that exhibit two characteristics: 1) the force acting on the object and the magnitude of the object's acceleration are always directly proportional to the displacement of the object from its equilibrium position, and 2) both the force vector and the acceleration vector are directed opposite to the displacement vector and therefore in toward the object's equilibrium position. The force acting on an object undergoing SHM follows Hooke's law $F = - k x$ where F is the restoring force acting on the object, k is a proportionality constant called the force constant, and x is the magnitude of the object's displacement from equilibrium. At the equilibrium position the net force on the object equals zero. The negative sign indicates that the force and displacement vectors are oppositely directed.

Amplitude, Frequency, and Period

The amplitude (A) of an object undergoing SHM refers to the maximum displacement of the object from the equilibrium position. The period (T) of the motion is the time required for the motion to repeat while the frequency (f) refers to the number of complete repetitions of the motion that occur each second. The frequency is inversely related to the period, $f = 1/T$.

Energy in the Simple Harmonic Oscillator

The potential energy stored in a vibrating system undergoing SHM is given by $PE = \frac{1}{2} k x^2$

and was previously discussed in Chapter 6. The total energy stored equals the sum of the kinetic energy and potential energy. Assuming that no energy is dissipated due to friction, this total energy remains constant.

$$E = KE + PE = \frac{1}{2} m v^2 + \frac{1}{2} k x^2$$

At maximum displacement from equilibrium, $v = 0$ and $x = A$. At this point $E = \frac{1}{2} k A^2$. When the displacement equals zero, i.e., $x = 0$, all of the energy is in the form of kinetic energy and $E = \frac{1}{2} m v^2$.

Using the energy principle, it is possible to show that the velocity of the object at any point in its motion can be determined from the equation

$$v = v_o (1 - x^2/A^2)^{\frac{1}{2}} \quad \text{or} \quad v = [(k/m)(A^2 - x^2)]^{\frac{1}{2}}$$

v is the velocity of the object at a displacement x from equilibrium, v_o is the maximum velocity of the object. Maximum velocity occurs as the object is passing through the equilibrium position. A is the amplitude of the motion.

The Period of SHM: the Reference Circle

Using the circle as a reference, an analogy can be drawn between the one-dimensional back and forth motion of an object undergoing SHM while attached to a spring with one component of the two-dimensional motion exhibited by an object traveling in a circle. As a result of this analogy, the following equation may be derived for the period of an object of mass m attached to a spring of force constant k and undergoing SHM:

$$T = 2 \pi (m/k)^{\frac{1}{2}}$$

The larger the mass of the attached object, the greater its inertia and the longer its period. The force constant k is related to the "stiffness" of the spring. The greater the "stiffness" of the spring, the greater the magnitude of the restoring force and acceleration at various displacements and the shorter the period.

SHM is Sinusoidal

Using the motion of a spring undergoing SHM as a reference, it is possible to show that the graph of the displacement of the object with time has the form of a sine or cosine wave, i.e., the position varies as a function of time. If the displacement of the object equals the amplitude (A) at $t = 0$, then the equation for the displacement, velocity, and acceleration as a function of time can be written as follows:

displacement $\quad x = A \cos \omega t \qquad$ where $\qquad \omega = 2 \pi f \quad$ or $\quad \omega = 2\pi/T$

velocity $\qquad v = - \omega A \sin \omega t \qquad$ acceleration $\quad a = - \omega^2 A \cos \omega t$

A graph of displacement vs. time for an object undergoing SHM at the end of a spring is shown below:

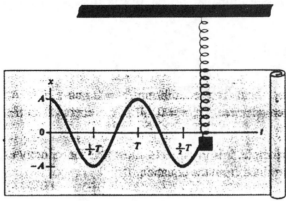

Simple Pendulum

A **simple pendulum** is assumed to have its entire mass concentrated at the end of a light string. The simple pendulum undergoes SHM if the maximum angle that it is displaced from equilibrium is small (approximately 15° or less). The formula for the period of motion is

$$T = 2 \pi \, (L/g)^{1/2}$$

L is the length of the pendulum and g is the gravitational acceleration. The period of a simple pendulum depends only on its length and the value of g; because of this it can be used as a timing device. The first pendulum clock was built by Christaan Huygens (1629-1695).

Damped Harmonic Motion

A system undergoing SHM will exhibit **damping**. Damping is the loss of mechanical energy as the amplitude of motion gradually decreases. In the mechanical systems studied in the previous sections, the losses are generally due to air resistance and internal friction and the energy is transformed into heat.

For the amplitude of the motion to remain constant, it is necessary to add enough mechanical energy each second to offset the energy losses due to damping. A simple example of this is a small child on a swing. If the parent stops giving a periodic push to the swing, its amplitude will gradually decrease until it stops at the equlibrium position.

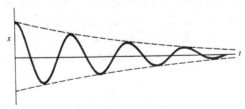

In many instances damping is a desired effect. For example, shock absorbers in a car remove unwanted vibration.

Forced Vibrations: Resonance

An object subjected to an external oscillatory force tends to vibrate. The vibrations that result are called **forced vibrations**. These vibrations have the same frequency as the external force and not the natural frequency of the object.

If the external forced vibrations have the same frequency as the natural frequency of the object, the amplitude of vibration increases and the object exhibits **resonance**. The natural frequency (or frequencies) at which resonance occurs is called the **resonant frequency**. Resonance can be beneficial, e.g., the small periodic additions of energy from the parent to the child on the swing can result in a large amplitude of swing. However, resonance can also be destructive. Proper damping must be included in the design of buildings, bridges, grandstands, etc., to ensure that structural damage does not occur.

Wave Motion

A wave is, in general, a disturbance which moves through a medium. It carries energy from one location to another without transporting the material of the medium. Examples of mechanical waves include water waves, waves on a string, and sound waves.

Transverse Waves

In a **transverse wave**, the particles of the medium move at right angles to the direction of motion of the wave. The top part of the wave is called the **crest** while the portion of the wave below the equilibrium position is called the **trough**. A wave on a rope, as shown below, approximates a transverse wave.

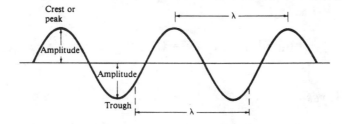

Longitudinal Waves

In a **longitudinal wave**, the particles of the medium move back and forth, parallel to the direction of motion of the wave. The result is a series of expansions alternating with compressions which travel along the length of the medium. The diagram below is a representation of a longitudinal wave traveling along a coiled spring, e.g. a slinky.

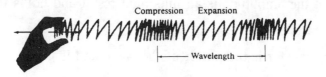

The region where the particles of the medium are close together is called a **compression** (or **condensation**) while the region where the particles are farther apart is called an **expansion** (or **rarefaction**). In analyzing longitudinal waves, it is convenient to compare a region of compression with the crest of a transverse wave and an expansion with the trough of a transverse wave.

Periodic Waves

The velocity (v) of a wave through a medium remains fixed as long as the medium does not change. For a given **periodic** wave, the velocity is given by

$$v = f \lambda$$

where the frequency (f) is the number of waves passing a particular point each second. The SI unit of frequency is the hertz (Hz) where 1 Hz = 1 wave per second. The wavelength (λ) is the distance between any two repeating points on the wave, e.g. the distance between adjacent crests or adjacent compressions.

The velocity of a wave depends on the properties of the medium through which it travels. For a transverse wave on a string, the velocity is given by the following equation:

$$v = [F_T/(m/L)]^{1/2}$$

The tension in the string is F_T, while m/L is the mass per unit length of the string.

For a longitudinal wave, the velocity depends on the elastic force factor and the density of the medium (ρ). For a longitudinal wave traveling through a solid rod, the elastic force factor is the **elastic modulus** (E). For a longitudinal wave traveling through a liquid or gas, the elastic force factor is the **bulk modulus** (B). The following equations are used to determine the velocity of the wave:

solid rod $\quad v = (E/\rho)^{1/2}$ $\qquad$ liquid or gas $\quad v = (B/\rho)^{1/2}$

Energy Transmitted by Waves

The energy (E) transmitted by a wave is given by

$$E = 2 \pi^2 \rho A v t f^2 x_o^2$$

where A is the cross-sectional area through which the wave travels and not the amplitude of the wave, vt is the distance the wave travels in time t, f is the frequency, and x_o is the amplitude of the wave. The fact that the energy transmitted by the wave is proportional to the square of the frequency and the square of the amplitude is an important result and holds for all types of waves.

The average rate of energy transfer is the power of the wave and is given by

$$\bar{P} = E/t = 2 \pi^2 \rho A v f^2 x_o^2$$

The **intensity** (I) of a wave is defined as the power transmitted across a unit area (A) perpendicular to the direction of energy flow and is given by

$$I = \bar{P}/A = 2 \, \pi^2 \, v \, \rho \, f^2 \, x_o^2$$

As waves travel outward from the center of a disturbance, e.g., an earthquake, the value of the area (A) increases and the intensity and amplitude of the waves decrease. If the wave is in an isotropic medium (same in all directions), the wave is spherical and both intensity and amplitude decrease with the square of the distance from the source. For a sphere, the surface area is $A = 4 \, \pi \, r^2$.

Behavior of Waves: *Reflection*

As shown in the diagram, the wave front of a wave striking a straight barrier follows the **law of reflection**. This law states that the **angle of incidence** equals the **angle of reflection**, $\theta_i = \theta_r$. The angle of incidence is the angle between the direction of the incoming wave and a normal drawn to the surface. The angle of reflection is the angle between the direction of the reflected wave and the normal to the surface.

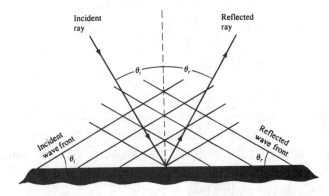

A transverse wave traveling in a rope undergoes reflection at the point in the rope where the medium changes. If the rope is fixed at one end, as shown in diagram a, a wave crest will undergo a 180° phase change upon reflecting from this point, i.e., a crest will reflect as a trough and vice versa. If the end is free as shown in diagram b, then no phase change occurs and a crest reflects as a crest and a trough reflects as a trough.

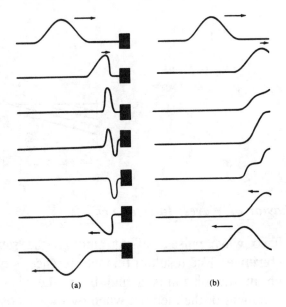

(a)   (b)

Behavior of Waves: Partial Transmission and Partial Reflection of a Wave Pulse at the Interface of Two Mediums

When a crest traveling through a light section of rope reaches a point where it enters a heavier section, then part of the crest is reflected as a trough and part is transmitted as a crest (diagram a). In diagram b, a crest is traveling through a heavy section of rope and reaches a point where it enters a lighter section of rope, then part of the crest is reflected as a crest and part of the crest is transmitted as a crest. A phase change occurs only for that part of the pulse that reflects upon entering a heavier medium.

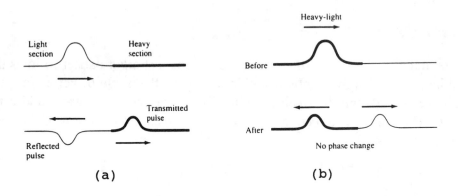

(a)　　　　(b)

Behavior of Waves: Refraction

A water wave traveling from deep water into shallow water changes direction. This phenomenon is known as **refraction**. The velocity is greater in deep water and, as can be seen in the diagram below, the change in direction is such that the angle of incidence is greater than the angle of refraction. The angle of refraction is defined as the angle between the direction of the refracted wave and the normal to the interface between the two sections.

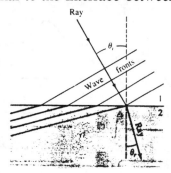

Behavior of Waves: Interference

Two wave pulses passing through the same region of space at the same time exhibit **interference**. The result of this interference is predicted by the **principle of superposition**. As shown in the diagrams a and b on the next page, this principle states that the resultant displacement of the medium when two waves overlap equals the algebraic sum of their separate displacements.

Destructive interference occurs if the amplitude of the resultant is smaller than either wave pulse. For example, as shown in diagram a, if a crest overlaps with a trough of equal amplitude, e.g., 1 meter, the resulting displacement equals zero, i.e., 1 m - 1 m = 0. After the momentary interaction, the pulses continue moving in their original directions with their original displacements.

Constructive interference occurs if the resultant is greater than the amplitude of either wave pulse. For example, as shown in diagram b, if two wave pulses, each 1 meter high and traveling in opposite directions, interfere, the result will be a momentary displacement of the medium of 2 meters, i.e., 1 m + 1 m = 2 m.

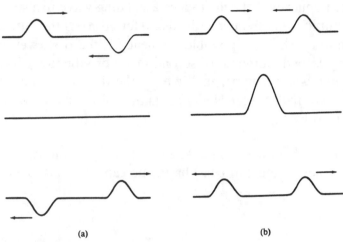

(a)   (b)

Behavior of Waves: Diffraction

Waves tend to bend as they pass an obstacle. For example, water waves passing a log in a river will tend to bend around the log after they pass. This bending is known as **diffraction**. The amount of diffraction depends on the wavelength of the waves and the size of the obstacle.

Standing Waves in Strings

As shown in the diagram below, a **standing wave** is produced by the superposition of two periodic waves having identical frequencies and amplitudes which are traveling in opposite directions. In stringed musical instruments, the standing wave is produced by waves reflecting off a fixed end and interfering with oncoming waves as they travel back through the medium.

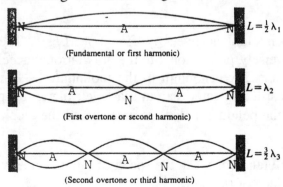

$L = \frac{1}{2}\lambda_1$

(Fundamental or first harmonic)

$L = \lambda_2$

(First overtone or second harmonic)

$L = \frac{3}{2}\lambda_3$

(Second overtone or third harmonic)

If a string is of fixed length and tension, the standing wave is produced only for certain wavelengths and resonant frequencies. A standing wave has certain points along its entire length where the displacement is always zero. These points are called **nodal points**. The nodal points are found at half wavelength ($\frac{1}{2}\lambda$) intervals along the length of the string.

Points of maximum amplitude, called **antinodal points**, are located halfway between adjacent nodal points. The displacement of the medium at the point fluctuates between a crest and a trough. The amplitude of the wave at the antinodal points equals the sum of the amplitudes of the two component waves which produce the standing wave pattern.

The lowest possible frequency that can produce a standing wave in a string fixed at both ends is called the **first harmonic**. This frequency is also referred to as the **fundamental frequency** or **first mode of vibration**. The next possible frequency is the first overtone. This frequency is also referred to as the second harmonic or second mode of vibration. The frequencies of the harmonics are consecutive whole number multiples of the first harmonic. The positions of the nodal points are designated by the letter N and the antinodal points by the letter A. The length of the string is represented by L.

In general, $\frac{1}{2}\lambda_n = L/n$, where n = 1, 2, 3, 4, etc., and n is the number of vibrating segments in the string. The frequency of the particular harmonic can be determined as follows:

$f = v/\lambda_n = v/(2L/n)$

SELECTED TEXT QUESTIONS WITH ANSWERS

QUESTION 2. Is the acceleration of a simple harmonic oscillator ever zero? If so, where?

ANSWER: One condition for SHM is a restoring force that is proportional to displacement from equilibrium. For a spring the restoring force follows Hooke's law ($F = -kx$). At the point where the displacement equals zero, the force exerted by the spring is zero. Based on Newton's second law $a = F/m$, if $F = 0$ then $a = 0$. As an object in SHM passes through the equilibrium position it has maximum speed but zero acceleration.

QUESTION 8. If a pendulum clock is accurate at sea level, will it gain or lose time when taken into the mountains?

ANSWER: The period of a pendulum is given by $T = 2\pi (L/g)^{\frac{1}{2}}$. The period is directly proportional to the length but inversely proportional to the gravitational acceleration. In section 5-6, it was shown that $g = Gm/r^2$. A pendulum in the mountains is in a region where the distance from the center of the earth is greater than at sea level; therefore, g is less than at sea level. If g is less the period of the pendulum clock is greater and the clock loses time.

QUESTION 16. Explain the difference between the speed of a transverse wave traveling down a cord and the speed of a tiny piece of the cord.

ANSWER: In a transverse wave, the tiny piece moves up and down while the wave moves in a direction perpendicular to the tiny piece. As shown in the diagram on page 11-7, if the wave is a periodic transverse wave, then the tiny piece is moving up and down in SHM. The velocity of the tiny piece is a minimum at the top and bottom of its motion and a maximum as it passes through the equilibrium position.

The wave travels along the cord in a direction perpendicular to the direction of motion of a tiny piece of cord. The speed of the wave along the length of the cord remains constant as long as the tension in the cord and the mass per unit length remain constant.

QUESTION 21. Two linear waves have the same amplitude and are otherwise identical, except one has half the wavelength of the other. Which transmits more energy? How much more energy does it transmit?

ANSWER: The relation between the energy transported by a wave and the frequency is given by equation $E = 2\pi^2\rho Avf^2x_o^2$, where ρ is the density of the medium, A is the cross-sectional area that the wave is passing through, v is the speed of the wave, f is the frequency of the wave, and x_o is the amplitude of the wave.

The frequency (f) of a wave is related to the wavelength by the equation $f = v/\lambda$. Based on this equation the wave that has half the wavelength must have twice the frequency. According to equation $E = 2\pi^2\rho Avf^2x_o^2$, the higher frequency wave carries the greater energy. Since the energy is related to the square of the frequency, the shorter wavelength wave has twice the frequency and therefore 2^2 or four times the energy.

PROBLEM SOLVING SKILLS

For problems involving a spring undergoing SHM:

1. Complete a data table listing the magnitude of the restoring force at maximum displacement, mass of object, amplitude of motion, etc.
2. Use Hooke's law to determine the force constant of the spring.
3. To determine the period, use the formula that relates the period to the force constant and the mass of the object.
4. Determine the total energy of the system and use the law of conservation of energy to determine the velocity of the object at any point in the motion.

For problems involving a simple pendulum undergoing SHM:

1. Use $T = 2\pi (L/g)^{1/2}$ to solve for the period, length, or the magnitude of the gravitational acceleration.

For problems involving the speed of a longitudinal waves passing through a solid, liquid, or gas:

1. If the medium is a long solid rod, note the density and the elastic modulus of the rod and solve for the velocity.
2. If the medium is a liquid or a gas, note the bulk modulus and the density and solve for the velocity.

For problems involving energy transported by a wave, the power of a wave, and wave intensity:

1. Complete a data table and list the velocity, frequency, and amplitude of the wave and the cross-sectional area through which the wave travels.
2. Apply the formula for energy, power, and intensity to solve the problem.

For problems involving standing waves in strings:

1. Complete a data table listing the tension, mass and length of string, and the frequency of the vibration causing the standing wave.
2. Draw an accurate diagram for the harmonic described in the problem. Locate the position of the nodes and antinodes. Determine the wavelength of the wave.
3. If the frequency is known, use $v = f \lambda$ to determine the velocity. If the tension and mass per unit length are known, use $v = [F_T/(m/L)]^{1/2}$ to solve for the velocity.
4. Solve for the unknown quantity.

PROGRAMMED PROBLEMS

PROBLEM 1. A 1.0 kg object is attached to a spring. The force constant of the spring is 35.0 N/m. The object is displaced 0.100 m from the equilibrium position. After release, the object exhibits SHM. Determine the a) period of the motion, b) total energy of the system, and c) object's speed when it is 0.0500 meter from equilibrium.

a. 1 Complete a data table using information both given and implied in the problem.	Solution: (Sections 11-1 and 11-2) $k = 35.0$ N/m $\qquad v_o = ?$ $m = 1.0$ kg $\qquad v$ (at 0.0500 m) $= ?$ $A = 0.100$ m $\qquad v$ (at 0.100 m) $= 0$
a. 2 Determine the period of the motion.	$T = 2 \pi (m/k)^{\frac{1}{2}} = 2 \pi (1.0 \text{ kg}/35.0 \text{ N/m})^{\frac{1}{2}} = 1.06$ s
b. 1 Determine the total energy in system.	The total energy of the system remains constant and equals the sum of the kinetic energy and potential energy at any point. The velocity, and therefore the kinetic energy, at maximum displacement of the object equals zero. The total energy equals the potential energy stored the spring at this point. $E = \frac{1}{2} m v^2 + \frac{1}{2} k A^2 = 0 \text{ J} + \frac{1}{2} (35.0 \text{ N/m})(0.100 \text{ m})^2$ $E = 0.175$ J
c. 1 Determine the object's speed at x = 0.0500 m.	Use the law of conservation of energy to solve for the velocity. $E = \frac{1}{2} m v^2 + \frac{1}{2} k x^2$ $0.175 \text{ J} = \frac{1}{2} (1.0 \text{ kg}) v^2 + \frac{1}{2} (35.0 \text{ N/m})(0.0500 \text{ m})^2$ $0.175 \text{ J} = 0.5 v^2 + 0.0438 \text{ J}$ $v = 0.513$ m/s An alternate approach would be to use the equation $v = v_o (1 - x^2/A^2)^{\frac{1}{2}}$ Use the energy method to determine the velocity at equilibrium (v_o). At equilibrium, x = 0. $E = \frac{1}{2} m v^2 + \frac{1}{2} k x^2$

$$0.175 \text{ J} = \tfrac{1}{2} (1.0 \text{ kg}) \, v^2 + \tfrac{1}{2} (35.0 \text{ N/m})(0 \text{ m})^2$$

$v_o = 0.592$ m/s,   then

$$v = v_o(1 - x^2/A^2)^{\frac{1}{2}}$$

$$= (0.592 \text{ m/s}) \, [1 - (0.0500 \text{ m})^2/(0.100 \text{ m})^2]^{\frac{1}{2}}$$

$$v = 0.513 \;\; \text{m/s}$$

Problem 2. An object oscillates with SHM according to the equation $x = 2.00 \cos \pi t$ meters, as shown in the diagram at the top of page 11-6. Determine the a) amplitude, frequency, and period of the motion and b) displacement, velocity, and acceleration of the object at $t = \tfrac{1}{3}$ s.

a. 1 Determine the amplitude, frequency, and period of the motion.	Solution: (Section 11-3) The equation follows the general form of a sinusoidal function in SHM, $x = A \cos \omega t$. Thus, the amplitude $A = 2.00$ m. $\omega = 2 \pi f$, then $2 \pi f = \pi$ rad/s $f = (\pi \text{ rad/s})/(2 \pi \text{ rad}) = 0.500$ vibrations/s $f = 0.500$ Hz The period may now be determined: $T = 1/f = 1/(0.500 \text{ vib/s}) = 2.00$ s
b. 1 Determine the displacement, velocity, and acceleration of the object at $t = \tfrac{1}{3}$ second.	Solve for the displacement at $t = \tfrac{1}{3}$ s by substituting this value into the displacement equation. $x = A \cos \omega t$ meters $= (2.00 \text{ m}) \cos (\pi \text{ rad/s})(\tfrac{1}{3} \text{ s})$ $x = (2.00 \text{ m})(\cos \pi/3 \text{ rad})$ but $(\pi/3 \text{ rad})(360°/2\pi \text{ rad}) = 60.0°$ $x = (2.00 \text{ m}) \cos 60.0° = (2.00 \text{ m})(0.500)$ $x = 1.00$ m Solve for the velocity and acceleration by substituting $t = \tfrac{1}{3}$ s into the appropriate equations.

$$v = - A\omega \sin \omega t$$

$$= - (2.00 \text{ m})(\pi \text{ rad/s}) [\sin (\pi \text{ rad/s})(\tfrac{1}{3} \text{ s})]$$

$$v = - (2.00 \pi \text{ m/s}) \sin 60.0° = - 5.44 \text{ m/s}$$

$$a = - A \omega^2 \cos \omega t$$

$$= - (2.00 \text{ m/s})(\pi \text{ rad/s})^2 [\cos (\pi \text{ rad/s})(\tfrac{1}{3} \text{ s})]$$

$$a = - (19.7 \text{ m/s}^2)(\cos 60.0°) = - 9.86 \text{ m/s}^2$$

PROBLEM 3. A simple pendulum is used in a physics laboratory experiment to obtain an experimental value for the gravitational acceleration. A particular student measures the length of the pendulum to be 0.705 meter, displaces it 10.0° from the equilibrium position, and releases it. Using a stopwatch, the student determines that the period of the pendulum is 1.68 s. Determine the experimental value of the gravitational acceleration.

a. 1

Rearrange the formula for the period of a simple pendulum and and solve for g.

Solution: (Section 11-4)

$T = 2 \pi (L/g)^{\frac{1}{2}}$ can be used because a 10° angle is considered to be a small angle.

$T = 2 \pi (L/g)^{\frac{1}{2}}$ and $T^2 = 4 \pi^2 L/g$,

solving for g gives

$$g = 4 \pi^2 L/T^2 = 4 \pi^2 (0.705 \text{ m})/(1.68 \text{ s})^2$$

$$g = 9.85 \text{ m/s}^2$$

PROBLEM 4. Determine the a) speed of sound through a long aluminum rod and b) wavelength of the sound waves produced by a 510 Hz vibration in the rod. The elastic modulus of aluminum is 70×10^9 N/m² and the density is 2.7×10^3 kg/m³.

a. 1

Determine the speed of sound in the aluminum rod

Solution: (Section 11-8)

The speed of sound in a metal rod depends on the elastic modulus (E) and the density (ρ). Substitute the values into the equation for the speed of sound in a metal rod.

$$v = (E/\rho)^{\frac{1}{2}}$$

$$= [(70 \times 10^9 \text{ N/m}^2)/(2.7 \times 10^3 \text{ kg/m}^3)]^{\frac{1}{2}}$$

$$v = 5100 \text{ m/s}$$

b. 1	The frequency and velocity are now known. Use the equation for the speed of a periodic wave to determine the wavelength.
Determine the wavelength of the sound waves in the rod.	$v = f \lambda$, then 5100 m/s = (510 Hz) λ
	λ = 10 m

PROBLEM 5. A sound wave has a frequency of 510 Hz. Determine the wavelength of the sound produced as it passes through a) air on a day when the speed of sound is 340 m/s and b) water. The bulk modulus of water is 2.0×10^9 N/m² and the density is 1.0×10^3 kg/m³.

a. 1	Solution: (Section 11-7 and 11-8)
Determine the wavelength.	$\lambda = v/f = (340 \text{ m/s})/(510 \text{ m/s})$
	λ = 0.67 m

| b. 1 | Both the bulk modulus and the density of the water are known; the velocity of the waves may be determined. |
| Determine the velocity of the waves as they travel through water. | $v = (B/\rho)^{\frac{1}{2}} = [(2.0 \times 10^9 \text{ N/m}^2)/(1.0 \times 10^3 \text{ kg/m}^2)]^{\frac{1}{2}}$ = 1400 m/s |

b. 2	$\lambda = v/f$
Determine the wavelength of the waves.	= (1400 m/s)/(510 Hz)
	λ = 2.75 m

PROBLEM 6. Spherical sound waves are emitted from a 10.0 watt source in an isotropic medium. Determine the wave intensity a) 1.0 m and b) 2.0 m from the source.

a. 1	Solution: (Section 11-9)
Determine the area of a sphere 1.0 m in diameter. Then determine the intensity at 1.0 m from the 10 watt source.	The intensity is related to the power by $I = \bar{P}/A$ where the area of a sphere is given by $A = 4 \pi r^2$.
	$I = \bar{P}/A = (10.0 \text{ W})/[4 \pi (1.0 \text{ m})^2]$ = 0.80 W/m²

| b. 1 | $I = \bar{P}/A = (10.0 \text{ W})/[4 \pi (2.0 \text{ m})^2]$ = 0.20 W/m² |
| Apply the intensity formula for r = 2.0 m. | Based on the results of parts a and b, note that the intensity is decreasing as the square of the distance from the source. |

a. 1	Solution: (Section 11-12)
Draw a diagram representing the third harmonic.	$L = \frac{3}{2}\lambda_3$ (Second overtone or third harmonic)

a. 2	The nodal points are $\frac{1}{2}\lambda$ apart; the wavelength of the waves can be determined.
Use the diagram to calculate the wavelength.	$\frac{1}{2}\lambda_3 = \frac{1}{3}L$ $\lambda_3 = \frac{2}{3}L = \frac{2}{3}(50.0 \text{ cm}) = 33.4 \text{ cm} = 0.334 \text{ m}$

a. 3	$v = f\,\lambda_3$
Determine the velocity of the periodic waves.	$= (200 \text{ Hz})(0.334 \text{ m})$ $v = 66.7 \text{ m/s}$

b. 1	$v = [F_T/(m/L)]^{\frac{1}{2}}$ and $v^2 = F_T/(m/L)$
Rearrange the equation that relates the tension in a string to the velocity of the waves and solve for the mass.	$F_T = v^2\, m/L$ and rearranging gives $m = (F_T\, L)/v^2$ $m = (50.0 \text{ N})(0.500 \text{ m})/(66.7 \text{ m/s})^2 = 5.62 \times 10^{-3} \text{ kg}$

a. 1	Solution: (Section 11-12)
Determine the velocity of the waves in the wire.	The tension, mass, and length of the wire are known. $v = [F_T/(m/L)]^{\frac{1}{2}} = [(6.00 \text{ N})/(0.00200 \text{ kg}/0.800 \text{ m})]^{\frac{1}{2}}$ $v = 49.0 \text{ m/s}$

b. 1

Draw a diagram of the standing wave produced by the second harmonic.

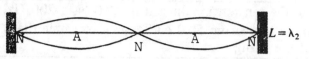

$L = \lambda_2$

(First overtone or second harmonic)

---

b. 2

Use the diagram to determine the wavelength.

$\frac{1}{2} \lambda_2 = \frac{1}{2} L$

$\lambda_2 = L = 0.800 \text{ m}$

---

b. 3

Solve for the frequency of the first harmonic.

$f = v/\lambda_1$

$\quad = (49.0 \text{ m/s})/(0.800 \text{ m})$

$f = 61.3 \text{ Hz}$

PRACTICE PROBLEMS

PROBLEM 1. A 0.20 kg mass is attached to the end of a spring of force constant 25.0 N/m. The spring is displaced 0.20 m and released. Determine the a) total energy stored in the spring, b) maximum velocity of the object, and c) object's velocity when it is 0.10 m from the equilibrium position.

ANS. a) 0.50 J, b) 2.2 m/s, c) 1.9 m/s

PROBLEM 2. A vertical spring stretches by 0.0500 m when a 0.100 kg mass is attached. The spring is then displaced 0.200 m below the new equilibrium position and released. Determine the a) force constant of the spring and b) period of the motion.

ANS. a) 19.6 N/m, b) 0.449 s

PROBLEM 3. Calculate the length of the pendulum in a grandfather's clock if it is set so that the frequency of vibration is 1.00 Hz. Assume that the value of the gravitational acceleration is 9.80 m/s^2.

ANS. 0.248 m

PROBLEM 4. A frequency of 120 Hz produces a third harmonic standing wave in a string 1.0 m long. The mass of the string is 5.0 x 10^{-4} kg. a) Draw an accurate diagram and determine the wavelength of the waves producing the standing wave. Determine the b) velocity of the waves and c) tension in the string.

ANS. a) 0.67 m, b) 80 m/s, c) 3.2 N

CHAPTER 12

SOUND

KEY TERMS AND PHRASES

sound is a longitudinal wave produced by a vibration which travels away from the source through solids, liquids, or gases, but not through a vacuum. Since a sound wave is a longitudinal wave there are regions of compression (condensation) and expansion (rarefaction) as the wave moves through the medium that transports it.

pitch refers to the frequency of a sound wave and is measured in hertz (Hz).

intensity level of sound is the energy transported by a wave per unit time per unit area. Intensity level of sound is measured in bels; however, the decibel (dB) is more commonly used.

open pipe is a wind instrument that is open to the air at both ends. The first harmonic standing wave that produces resonance has an antinode at near each end and one node in the middle of the air column.

closed pipe is a wind instrument that is open to the air at one end but closed at the other end. The first harmonic standing wave that produces resonance in the pipe has an antinode near the open end and a node at the closed end.

sound quality depends on the presence of overtones, their number, and relative amplitudes. The result gives each musical instrument its characteristic quality and timbre.

beats are regular pulsations in the loudness of a sound due to two waves of equal amplitude but slightly different frequencies traveling in the same direction. The number of beats per second is equal to the difference between the frequencies of the component waves and is known as the **beat frequency.**

Doppler effect refers to the perceived change of frequency of a wave when there is relative motion between the source and the listener.

shock waves and **sonic boom**s are produced when the speed of the source of sound exceeds the speed of sound. The sound waves in front of the source tend to overlap and constructively interfere. The superposition of the waves produce an extremely large amplitude, high energy wave called a shock wave. When the shock wave passes a listener, this energy is heard as a sonic boom.

SUMMARY OF MATHEMATICAL FORMULAS

speed of sound in air measured in meters per second	$v = 331 + 0.6\,T$ m/s	The speed of sound (v) in air changes with temperature. 331 m/s is the speed of sound in air at 0°C and T is the temperature of the air in centigrade.
sound intensity level	β (in dB) $= 10 \log I/I_o$	β is the intensity level in decibels. I is the intensity of the sound in W/m². The threshold of hearing $I_o = 1.0 \times 10^{-12}$ W/m².
harmonic frequency produced by an open pipe	$f_n = v/\lambda_n = v/(2L/n)$	The harmonic frequency produced by an open pipe depends on the speed of sound (v) and the length of the pipe (L). The frequencies of successive harmonics are consecutive whole number multiples of the first harmonic, i.e., n = 1, 2, 3, etc.
harmonic frequency produced by a pipe closed at one end	$f_n = v/\lambda_n = v/[4L/(2n - 1)]$	The harmonic frequency produced by an closed pipe depends on the speed of sound (v) and the length of the pipe (L). The frequencies of successive harmonics are consecutive odd whole number multiples of the first harmonic, i.e., n = 1, 3, 5, 7, etc.
sound interference pattern produced by two sources of sound which are in phase and have the same wavelength	constructive interference $\sin \theta = m\lambda/d$ destructive interference $\sin \theta = (m + \tfrac{1}{2})\lambda/d$	angular displacement (θ) of maxima and minima produced by two sources of sound which are in phase and have the same wavelength (λ). m = 1, 2, 3, etc.
beat frequency	$f_b = |f_1 - f_2|$	The beat frequency equals the difference between the two frequencies.

Doppler effect		The frequency (f ′) heard by the listener depends on the speed of sound (v), the frequency of the sound emitted by the source (f), speed of the source (v_s) or the speed of the listener (v_o), and the direction of motion of the source (or listener).
source moving toward a stationary listener	$f' = [1/(1 - v_s /v)] \, f$	
source moving away from a stationary listener	$f' = [1/(1 + v_s /v)] \, f$	
listener moving toward a stationary source	$f' = (1 + v_o/v) \, f$	
listener moving toward a stationary source	$f' = (1 - v_o/v) \, f$	

CONCEPT SUMMARY

Sound Waves

Sound is a longitudinal wave produced by a vibration which travels away from the source through solids, liquids, or gases, but not through a vacuum. Since a sound wave is a longitudinal wave there are regions of compression (condensation) and expansion (rarefaction) as the wave moves through the medium that transports it.

The speed is independent of the barometric pressure, frequency, and wavelength of the sound. However, the speed of sound in a gas is proportional to the temperature. The following equation is useful in determining the speed of sound in air

$v = 331 + 0.6 \, T$ m/s

where v is the speed of sound in meters per second, 331 is the speed of sound in m/s at 0°C, and T is the temperature in degrees centigrade.

Pitch

Pitch refers to the frequency of a sound wave and is measured in hertz (Hz). The range of human hearing is from 20 Hz to 20,000 Hz. This is called the audible range. This range is considered to be the limits of human hearing. The actual range that can be heard varies from person to person and the ear becomes less responsive to higher frequencies with age.

Sound waves above 20,000 Hz are called ultrasonic and have found application in medicine and other fields. Those below 20 Hz are called infrasonic waves.

Sound Intensity

The ear transforms the energy of sound waves into electrical signals which are carried to the brain by the nerves. The ear is not equally responsive to all frequencies of sound. It is most sensitive to sounds between 2000 Hz and 3000 Hz.

Loudness is a subjective physiological sensation in a human being that increases with the intensity of a sound. Our subjective sensation of loudness depends not only on **intensity** but also on frequency. A person easily hearing a sound at 1000 Hz may not be able to hear a sound of equal intensity at 50 Hz.

The loudness of a sound is approximately proportional to the common logarithm of the intensity. The intensity level is measured in **bels**, named after Alexander Graham Bell; however, the **decibel** (dB) is more commonly used. The decibel equals one-tenth bel, i.e. 1 dB = 0.1 bel. The formula for the intensity level is

$$\beta \text{(in dB)} = 10 \log I/I_o$$

where β is the intensity level in decibels, I is the intensity of the sound in watt/m^2, and I_o is the minimum intensity audible to the average person and is called the threshold of hearing.

$$I_o = 1.0 \times 10^{-12} \text{ watt/m}^2$$

The sound levels in dB that are common in everyday life extend from 0 dB (the threshold of hearing) to 140 dB for a jet plane 30 meters away. Ordinary conversation is approximately 65 dB while an indoor rock concert may be 120 dB, which is at the threshold of pain. Exposure to sound levels above 85 dB over an extended period of time may lead to permanent damage to a person's hearing. A table listing intensity levels and intensities of various sounds is given in the text.

Sources of Sounds: Musical Instruments

String Instruments

While the theory behind standing waves in stringed instruments was discussed in Chapter 11, it should be noted that the sounds produced by vibrating strings are not very loud. Thus, many stringed instruments make use of a sounding board or box, sometimes called a resonator, to amplify the sounds produced. The strings on a piano are attached to a sounding board while for guitar strings a sound box is used. When the string is plucked and begins to vibrate, the sounding board or box begins to vibrate as well. Since the board or box has a greater area in contact with the air, it tends to amplify the sounds.

Wind Instruments

The sounds that are produced by a **wind instrument** are the result of standing waves produced in the air contained within the instrument. In some wind instruments, such as woodwinds or brasses, the air is set into vibration by a vibrating reed or the vibrating lip of the musician. In other cases, e.g., the flute or organ, a stream of air is directed against one edge of an opening or mouthpiece. The resulting turbulence produces vibrations within the instrument. The vibrations cover a range of frequencies which are due to longitudinal standing waves which are created in the air column.

Open Pipe

A wind instrument that is open to the air at both ends is known as an **open tube** or **pipe**. The longitudinal standing wave that produces the sound has an antinode at each end and at least one node in the air column. Assuming that the speed of sound is constant, and $v = f \lambda$ for a periodic wave, the frequency produced depends on the length of the tube.

The possible modes of vibration, called harmonics, are similar to those produced in strings. In order to simplify the discussion, the diagrams below show transverse standing waves rather than longitudinal standing waves. The antinodal point in a longitudinal standing wave would actually be a region of alternating compressions and expansions and high pressure variation while at the nodal points the air pressure would remain relatively constant.

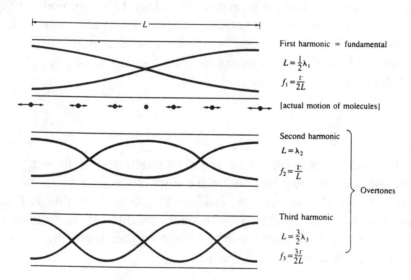

First harmonic = fundamental

$$L = \tfrac{1}{2}\lambda_1$$

$$f_1 = \tfrac{v}{2L}$$

[actual motion of molecules]

Second harmonic

$$L = \lambda_2$$

$$f_2 = \tfrac{v}{L}$$

Overtones

Third harmonic

$$L = \tfrac{3}{2}\lambda_3$$

$$f_3 = \tfrac{3v}{2L}$$

In general, $\tfrac{1}{2}\lambda_n = L/n$, where n is an integer and refers to the mode of vibration, e.g., for the 3rd harmonic n = 3. The frequency of the particular harmonic can then be determined since $f = v/\lambda_n = v/(2L/n)$. The frequencies of successive harmonics are consecutive whole number multiples of the first harmonic, i.e., 2, 3, 4, etc., times the frequency of the first harmonic.

Closed Pipe

A wind instrument that is open to the air at one end but closed at the other end is known as a **closed tube** or **closed pipe**. The longitudinal standing wave produced in the pipe has an antinode at the open end but a node at the closed end. Assuming the speed of sound is constant, and $v = f \lambda$ for a periodic wave, then the frequency produced depends on the length of the tube.

The diagrams shown below represent the first three modes of vibration in a closed pipe.

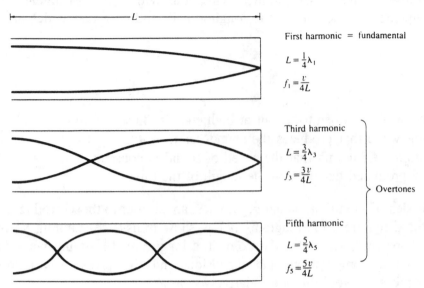

First harmonic = fundamental

$L = \frac{1}{4}\lambda_1$

$f_1 = \frac{v}{4L}$

Third harmonic

$L = \frac{3}{4}\lambda_3$

$f_3 = \frac{3v}{4L}$

Fifth harmonic

$L = \frac{5}{4}\lambda_5$

$f_5 = \frac{5v}{4L}$

Overtones

In general, $\frac{1}{4}\lambda_n = L/(2n - 1)$, where n is an integer and refers to the mode of vibration, e.g., for the third harmonic, n = 3. The frequency of the particular harmonic can then be determined, since $f_n = v/\lambda_n = v/[4L/(2n - 1)]$. The frequencies of successive harmonics for a pipe closed at one end are consecutive odd number multiples of the first harmonic, i.e., 3, 5, 7, etc., times the frequency of the first harmonic.

Quality

The **quality** of a sound depends on the presence of overtones, their number, and relative amplitudes. A piano and a guitar may be playing the same note, with the same frequency and amplitude, yet the sounds are clearly distinguishable. The reason for this is that the relative amplitudes of the harmonics that are produced are different for different instruments and the note which is produced is the result of the superposition of the various harmonics. The result gives each instrument its characteristic quality and timbre.

Even two guitars will sound different because of certain characteristics in the construction of the instrument. These characteristics determine the relative amplitudes of the harmonics produced.

Music Versus Noise

Our minds interpret sounds that include frequencies that are simple multiples of one another as harmonious or pleasing to the ear. Noise is the result of many vibrations of different frequencies and amplitudes with no particular relationship to one another. Oftentimes, the distinction between music and noise is not sharp. The determination must be made by the individual. Thus what is music to the ears of the typical teenager might be considered noise by the typical parent.

Interference of Sound Waves

An **interference pattern** will be produced by two sources of sound waves separated by a certain distance (d) if the sounds produced are of the same frequency and amplitude. The interference pattern shown below is the result of two sources, source 1 (s_1) and source 2 (s_2), both of which are in phase. The term "in phase" means that both sources produce compressions at the same moment of time and expansions at the same moment of time.

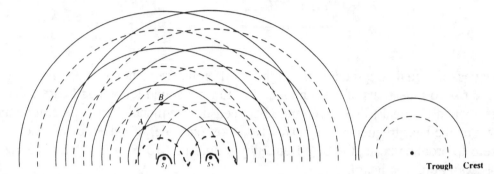

Trough   Crest

At locations where the path difference is a whole number multiple of the wavelength, the waves will arrive in phase and constructive interference occurs. Point A in the diagram is such a point. The distance from source 1 to point A is 1λ, while from source 2 the distance is 2λ. The path difference is therefore $2\lambda - 1\lambda = 1\lambda$.

At locations where the path difference is odd multiples of $\frac{1}{2}\lambda$, the waves arrive out of phase, i.e., a compression is superimposed on a expansion, and destructive interference occurs. Point B in the diagram is such a point. The distance from source 1 to point B is $3/2\ \lambda$ and from source 2 is 2λ. The path difference is therefore $2\lambda - 3/2\ \lambda = \frac{1}{2}\lambda$.

In the diagram at the top of the next page, the path difference is represented by the distance from source 1 to E. This distance is $m\lambda$ for constructive interference and $(m + \frac{1}{2})\lambda$ for destructive interference, where m = 0, 1, 2, 3, 4, etc.

The following formulas can be used to determine the positions of constructive and destructive interference if the angle θ is relatively small (less than 15°). For small angles, $\tan\theta \approx \sin\theta \approx \theta$, where θ is in radians.

For constructive interference: $\tan\theta = \Delta X/L$ but $\sin\theta = m\lambda/d$ and $\Delta X/L = m\lambda/d$

For destructive interference: $\tan\theta = \Delta X/L$ but $\sin\theta = (m + \frac{1}{2})\lambda/d$ and $\Delta X/L = (m + \frac{1}{2})\lambda/d$

ΔX is the perpendicular distance from the center line to the point in question. L is the length of the center line, $m\lambda$ or $(m + \frac{1}{2})\lambda$ is the path difference, and d is the distance between the sources of the sound waves.

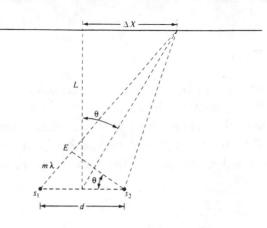

Beats

Two waves of equal amplitude but slightly different frequencies traveling in the same direction give rise to pulsations of maximum and minimum sound known as **beats**. The number of beats per second is equal to the difference between the frequencies of the component waves and is known as the beat frequency, $f_b = |f_1 - f_2|$. For example, if waves of 600 hertz and 610 hertz interfere to produce beats, the beat frequency would be 610 Hz - 600 Hz = 10 Hz, i.e., 10 beats per second would be heard.

Doppler Effect

When a source of sound waves and a listener approach one another, the pitch of the sound is increased as compared to the frequency heard if they remain at rest. If the source and the listener recede from one another, the frequency is decreased. This phenomena is known as the **Doppler effect**. The pitch heard by the listener is given by the following equations:

$$f' = [1/(1 - v_s/v)] f \qquad\qquad f' = [1/(1 + v_s/v)] f$$

source moving toward   source moving away from
a stationary listener   a stationary listener

$$f' = (1 + v_o/v) f \qquad\qquad f' = (1 - v_o/v) f$$

listener moving toward   listener moving away from
a stationary source   a stationary source

f' is the frequency of the sound heard by the listener (observer), f is the frequency of the sound emitted by the source, v is the speed of sound in air, v_s is the velocity of the source, and v_o is the velocity of the listener (observer).

Common examples of the Doppler effect include the change in pitch of the siren of a police car or ambulance as it passes at high speed, as well as the change in pitch of the whistle of a passing train.

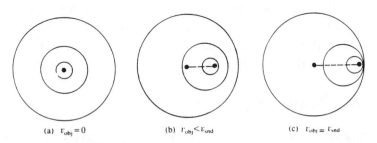

(a) $v_{obj} = 0$ (b) $v_{obj} < v_{snd}$ (c) $v_{obj} = v_{snd}$

Light waves also exhibit the Doppler effect. The spectra of stars that are receding from us is shifted toward the longer wavelengths of light. This is known as the **red shift**. Measurement of the red shift allows astronomers to calculate the speed at which stars are moving away. Since almost all stars and galaxies exhibit a red shift, it is believed that the universe is expanding.

Shock Waves and the Sonic Boom

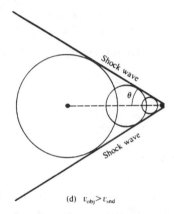

(d) $v_{obj} > v_{snd}$

When the speed of a source of sound exceeds the speed of sound, the sound waves in front of the source tend to overlap and constructively interfere. The superposition of the waves produce an extremely large amplitude wave called a **shock wave**.

The shock wave contains a great deal of energy. When the shock wave passes a listener, this energy is heard as a **sonic boom**. The sonic boom is heard only for a fraction of a second; however, it sounds as if an explosion has occurred and can cause damage.

SELECTED TEXT QUESTIONS WITH ANSWERS

QUESTION 8. What is the reason catgut strings on some musical instruments are wrapped with fine wire?

ANSWER: The velocity of waves on a string is given by $v = [F_T/(m/L)]^{1/2}$ where m/L is the mass per unit length of the string. Based on this equation, the fine wire increases the mass per unit length of the string and decreases the velocity of the wave.

The lower velocity results in a lower frequency for the same wavelength. As a result, it is possible to obtain lower frequency notes without changing the length of the string. Note that the velocity is also a function of the tension (F_T) in the string. Therefore, both the tension and the mass per unit length can changed in order to produce the desired frequency.

QUESTION 10. How will the temperature in a room affect the pitch of organ pipes?

ANSWER: The frequency the first harmonic of an open pipe is given by $f_1 = v/2L$ while for a closed pipe $f_1 = v/4L$. Therefore, the frequency is directly proportional to the velocity of sound in air. However, the velocity of sound in air changes with temperature, i.e., $v = (331 + 0.60\ T)$ m/s where the temperature T is measured in degrees Celsius. As the temperature increases, the velocity increases and the pitch produced by the organ pipe increases.

Question 19. A sonic boom sounds much like an explosion. Explain the similarity of the two.

ANSWER: When the speed of a source of sound exceeds the speed of sound, the sound waves in front of the source tend to overlap and constructively interfere. This results in a burst of high pressure air, sonic boom, passing the listener followed by a sudden return to normal atmospheric pressure. This sudden change in pressure and return to normal pressure is very similar to the change in air pressure when an explosion occurs.

PROBLEM SOLVING SKILLS

For problems involving the sound intensity:

1. Complete a data table noting the intensity level of the sound, the intensity of the threshold of hearing, frequency of the sound, the density of air, and the speed of sound.
2. Solve for the sound intensity in decibels and the amplitude of the sound wave.

For problems involving harmonics produced in pipes:

1. If necessary, solve for the speed of sound in the pipe.
2. Note whether the problem involves an open or closed pipe. Draw an accurate diagram locating the nodes and antinodes for the harmonic(s) requested.
3. Determine the wavelength of the waves producing the particular harmonic.
4. Solve for the frequency of the particular harmonic. If the frequency is given use the above steps to solve for the length of the pipe.

For problems involving an interference pattern produced by two sources of sound of the same frequency and amplitude which are separated by a distance d:

1. Draw an accurate diagram labeling d, ΔX, L, m, $m\lambda$, and λ.
2. Complete a data table listing the information given in the problem and the information requested in the solution.

3. Note whether the problem involves constructive interference or destructive interference.
4. Choose the appropriate formula(s) and solve the problem.

For problems involving the Doppler effect:

1. Complete a data table listing information both given and implied in the problem.
2. Note whether the source or the listener is moving; also note whether the source and listener are approaching each other or moving away from each other.
3. Select the appropriate formula and solve for frequency heard by the listener.
4. Solve for the wavelength of the sound between the source and the listener.

PROGRAMMED PROBLEMS

a. 1	Solution: (Section 12-1)
Solve by using the formula for the speed of sound in air.	$v = 331 + 0.6\ T$ m/s
	at 0°C $v = 331 + (0.6)(10°C) = 337$ m/s
	at 30°C $v = 331 + (0.6)(30°C) = 349$ m/s

a. 1	Solution: (Sections 12-2 and 12-3)
Apply the formula for the intensity level in decibels.	$I_o = 1.0 \times 10^{-12}$ W/m^2
	$\beta = 10 \log (I/I_o)$
	$65\ db = 10 \log (I/1.0 \times 10^{-12}$ W/m$^2)$
	$6.5 = \log (I/1.0 \times 10^{-12}$ W/m$^2)$
	$10^{6.5} = I/1.0 \times 10^{-12}$ W/m^2
	$3.16 \times 10^6 = I/1.0 \times 10^{-12}$ W/m^2
	$I = (3.16 \times 10^6)(1.0 \times 10^{-12}$ W/m$^2)$
	$I = 3.16 \times 10^{-6}$ W/m^2
b. 1.	$I = 2\ \pi^2\ v\ \rho\ f^2\ x_o^2$
Use the equation introduced in chapter 11 that relates intensity, frequency and velocity to solve for the amplitude.	3.16×10^{-6} W/m$^2 = 2\ \pi^2\ (340$ m/s$)\ (1.29$ kg/m$^3)\ (600$ Hz$)^2\ x_o^2$
	$x_o^2 = 1.01 \times 10^{-15}$ m^2
	$x_o = 3.1 \times 10^{-8}$ m

a. 1	Solution: (Section 12-5)
Determine the speed of sound.	$v = [331 + (0.6)(30)]$ m/s $= 349$ m/s

a. 2	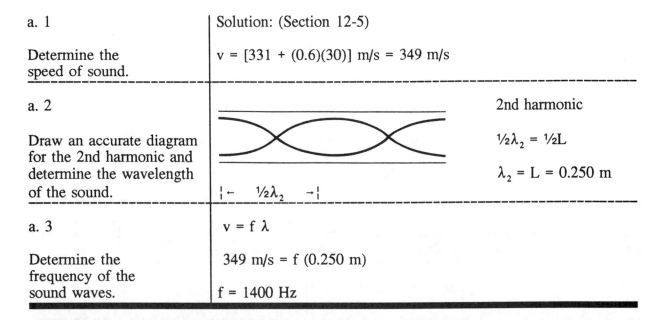 2nd harmonic $\frac{1}{2}\lambda_2 = \frac{1}{2}L$ $\lambda_2 = L = 0.250$ m
Draw an accurate diagram for the 2nd harmonic and determine the wavelength of the sound.	

a. 3	$v = f\lambda$
Determine the frequency of the sound waves.	349 m/s $= f(0.250$ m$)$ $f = 1400$ Hz

PROBLEM 4. A pipe 0.500 m long is closed at one end. A guitar is placed near the open end of the tube and the string is plucked. The guitar string is 0.750 m long and has a mass of 0.0010 kg. The string vibrates in its second mode (2nd harmonic) and produces a third harmonic standing wave in the closed pipe. If the speed of sound is 340 m/s, determine the a) frequency of the sound produced by the air column, b) tension in the guitar string. Hint: the frequency of the vibrating guitar string is the same as the frequency produced in the closed pipe.

a. 1	Solution: (Section 12-5)
Draw an accurate diagram for the 3rd harmonic for a closed pipe and determine the wavelength of the sound waves.	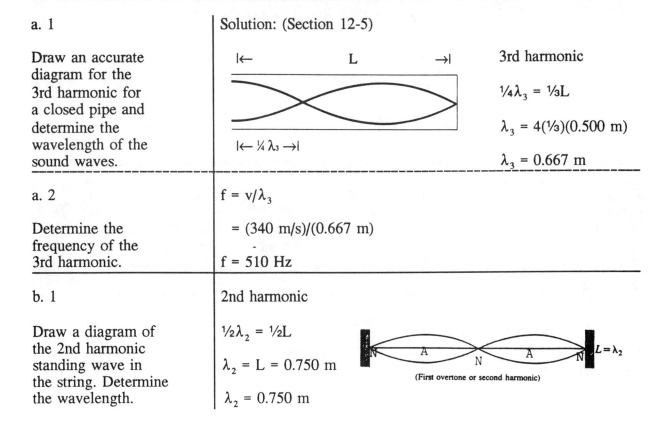 3rd harmonic $\frac{1}{4}\lambda_3 = \frac{1}{3}L$ $\lambda_3 = 4(\frac{1}{3})(0.500$ m$)$ $\lambda_3 = 0.667$ m

a. 2	$f = v/\lambda_3$
Determine the frequency of the 3rd harmonic.	$= (340$ m/s$)/(0.667$ m$)$ $f = 510$ Hz

b. 1	2nd harmonic
Draw a diagram of the 2nd harmonic standing wave in the string. Determine the wavelength.	$\frac{1}{2}\lambda_2 = \frac{1}{2}L$ $\lambda_2 = L = 0.750$ m $\lambda_2 = 0.750$ m (First overtone or second harmonic)

b. 2	$v = f \lambda_2$
Determine the velocity of the waves in the string.	$v = (510 \text{ Hz})(0.750 \text{ m}) = 382 \text{ m/s}$
b. 3	$v = [F_T/(m/L)]^{\frac{1}{2}}$ and rearranging gives
Solve for the tension in the string.	$F_T{}^2 = v^2 m/L = (382 \text{ m/s})^2 (0.00300 \text{ kg})/(0.750 \text{ m})$
	$F_T = 584 \text{ N}$

PROBLEM 5. A student strikes two tuning forks and hears 2 beats per second. He notes that 440 Hz is printed on one tuning fork. Determine the frequency of the other fork.

a. 1	Solution: (Section 12-7)
Determine the frequency of the second tuning fork.	This problem has two possible answers. The difference between the two frequencies must be 2 Hz; however, we cannot be sure which tuning fork has the higher frequency.
	Let $f_1 = 440 \text{ Hz}$ and $f_2 = ?$,
	then either $f_1 - f_2 = 2 \text{ Hz}$
	$440 \text{ Hz} - f_2 = 2 \text{ Hz}$ and $f_2 = 438 \text{ Hz}$
	or $f_2 - f_1 = 2 \text{ Hz}$
	$f_2 - 440 \text{ Hz} = 2 \text{ Hz}$ and $f_2 = 442 \text{ Hz}$
	Therefore the frequency of the second tuning fork may either be 438 Hz or 442 Hz.

PROBLEM 6. Two point sources of sound are in phase and separated by a distance (d) of 0.300 m. The frequency of the sound is 600 Hz. A listener stands at a point that is 1.00 m along a center line (L) that bisects the line which connects the two speakers. The listener then walks perpendicular to the center line. Determine the distance (Δ X) from the center line to the first two a) nodal points, b) antinodal points that are to the left of the center line. Assume that the speed of sound is 340 m/s.

a. 1	Solution: (Section 12-7)
Determine the wavelength of the sound waves.	$\lambda = v/f = (340 \text{ m/s})/(600 \text{ Hz})$
	$\lambda = 0.567 \text{ m}$

a. 2 Draw an accurate diagram and list each of the quantities given.	d = 0.300 m L = 1.00 m λ = 0.567 m ΔX = ? 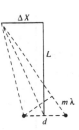

---

a. 3 Apply the formula for destructive interference to determine the positions of the first two nodal points, m = 0 and m = 1.	1st nodal point (m = 0): $\Delta X/L$ = $(0 + \frac{1}{2})\lambda/d$ $\Delta X/(1.00 \text{ m})$ = $(\frac{1}{2})(0.567 \text{ m})/(0.300 \text{ m})$ ΔX = 0.945 m 2nd nodal point (m = 1): $\Delta X/L$ = $(1 + \frac{1}{2})\lambda/d$ $\Delta X/(1.00 \text{ m})$ = $(3/2)(0.567 \text{ m})/(0.300 \text{ m})$ ΔX = 2.84 m
b. 1 Apply the formula for constructive interference to determine the positions of the first two antinodal points to the left of the center line, m = 1 and m = 2.	The sources are in phase and therefore an antinodal point occurs along the center line (m = 0). The first two antinodal points to the the left of the center line occur at m = 1 and m = 2. 1st nodal point: $\Delta X/L$ = $m\lambda/d$ $\Delta X/(1.00 \text{ m})$ = $1(0.567 \text{ m})/(0.300 \text{ m})$ ΔX = 1.89 m 2nd nodal point: $\Delta X/(1.00 \text{ m})$ = $2(0.567 \text{ m})/(0.300 \text{ m})$ ΔX = 3.78 m

PROBLEM 7. A stationary source emits sound of frequency 600 Hz on a day when the speed of sound is 340 m/s. A listener moves toward the source of sound at 20.0 m/s. Determine the a) frequency heard by the listener, b) wavelength of the sound between the source and the listener, and c) answer parts a and b for the listener moving away from the source.

a. 1 Determine the frequency heard by the listener.	Solution: (Section 12-8) This problem involves the Doppler effect. Apply the formula for a moving listener approaching a stationary source. $f' = (1 + v_o/v)f$ $f' = [1 + (20.0 \text{ m/s})/(340 \text{ m/s})](600\text{Hz})$ = 635 Hz

b. 1 Determine the wavelength of the sound between the source and listener.	The source of the sound is not moving; therefore, the wavelength of the sound remains the same regardless of the listener's speed. $\lambda = v/f$ $\lambda = (340 \text{ m/s})/(600 \text{ Hz}) = 0.567 \text{ m}$
c. 1 Answer parts a and b for the listener moving away from the source.	The wavelength of the sound will not change whether the listener is moving toward or away from the source; however, the frequency heard will change. Apply the formula for a listener moving away from the source in order to determine the frequency heard by the listener. $f' = (1 - v_o/v) f = [1 - (20.0 \text{ m/s})/(340 \text{ m/s})](600\text{Hz})$ $f' = 565 \text{ Hz}$

PRACTICE PROBLEMS

PROBLEM 1. During an indoor rock concert, sound intensities of 0.10 W/m^2 are produced. Determine the a) intensity levels of these sounds in decibels, b) amplitude of the sound wave if its frequency is 500 hertz. The density of air is 1.29 kg/m^3 and the speed of sound is 340 m/s.

ANS. a) 110 dB,   b) x_o = 6.68 x 10^{-6} m or 0.0068 mm

PROBLEM 2. a) Determine the frequencies of the fourth harmonic produced by an organ pipe 0.125 m long which is open at both ends on a day when the speed of sound is 340 m/s. b) Determine the frequency of the fifth harmonic of the same organ pipe if it is closed at one end.

ANS. a) 5440 Hz, b) 3400 Hz

PROBLEM 3. Two point sources of sound are in phase and separated by a distance of 2.0 m. The speed of sound is 340 m/s. A listener stands at a point 5.0 m along a center line which bisects a line connecting the two speakers. The listener then walks 0.50 m perpendicular to the center line before locating a nodal point. Determine the wavelength and frequency of the sound.

ANS. 0.40 m, 850 Hz

PROBLEM 4. A child whirls a toy at the end of a string in a circle of radius 2.0 m at 2.0 rev/s. As the toy revolves, it makes a whistling sound of frequency 200 Hz. What are the maximum and minimum frequencies heard by a stationary listener standing some distance away. Assume that the speed of sound is 340 m/s.

ANS. 216 Hz, 186 Hz

CHAPTER 13

TEMPERATURE AND THE KINETIC THEORY

KEY TERMS AND PHRASES

temperature is a measure of relative hotness or coldness. In kinetic theory, temperature is a measure of the average kinetic energy of the particles of a substance.

thermometer is a device used to measure the temperature of a substance. The thermometer is usually calibrated using either the Fahrenheit temperature scale (°F) or the Celsius (°C). Water is the reference used by both temperature scales. At 1 atmosphere of pressure, the freezing point of water is 32°F or 0°C, while the boiling point is 212°F or 100°C.

coefficient of linear expansion is the ratio between the change in length of a solid rod to its original length per one degree change in temperature.

coefficient of volume expansion is the ratio between the change in volume of a solid or a liquid its original volume per one degree change in temperature.

Boyle's law states that, if the temperature of a gas is held constant, the volume occupied by an enclosed gas varies inversely with the pressure exerted on it.

Charles' law states that the volume of an enclosed gas held at constant pressure is directly proportional to the absolute temperature.

Gay-Lussac's law states that the pressure exerted by a gas held at constant volume is directly proportional to the absolute temperature.

ideal gas refers to a monatomic, low density and low pressure gas. The idea of an 'ideal' or perfect gas is used to create a model which can then be used to study real gases.

ideal gas law states that the three gas laws, Boyle's, Charles, and Gay-Lussac's, can be combined into a single more general law that relates the pressure (P), volume (V), and temperature (T) of a fixed quantity of gas (n).

Avogadro's hypothesis states that equal volumes of gas at the same pressure and temperature contain equal numbers of molecules.

Avogadro's number is the number of atoms (or molecules) per mole. The modern value of Avogadro's number is 6.02×10^{23} molecules.

postulates of the kinetic theory of gases are 1) A gas consists of a large number of molecules moving in random directions with a variety of speeds. 2) The average distance between any two molecules in a gas is large compared to the size of an individual molecule. 3) The molecules obey the laws of classical mechanics and are presumed to interact with one another only when they collide. 4) Collisions between molecules or between a molecule and the side walls of the container are perfectly elastic.

root-mean-square velocity (v_{rms}) refers to the square root of the average of the squares of the magnitudes of the velocities of the molecules in a gas.

phase diagram is a pressure-temperature diagram which allows the reader to determine the temperatures and pressures at which the various phases (solid, liquid, gas) for a particular substance exist. The diagram also indicates those conditions under which equilibrium occurs.

triple point is the point at which all three states of matter (solid, liquid, gas) coexist in equilibrium. For water, the triple point is at 0.01°C and 0.006 atm.

sublimation is the direct conversion of a solid to a vapor without passing through the liquid state.

evaporation is the process by which molecules escape from a liquid and enter the gas or vapor phase.

vapor pressure refers to the partial pressure contributed by the molecules that have evaporated from a liquid.

relative humidity is a measure of the water vapor content of the air compared to the maximum amount the air can hold at a particular temperature.

dew point is reached when the relative humidity reaches the saturation point and the air can hold no more water. If the temperature drops below this point, condensation or dew occurs.

boiling occurs when the temperature of the liquid rises to the point where the vapor pressure inside the bubbles that form in the liquid equals or exceeds the external air pressure.

diffusion is the spontaneous movement of molecules from a region of high concentration to a region of low concentration due to random molecular movement. The rate of diffusion of one substance into another is described by the **Fick's law**.

Graham's law describes the rate of diffusion of gas molecules. The law states that the rate of diffusion of molecules of a gas is inversely proportional to the square root of the molecular mass.

SUMMARY OF MATHEMATICAL FORMULAS

Celsius-Fahrenheit temperature conversion	$°F = 9/5 \ °C + 32°$ or $°C = 5/9 \ (°F - 32°)$	These formulas allow the conversion of temperatures from Fahrenheit to Celsius and vice versa.
linear expansion	$\Delta L = L_o \ \alpha \ \Delta T$ and $L = L_o \ (1 + \alpha \ \Delta T)$	The linear expansion (ΔL) of a solid rod is related to the rod's original length (L_o), coefficient of linear expansion (α) and the change in temperature (ΔT).
volume expansion	$\Delta V = V_o \ \beta \ \Delta T$ and $V = V_o \ (1 + \beta \ \Delta T)$	The volume expansion (ΔV) of a substance is related to the original volume (V_o), coefficient of volume expansion (β), and the change in temperature (ΔT).
Boyle's law	$P \ V = constant$ $P_1 \ V_1 = P_2 \ V_2$	If the temperature of a gas is held constant, the volume occupied by an enclosed gas varies inversely with the pressure exerted on it.
Charles' law	$V/T = constant$ $V_1 \ /T_1 = V_2 \ /T_2$	The volume of an enclosed gas held at constant pressure is directly proportional to the absolute temperature.
Gay-Lussac's law	$P/T = constant$ $P_1 \ /T_1 = P_2 \ /T_2$	The pressure exerted by a gas held at constant volume is directly proportional to the absolute temperature.
ideal gas law	$P \ V = n R T$	The ideal gas law relates the pressure (P), volume (V), and temperature (T) of a fixed quantity of gas (n). R is a constant of proportionality called the universal gas constant.
equations from kinetic theory of gases	$P \ V = ⅓ N m v^2$ $v_{rms} = (\overline{v^2})^{½}$ $\overline{KE} = 3/2 \ k T$	N is the total number of molecules in the gas. m is the mass of an individual molecule. v_{rms} is the square root of average value of the squares of the velocities of the molecules. T is the temperature in degrees Kelvin.

relative humidity	R.H. = $(\rho_{wv})/(\rho_{swv})$ x 100% R.H. = $(P_{vp})/(P_{svp})$ x 100%	The relative humidity (R.H.) is the ratio of the density of water vapor in the air (ρ_{wv}) as compared to the density of water vapor in the air when the air is saturated(ρ_{swv}). or The relative humidity is the ratio of the partial pressure of the water vapor in the air (P_{vp}) to the vapor pressure of water when the air is saturated (P_{svp}).
Diffusion Fick's law	$J = D\, A\, (C_1 - C_2)/d$	The rate of diffusion (J) of one substance into another is related to the diffusion constant (D), the cross-sectional area (A) of the region through which the diffusion occurs, and the concentration of the diffusing substance (C). C_1 is the region of higher concentration and C_2 is the region of lower concentration. Δx is the unit of distance .
gaseous diffusion Graham's law	$J \propto 1/(m)^{1/2}$	The rate of diffusion (J) of gas molecules is inversely proportional to the square root of the molecular mass.

CONCEPTS AND EQUATIONS

Temperature

Temperature is a measure of relative hotness or coldness. The temperature of a substance is measured with a device called a **thermometer**. The thermometer is usually calibrated in terms of degrees **Fahrenheit** (°F) or degrees **Celsius** (°C). Water is the reference used by both temperature scales. At 1 atmosphere of pressure, the freezing point of water is 32°F or 0°C, while the boiling point is 212°F or 100°C. Although the Fahrenheit temperature scale shall not be used in this study guide, it is useful to memorize the conversion formulas for Fahrenheit to centigrade or vice versa:

°F = 9/5 °C + 32°   or   °C = 5/9 (°F - 32°)

In many instances it will be necessary to use the Kelvin temperature scale, where K = °C + 273°. The Kelvin temperature is also known as the absolute temperature scale.

Linear Expansion

The change in length (ΔL) of a solid when it is heated or cooled depends on the original length (L_o), coefficient of linear expansion (α), and the change in temperature (ΔT). The unit used for the coefficient of linear expansion is ($/C^\circ$) or (C°)$^{-1}$. The change in length is given by $\Delta L = L_o \, \alpha \, \Delta T$. The final length ($L$) of the solid after the temperature change is given by

$$L = L_o + \Delta L = L_o (1 + \alpha \, \Delta T)$$

Volume Expansion

The change in volume (ΔV) of a solid or liquid when it is heated or cooled depends on the original volume (V_o), coefficient of volume expansion (β), and the change in temperature (ΔT). The volume and/or change of volume of a substance is given in cm^3, m^3, or liters. The unit used for the coefficient of volume expansion is ($/C^\circ$) or (C°)$^{-1}$. The change in volume (ΔV) of a solid or liquid that undergoes a temperature change is given by $\Delta V = V_o \, \beta \, \Delta T$ while the volume (V) at the final temperature is given by

$$V = V_o + \Delta V = V_o (1 + \beta \, \Delta T)$$

The Ideal Gas

The changes in volume of a gas cannot be described by the equation given for a solid or liquid. However, within a certain range of temperatures and pressures many gases have been found to follow three simple laws: Boyle's law, Charles' law and Gay-Lussac's law. A gas that follows these laws completely is an idealization called an **ideal gas**.

The variables involved in these laws are the pressure (P), the volume (V), the temperature (T), and the number of moles of gas (n). The Kelvin temperature scale is always used. One mole contains 6.02×10^{23} molecules and equals the molecular weight of the substance expressed in grams.

Boyle's Law

If the temperature of a gas is held constant, the volume occupied by an enclosed gas varies inversely with the pressure exerted on it. **Boyle's law** can be written as follows:

$$V \propto 1/P \quad \text{and} \quad P \, V = \text{constant} \quad \text{(at constant temperature)}$$

Charles' Law

The volume of an enclosed gas held at constant pressure is directly proportional to the absolute temperature:

$$V \propto T \quad \text{and} \quad V/T = \text{constant} \quad \text{(at constant pressure)}$$

Gay-Lussac's Law

The pressure exerted by a gas held at constant volume is directly proportional to the absolute temperature:

$P \propto T$ and P/T = constant   (at constant volume)

Ideal Gas Law

The three gas laws can be combined into a single more general law that relates the pressure (P), volume (V), and temperature (T) of a fixed quantity of gas (n):

$P V = n R T$

R is a constant of proportionality called the universal gas constant. Depending on the situation, R may be expressed in different sets of units. Typical values of R are

$R = 8.31$ joules/mole·K $= 0.0821$ liter atm/mole·K $= 1.99$ cal/mole·K $= 62.4$ liter torr/mole·K

Avogadro's Number and the Ideal Gas Law

Avogadro's hypothesis formulated in 1811 states that "equal volumes of gas at the same pressure and temperature contain equal numbers of molecules." The number of molecules per mole is known as **Avogadro's number** (N_A). The modern value of N_A is 6.02×10^{23} molecules.

The total number of molecules in a gas (N) equals the product of the number of moles of gas (n) and the number of molecules per mole (N_A); thus $N = n N_A$ and the ideal gas equation can be written as follows:

$P V = n R T = (N/N_A) R T$ but $(N/N_A)R = k$

where $k = 1.38 \times 10^{-23}$ J/K and k is known as **Boltzmann's constant**.

Postulates of the Kinetic Theory of Gases

1) A gas consists of a large number of molecules moving in random directions with a variety of speeds.
2) The average distance between any two molecules in a gas is large compared to the size of an individual molecule.
3) The molecules obey the laws of classical mechanics and are presumed to interact with one another only when they collide.
4) Collisions between molecules or between a molecule and the side walls of the container are perfectly elastic.

The pressure exerted by an enclosed gas may be rewritten in light of these postulates and is given by the equation $P V = \frac{1}{3} N m v^2$ where N is the total number of molecules in the gas, m is the mass of an individual molecule, and v^2 is the average value of the squares of the velocities of the molecules of the gas.

Root-Mean-Square Velocity (v_{rms})

The root-mean-square velocity is often confused with average velocity. The root-mean-square velocity refers to the square root of the average of the squares of the magnitudes of the velocities of the molecules in a gas.

$$v_{rms} = (\overline{v^2})^{1/2} = [(v_1^2 + v_2^2 + ... + v_N^2)/N]^{1/2}$$

The average or the mean speed ($\overline{v}$) is equal to the sum of the speeds of the molecules divided by the number of molecules.

$$\overline{v} = (v_1 + v_2 + ... + v_N)/N$$

Relationship between Average Kinetic Energy and the Absolute Temperature

From the kinetic theory, $P\ V = \frac{1}{3}\ N\ m\ \overline{v^2}$ and from the ideal gas law, $P\ V = n\ R\ T$. If we express both equations in the same units, then

$\frac{1}{3}\ N\ m\ \overline{v^2} = n\ R\ T$ and rearranging we find

$\frac{1}{2}\ m\ \overline{v^2} = \frac{1}{2}\ (3\ n\ R\ T/N) = 3/2\ (nR/N)\ T$ but $nR/N = k = $ constant, therefore

$$\overline{KE} = 3/2\ k\ T$$

From this equation we can see that the average kinetic energy of a molecule in a gas is directly proportional to the absolute temperature of the gas.

Real Gases and Changes of Phase

In section 13-11 of the text, the second postulate of the kinetic theory assumes that the volume occupied by the gas molecules is negligible compared to the volume of the container. However, when dealing with a real gas a correction factor must be included. Especially at high pressures, when the volume occupied by the gas is small, the behavior of a real gas deviates significantly from the predictions of the ideal gas law.

The third postulate states that "molecules of a gas are presumed to interact with one another only when they collide." However, contrary to this postulate, all molecules, even supposedly inert molecules such as helium, exhibit weak, short-range attractive forces for one another. These forces, called Van der Waal forces, are ignored when discussing the ideal gas law. If sufficient pressure is applied, it is found that as a gas cools the molecules begin to cling together and the gas will become a liquid below a certain **critical temperature**. For helium, this temperature is about -267°C (6 K). Above the critical temperature, the gas cannot be liquified, no matter what the pressure. The diagram A at the top of the next page shows a possible variation of volume with temperature for a real gas as compared to the predictions from the ideal gas law.

A pressure-temperature diagram (**phase diagram**) allows us to determine the temperatures and pressures at which the various phases (solid, liquid, gas) exist, as well as those conditions under which equilibrium occurs.

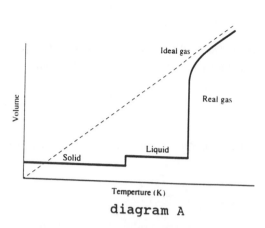

diagram A

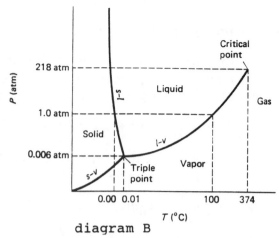

diagram B

Diagram B is the phase diagram for water. The point where the three lines intersect is called the **triple point**. At this point, all three states of matter (solid, liquid, gas) coexist in equilibrium. For water, the triple point is at 0.01°C and 0.006 atm.

Sublimation

Sublimation is the direct conversion of a solid to a vapor without passing through the liquid state. As shown in the phase diagram, the sublimation of ice to the vapor phase does not occur at ordinary temperatures and pressures. For water, the temperature must be below 0.01°C and the pressure must be below 0.006 atm. However, common substances that exhibit sublimation at room temperature and pressure are moth balls (naphthalene) and dry ice (CO_2). They go directly from the solid phase to the vapor phase.

Evaporation

Evaporation is the process by which molecules escape from a liquid and enter the gas or vapor phase. The molecules that escape overcome forces which hold them in the liquid phase. It is the molecules of high kinetic energy that leave the liquid, with the result that the average kinetic energy of the remaining molecules decreases. Since temperature is proportional to the average kinetic energy of the molecules, the temperature of the liquid decreases. Thus evaporation is a cooling process.

If the temperature of the liquid increases, it is observed that the evaporation rate increases. The kinetic theory predicts that at the higher temperature, more molecules will have kinetic energy in excess of that required to escape from the liquid and is therefore in agreement with observation.

Vapor Pressure

The molecules that have evaporated from the liquid travel in random directions. Some of these molecules return to the liquid but evaporation will continue until the number of molecules leaving the liquid per second equals the number that are reentering the liquid per second. This point is the equilibrium point and the air is then said to be saturated.

The evaporated molecules contribute to the total pressure on the liquid from which they escaped. The **partial pressure** due to these molecules after (saturation) equilibrium has been reached is called the **vapor pressure**. At higher temperatures the vapor pressure is higher due to the greater number of molecules that have escaped from the liquid. At the boiling point of the liquid, the vapor pressure equals the atmospheric pressure. Table 13-4 in the text lists the saturated vapor pressure of ice and water at various temperatures. Ice is included because solids also have a very small, but measurable, vapor pressure.

Humidity

Humidity is a measure of the water vapor content of the air. **Absolute humidity** refers to the mass of water vapor present per unit volume of gas. When discussing atmospheric conditions, the term **relative humidity** (R.H.) is used. Relative humidity is determined as follows:

$$\text{R.H.} = \frac{\text{density of water vapor in the air}}{\text{density of saturated air at the same temperature}} \times 100\%$$

$$\text{or} \quad \text{R.H.} = \frac{\text{partial pressure of water vapor}}{\text{saturated vapor pressure of water}} \times 100\%$$

If the temperature of the air begins to drop, e.g., after sundown, the relative humidity may reach the saturation point and the air can hold no more water. If the temperature begins to drop below this point, called the **dew point, condensation** will occur in the form of dew on grass, cars, etc. or possibly as fog.

Boiling

Boiling occurs when the temperature of the liquid rises to the point where the vapor pressure inside the bubbles that form in the liquid equals or exceeds the external air pressure. Archimedes principle predicts that when the buoyant force exerted on the bubble by the liquid exceeds the weight of the bubble, the bubble will rise to the surface.

The boiling point of a liquid depends on the atmospheric (barometric) pressure. If the atmospheric pressure equals one atmosphere, boiling occurs at 100°C. Under normal conditions the barometric pressure varies from day to day and as a result the boiling point also varies slightly from 100°C. The barometric pressure decreases with increasing altitude and because of this it may be necessary to cook foods for a longer period of time or use a pressure cooker to reduce cooking time. The lid on the pressure cooker forms a tight seal on the pot and is equipped with a pressure relief valve that prevents the pot from exploding. The heat supplied to the water causes the liquid water to evaporate until the pressure inside the cooker reaches the desired temperature and steam escapes via the pressure relief valve. Because the pressure inside the cooker is greater than one atmosphere, the water boils at a higher temperature and the food cooks faster.

Diffusion

Diffusion is the spontaneous movement of molecules from a region of high concentration to a region of low concentration due to random molecular movement. The spreading of a drop of food color in a glass of water and cigarette smoke in a room are examples of diffusion.

Diffusion occurs because molecules of all of the substances present are able to move throughout the container. However, in a liquid the molecules are much closer together than in a gas. Thus the molecules of one liquid diffusing into a second liquid undergo many more collisions per cm of distance than in the diffusion of one gas into another. The average distance between collisions, called the mean free path, is much shorter in a liquid. As a result, the rate of diffusion of one liquid into another is much slower for one liquid into another as compared to one gas into another.

Fick's Law of Diffusion

The rate of diffusion of one substance into another is described by the diffusion equation, or **Fick's law**.

$$J = D\ A\ (C_1 - C_2)/d$$

J is the rate of diffusion past a certain point each second and D is a constant of proportionality known as the diffusion constant or diffusion coefficient. D depends on the properties of the substances involved in the diffusion process. A is the cross-sectional area of the region through which the diffusion occurs.

C is the concentration of the diffusing substance in mol/m^3, where C_1 is the region of higher concentration and C_2 is the region of lower concentration. $(C_1 - C_2)/d$ is the concentration gradient. The concentration gradient is the change in concentration per unit distance (d).

The diffusion equation can also be written in terms of partial pressures:

$$J_i = (D\ A/R\ T)\ (\Delta P_i\ /d)$$

i is the particular component being considered, R is the universal gas constant, T is the temperature in K, and ΔP_i is the change in partial pressure over the distance (d) being considered.

The rate of diffusion of gas molecules can also be described by **Graham's law**. Graham's law states that the rate of diffusion of molecules (J) of a gas is inversely proportional to the square root of the molecular mass: $J \propto 1/(m)^{1/2}$.

SELECTED TEXT QUESTIONS WITH ANSWERS

QUESTION 8. Why is it sometimes easier to remove the lid from a tightly closed jar after warming it under hot running water?

ANSWER: Hot water will raise the temperature of the metal lid and the glass jar. The coefficient of linear expansion is greater for the metal and the circumference of the lid becomes greater than that of the jar. As a result the lid loosens.

QUESTION 16. Will a grandfather's clock, accurate at 20°C, run fast or slow on a hot day (30°C)? The clock uses a pendulum supported on a long thin brass rod.

ANSWER: As the temperature increases the length of the pendulum increases ($\Delta L = L_o \, \alpha \, \Delta T$). As the length increases the period (T) of the pendulum increases, i.e., $T = 2\pi(L/g)^{\frac{1}{2}}$. As a result, it will take longer for the pendulum's motion to repeat and the clock will run slow.

QUESTION 28. Explain why a hot, humid day is far more uncomfortable than a hot, dry day at the same temperature.

ANSWER: On a hot day the human body perspires and uses evaporative cooling to keep cool. However, the partial pressure of water in the air on a hot, humid day is much higher than on a hot, dry day. Because of the higher amount of water in the air on a hot, humid day, the rate of evaporative cooling of perspiration is lower and droplets of perspiration tend to form and the person feels warm, "sticky," and uncomfortable. On a hot, dry day perspiration evaporates faster and the person feels cooler.

PROBLEM SOLVING SKILLS

For problems involving linear or volume expansion:

1. Complete a data table listing the initial length (or volume), coefficient of expansion, and change in temperature.
2. Determine the change in length (or volume). The final length (or volume) is the sum of the original plus the change.
3. If the problem involves a liquid expanding in a solid container, then the expansion of both the liquid and the solid must be taken into account. The resulting coefficient of volume expansion equals the difference between the coefficient for the liquid and the coefficient for the solid.

For problems involving the ideal gas equation:

1. Complete a data table listing the absolute pressure, volume, temperature in degrees Kelvin, and the number of moles of gas present.
2. The absolute pressure equals the gauge pressure plus the atmospheric pressure.
3. Use the ideal gas equation to solve the problem.
4. If more than one gas is present, it may be necessary to determine the partial pressure exerted by each gas. The total pressure equals the sum of the partial pressures.

For problems involving the kinetic theory of gases:

1. Complete a data table listing the absolute pressure, volume, temperature in degrees Kelvin, and the number of moles of gas.
2. Solve for the average kinetic energy of the gas molecules.
3. Solve for the root-mean-square velocity of the molecules.

For problems involving rate of diffusion:

1. Complete a data table noting the concentration in the two regions, the distance over which the diffusion occurs, and the diffusion constant.
2. Use Fick's law to solve for the rate of diffusion.

For problems involving gaseous diffusion:

1. Complete a data table noting the molecular weight of each gas.
2. Use Graham's law to determine the rate of diffusion.

PROGRAMMED PROBLEMS

> **PROBLEM 1.** An engineer is assigned the task of designing and calibrating an oral thermometer. The volume of mercury contained in the thermometer is 0.400 cm^3 at 35.0°C. The radius of the tube along which the mercury travels is 0.00500 cm. Determine the change of temperature required to cause the thread to travel 3.70 cm along the length of the tube and the actual temperature at this point. Ignore the expansion of the glass that contains the mercury. Note: the coefficient of volume expansion of mercury is 180 x 10^{-6}/C°.

a. 1	Solution: (Section 13-4)
Determine the increase in volume of the mercury as a result of the expansion along the tube.	The mercury is expanding along the length of a cylindrical tube. The increase in volume can be found as follows: $$\Delta V = \pi\, r^2\, h = \pi\ (0.00500\ cm)^2\ (3.70\ cm)$$ $$\Delta V = 2.90 \times 10^{-4}\ cm^3$$
a. 2	$$\Delta V = V_o\ \beta\ \Delta T$$
Determine the change in temperature required to cause the expansion. Also, determine the final temperature.	$$2.90 \times 10^{-4}\ cm^3 = (0.400\ cm^3)(180 \times 10^{-6}/C°)\Delta t$$ $$\Delta T = 4.0\ C°$$ $$\Delta T = T_f - T_i$$ $$4\ C° = T_f - 35°C$$ $$T_f = 39°C$$

Note: the assumption that the expansion of the glass is negligible is an oversimplification. The coefficient of volume expansion for glass is 27 x 10^{-6}/C°. The "apparent" coefficient of volume expansion for mercury in glass would be (180 - 27) x 10^{-6}/C° or 153 x 10^{-6}/C°.

A fever thermometer has a constriction just above the bulb. When the thermometer is removed from the person's mouth the volume of the mercury tends to decrease because of the lower air temperature outside of the mouth. The thread of mercury is broken at the constriction before the mercury can begin to flow back into the bulb. Because of this, an accurate measurement of the person's temperature may be obtained. In order to repeat the reading it is necessary to shake the thermometer in such a manner that some of the mercury in the stem passes through the constriction and into the bulb.

> **PROBLEM 2.** During a classroom demonstration the instructor clumsily fails in an attempt to push a 2.001 cm diameter brass sphere though a 1.999 cm opening in a steel ring. The instructor then heats the steel ring until the sphere "just" slips through the opening. If the initial temperature of the ring is 20.0°C, calculate the minimum temperature of the ring at which the sphere slips through the hole. Note: the coefficient of linear expansion of steel is 12 x 10^{-6}/°C.

a. 1	Solution: (Section 13-4)
Assume that the diameter of the sphere equals the final inside diameter of the ring.	$L_{final} = L_{original} + \Delta L = L_o + L_o \ \alpha \ \Delta T = L_o(1 + \alpha \ \Delta T)$
	$2.001 \ cm = (1.999 \ cm)[1 + (12 \ x \ 10^{-6}/°C)(T_f - 20.0°C)]$
	$1.001 = 1 + (12 \ x \ 10^{-6}/°C)(T_f - 20.0°C)$
	$0.001 = (12 \ x \ 10^{-6}/°C)(T_f - 20.0°C)$
	$83.3°C = T_f - 20.0°C$
	$T_f = 103°C$

PROBLEM 3. A 10.0 liter vessel contains 0.020 kg of an ideal gas at 50.0°C and a pressure of 3.00 atm. a) How many moles of gas are in the vessel? b) Determine the root-mean-square speed of the molecules in the gas.

a. 1	Solution: (Sections 13-8 through 13-11)
Apply the ideal gas equation and solve for the number of moles of gas present.	Express the temperature in degrees Kelvin:
	$50.0°C + 273°C = 323 \ K$
	$P \ V = n \ R \ T$
	$(3.00 \ atm)(10.0 \ liter) = n \ (0.0821 \ liter \ atm/mole \ K)(323 \ K)$
	$n = 1.13 \ moles$
b. 1	$n = (1.13 \ mole)(6.02 \ x \ 10^{23} \ molecules/mole)$
Determine the number of molecules present.	$n = 6.81 \ x \ 10^{23} \ molecules$
b. 2	$m = (0.020 \ kg)/(6.81 \ x \ 10^{23} \ molecules)$
Determine the mass of a single molecule of the ideal gas.	$m = 2.94 \ x \ 10^{-26} \ kg/molecule$
b. 3	$\overline{KE} = 3/2 \ k \ T \quad and \quad ½ \ m \ \overline{v^2} = 3/2 \ k \ T$
Determine the rms speed.	$v_{rms} = (3 \ k \ T/m)^{½}$
	$= [3 \ (1.38 \ x \ 10^{-23} \ J/molecule \ K)(323 \ K)/(2.94 \ x \ 10^{-26} \ kg)]^{½}$
	$v_{rms} = 674 \ m/s$

PROBLEM 4. An ideal gas occupies a volume of 5.00 liters at 30.0°C and gauge pressure of 2.00 atm. The gas is heated until its volume is 15.0 liters and gauge pressure is 2.80 atm. Determine the a) number of moles contained in the gas and b) final temperature of the gas.

a. 1 Convert from gauge pressure to absolute pressure.	Solution: (Sections 13-8 and 13-9) absolute pressure = gauge pressure + 1.00 atm $\quad\quad\quad\quad\quad\quad$ = 2.00 atm + 1.00 atm absolute pressure = 3.00 atm
a. 2 Determine the number of moles in the gas.	Apply the ideal gas law using the initial conditions to determine the number of moles in the gas. $P_i V_i = n R T_i$ (3.00 atm)(5.00 liter) = n (0.0821 liter atm/mole K)(303 K) n = 0.603 moles
b. 1 Determine the absolute pressure after the change.	The absolute pressure after the change has occurred is 2.80 atm + 1.00 atm = 3.80 atm.
b. 2 Use the ideal gas equation and solve for the final temperature.	Again apply the ideal gas formula and solve for the final temperature. $P_f V_f = n R T_f$ (3.80 atm)(15.0 liters) = (0.603 mole)(0.0821 liter atm/mole K)T_f $T_f = 1150$ K

PROBLEM 5. Five molecules have the following speeds given in m/s: 1.0, 2.0, 3.0, 4.0, 5.0. Determine the a) mean speed and b) rms speed.

a. 1 Determine the mean speed.	Solution: (Section 13-11) The mean speed ($\bar{v}$) equals the sum of the individual speeds divided by the total number of molecules. $\bar{v} = \dfrac{(1.0 + 2.0 + 3.0 + 4.0 + 5.0)\ m/s}{5}$ $\bar{v} = 3.0$ m/s

b. 1	The rms speed is equal to the square root of the mean of the square of the speeds of the molecules of the gas.
Determine the rms speed.	$v_{rms} = [(1.0^2 + 2.0^2 + 3.0^2 + 4.0^2 + 5.0^2)/5]^{1/2}$
	$v_{rms} = 3.3$ m/s

PROBLEM 6. Calculate the a) average translational kinetic energy and b) rms speed of an argon-40 atom when the temperature is 27.0°C. Note: argon is a monatomic gas and one mole of Ar-40 has a mass of 40.0 grams, or expressed in atomic mass units, the mass of one Ar-40 atom the mass is 40 u.

a. 1	Solution: (Section 13-11)
Convert 27.0°C to degrees Kelvin.	27.0°C + 273°C = 300 K
a. 2	$\overline{KE} = 3/2 \ k \ T = (3/2)(1.38 \times 10^{-23} \ J/K)(300 \ K)$
Determine the translational KE of an argon-40 atom.	$\overline{KE} = 6.21 \times 10^{-21}$ J
b. 1	$(40 \ g/mole)(1.0 \ kg/1000 \ g)(1 \ mole/6.02 \times 10^{23} \ atoms) =$ $\qquad\qquad\qquad\qquad\qquad\qquad 6.64 \times 10^{-26}$ kg
Determine the mass in kg of an Ar-40 atom.	alternate method: $(40.0 \ u)(1.66 \times 10^{-27} \ kg/u) = 6.64 \times 10^{-26}$ kg
b. 2	$\overline{KE} = \frac{1}{2} \ m \ v^2 \quad$ but $\quad \overline{KE} = 3/2 \ k \ T$
Calculate the rms speed of an Ar-40 atom.	$\frac{1}{2}(6.64 \times 10^{-26} \ kg) \ \overline{v^2} = 3/2 \ (1.38 \times 10^{-23} \ J/K)(300 \ K)$
	$\overline{v^2} = 1.87 \times 10^5 \ m^2/s^2$
	$v_{rms} = 432$ m/s

PROBLEM 7. In a laboratory experiment, oxygen is collected through displacement of water as shown in the diagram below. The gas is collected at 23°C until 0.300 liters of gas is obtained. The atmospheric pressure is 760 torr and the vapor pressure of water at 23°C is 21.1 torr. Determine the a) partial pressure of the O_2 in the container and b) number of moles of O_2 collected.

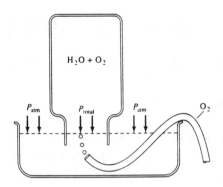

H₂O + O₂

P_{atm} P_{total} P_{atm} O₂

a. 1	Solution: (Section 13-9)
Determine the partial pressure exerted by the oxygen.	The partial pressure of the oxygen gas plus the water vapor must equal the external atmospheric pressure, thus
	$P_{atm} = P_{Oxygen\ gas} + P_{water\ vapor}$
	760 torr = $P_{Oxygen\ gas}$ + 21.1 torr
	$P_{Oxygen\ gas}$ = 739 torr
b. 1	The pressure is given in torr; therefore, use 62.4 liter torr/mole K for R. Also, remember that the temperature must be expressed in degrees Kelvin.
Solve for the number of moles of oxygen.	$P\,V = n\,R\,T$
	(739 torr) (0.300 liters) = n (62.4 liter torr/mole K)(296 K)
	n = 0.012 moles of O_2

PROBLEM 8. A cool mist vaporizer is used in the bedroom of a child who has a common cold. The relative humidity of the room is 20.0% at 20.0°C. The room is 3.00 m x 3.00 m x 2.70 m. Determine the amount of water vapor in grams and liters that must be added to the air of the room in order to raise the relative humidity to 60.0%.

a. 1	Solution: (Section 13-14)
Determine the initial partial pressure of the water vapor in torr.	According to Table 13-4 the saturated vapor pressure of water at 20.0°C is 17.5 torr.
	R.H. = P_p /P_v x 100%

	$20.0\% = P_p /17.5 \text{ torr} \times 100\%$
	$P_p = (20.0\%)/(100\%) \times 17.5 \text{ torr} = 3.50 \text{ torr}$
a. 2 Determine the number of moles of water vapor present at 20.0°C. Note: 20.0°C = 293 K	$V = 3.00 \text{ m} \times 3.00 \text{ m} \times 2.70 \text{ m} = 24.3 \text{ m}^3$ $P V = n R T$ $(3.50 \text{ torr})(24.3 \text{ m}^3)(10^3 \text{ liter}/1 \text{ m}^3) =$ $\qquad\qquad n (62.4 \text{ liter torr/mole·K})(293 \text{ K})$ $n = 4.65 \text{ moles}$
a. 3 Determine the number of grams of water in 4.65 moles.	A water molecule contains 2 atoms of hydrogen and 1 atom of oxygen. The molecular weight of water is $(2 \times 1) + (1 \times 16) =$ 18.0 g/mole. The number of grams of water added is 4.65 moles x 18.0 grams/mole = 83.7 grams
a. 4 Determine the volume of water initially present in the vapor phase.	The density of water is 1.00 g/ml and there are 1000 ml in 1.00 liter. $V = (83.7 \text{ g})(1 \text{ ml/g})(1 \text{ liter}/1000 \text{ ml}) = 0.0837 \text{ liters}$
a. 5 Determine the partial pressure of the water vapor in torr when the relative humidity is 60.0%.	$R.H. = P_p /P_v \times 100\%$ $60.0\% = P_p /17.5 \text{ torr} \times 100\%$ $P_p = (60.0\%)/(100\%) \times 17.5 \text{ torr}$ $P_p = 10.5 \text{ torr}$
a. 6 Determine the number of moles of water vapor present at 20.0°C when the R.H. = 60.0%.	$P V = n R T$ $(10.5 \text{ torr})(24.3 \text{ m}^3)(10^3 \text{ liter}/1 \text{ m}^3) =$ $\qquad\qquad n (62.4 \text{ liter torr/mole K})(293 \text{ K})$ $n = 14.0 \text{ moles}$
a. 7 Determine the number of grams of water in 14.0 moles.	14.0 moles x 18.0 grams/mole = 252 grams

a. 8	252 g required - 83.7 g initially present = 168 g to be added
Calculate the number of grams of water that must be added to raise the R.H. from 20.0% to 60.0 %.	

a. 9	The density of water is 1.00 g/ml and there are 1000 ml in 1.00 liter.
Determine the volume of water added to the vapor phase.	V = (168 g)(1 ml/gram)(1.00 liter/1000 ml) = 0.168 liters

PROBLEM 9. Estimate the boiling point of water at a place where the atmospheric pressure is 500 torr.

a. 1	Solution: (Section 13-14)
Use Table 13-4 to estimate the boiling point.	At the boiling point, the vapor pressure of water equals the external pressure. Therefore, the vapor pressure of the bubbles that form must be 500 torr. From Table 13-4, the vapor pressure of water at 90° is 526 torr. Therefore, the boiling point of water at a place where the atmospheric pressure is 500 torr is approximately 89°C.

PROBLEM 10. Determine the rate of diffusion per m^2 of a tobacco mosaic virus through water over a distance of 1.0 cm from a region where the concentration is 2.0 mole/m^3 to a region where the concentration is zero. The diffusion coefficient is 0.30 x 10^{-11} m^2/s.

a. 1	Solution: (Section 13-15)
Apply the formula for the rate of diffusion.	$J = D A (C_1 - C_2)/\Delta x$ and
	$J/A = D (C_1 - C_2)/\Delta x$
	$= [(0.30 \times 10^{-11} m^2/s)(2.0 \text{ mole}/m^3 - 0 \text{ mole}/m^3)]/(0.01 \text{ m})$
	$J/A = 6.0 \times 10^{-10} \text{ mole}/m^2\cdot s$

PROBLEM 11. The rate of diffusion of an unknown gas (x) is determined to be 2.92 times faster than that of ammonia (NH_3). Determine the a) mass of the individual molecules of the unknown gas and b) molecular weight of the gas. The molecular weight of ammonia is 17.0 g/mole. Hint: use Graham's law of diffusion, which states that the rate of diffusion of a gas is inversely proportional to the square root of the molecular mass.

a. 1	Solution: (Section 13-15)
Determine the mass of an ammonia molecule.	$m = 17.0$ g/mole x 1 mole/6.02×10^{23} molecules $m = 2.82 \times 10^{-23}$ g/molecule
a. 2 Use Graham's law of diffusion to determine the mass of a molecule of the unknown gas.	From Graham's law of diffusion, $J \propto 1/(m)^{1/2}$; therefore the ratio of the rates of diffusion of unknown gas to ammonia should be inversely related to the ratio of the square roots of their molecular masses, i.e., $$\frac{J_x}{J_{NH3}} = \frac{1/(m_x)^{1/2}}{1/(m_{NH3})^{1/2}} = \frac{(m_{NH3})^{1/2}}{(m_x)^{1/2}}$$ $J_x/J_{NH3} = 2.92 = (m_{NH3}/m_x)^{1/2}$ Using algebra to rearrange the equation: $m_x = m_{NH3}/(2.92)^2$ $m_x = (2.82 \times 10^{-23}$ g/molecule$)/(8.53)$ $m_x = 3.31 \times 10^{-24}$ g/mole
b. 1 Determine the molecular weight of the unknown gas.	The molecular weight of a gas is the mass of 1 mole of the gas expressed in grams. The mass of 1 molecule is now known and 1 mole of a gas contains 6.02×10^{23} molecules. The molecular weight may now be determined. $MW = (3.31 \times 10^{-24}$ g/molecule$)$ x $(6.02 \times 10^{23}$ molecules/mole$)$ $MW = 1.99$ gram/mole The molecular weight of diatomic hydrogen (H_2) is 2.00 g/mole; thus the unknown gas is comprised of hydrogen molecules.

PRACTICE PROBLEMS

PROBLEM 1. Determine the density of neon gas at 100°C and 2.00 atm. The molecular weight of neon is 20.2 grams per mole.

ANS. 1.32 gram/liter

PROBLEM 2. Calculate the rms speed of molecules in a tank of volume 2.00 m^3 which contains 20.0 kg of an ideal gas at a pressure of 5.00 x 10^5 N/m^2.

ANS. 387 m/s

PROBLEM 3. A sample of oxygen is collected over water at 40°C. The volume of the sample is 200 ml and the total pressure in the gas is 800 torr. Determine the temperature at which the same sample would occupy a volume of 300 ml if the water vapor is removed and the pressure in the gas is 700 torr. Note: the vapor pressure of water is 55.3 torr at 40°C.

ANS. 367 K or 94.0°C

PROBLEM 4. Determine the ratio of the rate of diffusion of He to Ne. The molecular weight of He is 4.0 g/mole and the molecular weight of Ne is 28.0 g/mole.

ANS. J_{He}/J_{Ne} = 2.3. He diffuses 2.3 times faster than Ne.

CHAPTER 14

HEAT

KEY TERMS AND PHRASES

SI unit of heat is the joule (J). Also used are units related to the joule, namely, the **calorie** and **kilocalorie** where 4.184 J = 1 calorie (exactly) and 1000 cal = 1 kilocalorie.

internal energy consists of the total potential and kinetic energy of all of the atoms in the substance.

heat is the transfer of energy (usually thermal energy) from an object at higher temperature to one at a lower temperature.

specific heat is the amount of heat required to raise the temperature of 1 kg of a substance by 1°C. For water at 15°C, the specific heat is 1.00 kcal/kg°C or 4180 J/kg°C.

calorimetry is the quantitative measurement of the heat exchanged between two substances that are initially at different temperatures. Calorimetry is based on the law of conservation of energy and assumes that the heat lost by an object(s) at the higher temperature equals the heat gained by the object(s) at the lower temperature, i.e., heat lost + heat gained = 0.

change of phase refers to an object changing from the solid state to the liquid state and vice versa or from the liquid state to the gaseous state and vice versa.

latent heat of fusion refers to the energy that must be added or removed from a substance in order for the substance to change from the solid state to the liquid state or vice versa. The change occurs at the melting (freezing) point of the substance and no change of temperature occurs during the change of phase.

latent heat of vaporization refers to the energy that must be added or removed from a substance in order for the substance to change from the liquid state to the gaseous state or vice versa. The change occurs at the boiling (condensation) point of the substance and no change of temperature occurs during the change of phase.

conduction of heat is the result of collisions between molecules in a material. In a solid, the molecules are not free to move throughout the volume of the solid. The increase in kinetic energy of the molecules is in the form of vibrational kinetic energy.

convection is transfer of heat due to mass movement of warm material from one region to another. In the process of moving, the warm material displaces cold material.

radiation is the transfer of energy due to electromagnetic waves. Electromagnetic waves require no substance to transfer the energy.

thermal equilibrium between an object and its surroundings is reached when they reach the same temperature, i.e., $T_{1f} = T_{2f}$.

SUMMARY OF MATHEMATICAL FORMULAS

internal energy of an ideal gas	$U = 3/2\ n\ R\ T$	The internal energy (U) of an ideal gas depends on the number of moles (n) and the temperature in degrees Kelvin (T). R represents the universal gas constant.
heat transfer and specific heat	$Q = m\ c\ \Delta T$	The heat (Q) required to raise the temperature of a substance is related to the mass (m), specific heat (c) of the substance and the change in temperature (ΔT).
latent heat of fusion or latent heat of vaporization	$Q = m\ \ell$	The heat (Q) added (or removed) in order to cause a substance to undergo a change of phase depends on the object's mass (m) and the latent heat (ℓ) where ℓ is either the heat of fusion (ℓ_F) or heat of vaporization (ℓ_V).
thermal conduction	$\Delta Q/\Delta t = K\ A\ (T_1 - T_2)/\ell$	The rate of heat flow ($\Delta Q/\Delta t$) through a substance due to conduction is related to the cross-sectional area of the object (A), the object's thickness (ℓ), and the temperature difference between the ends of the object ($T_1 - T_2$) in °C. K is the thermal conductivity of the material.

Stefan-Boltzmann's equation	$\Delta Q/\Delta t = e \, \sigma \, A \, T^4$	The rate at which an object radiates electromagnetic energy ($\Delta Q/\Delta t$) is related to the object's surface area(A), the object's temperature (T) in degrees Kelvin, and the emissivity (e) of the material. $\sigma = 5.67 \times 10^{-8} \; J/(s \, m^2 \, K^4)$ and is Stefan-Boltzmann's constant.
net rate of radiant energy transferred	$\Delta Q/\Delta t = e \, \sigma \, A \, (T_1^4 - T_2^4)$	The net rate of radiant energy flow be tween an object and its surroundings is related to the surface area, emissivity, and the difference in the temperatures between the object (T_1) and its surroundings (T_2).

EQUATIONS AND CONCEPTS

Units of Heat

The **SI** unit of heat is the joule (J). Also used are units related to the joule, namely, the **calorie** and **kilocalorie** where 4.184 J = 1 calorie (exactly) and 1000 cal = 1 kilocalorie. The kilocalorie is also known as the "large" calorie and is the unit associated with the energy content of foods.

Temperature, Heat, and Internal Energy

Temperature is the measure of the average kinetic energy of the individual molecules in a substance. All objects, whether solid, liquid, or gas, consist of atoms that are in motion. The **internal energy** consists of the total potential and kinetic energy of all of the atoms in the substance.

In an ideal monatomic gas we do not consider attractive or repulsive forces between atoms and so the internal energy (U) would be the total number of molecules in a gas times the average kinetic energy of the molecules.

$$U = N \, \overline{KE}$$

$$U = N \, (\tfrac{1}{2} \, m \, \overline{v^2}) = (3/2 \, k \, T) = 3/2 \, n \, R \, T$$

Heat is the transfer of energy (usually thermal energy) from an object at higher temperature to one at a lower temperature.

Specific Heat

The amount of heat (Q) required to raise the temperature of a substance depends on the quantity of matter or mass (m), the **specific heat** (c) of the substance involved, and the change in temperature (ΔT). This is expressed mathematically as

$$Q = m \; c \; \Delta T$$

c is the specific heat of the substance. This is the amount of heat required to raise the temperature of 1 kg of a substance by 1°C. For water at 15°C, c = 1.00 kcal/kg°C or 4180 J/kg°C.

Calorimetry

Calorimetry is the quantitative measurement of the heat exchanged between two substances that are initially at different temperatures. For example, if hot water is added to a container of cold water, some of the heat energy of the hot water will be transferred to the cold water and also to the container. In calorimetry, this heat exchange is treated quantitatively.

Calorimetry is based on the law of conservation of energy and assumes that the heat lost by an object(s) at the higher temperature equals the heat gained by the object(s) at the lower temperature, i.e., heat lost + heat gained = 0   or   heat gained = - heat lost.

Latent Heat

Latent Heat is the energy that must be added or removed from a substance in order for a **change of phase** to occur. For changes from the solid to the liquid state or vice versa the heat energy required is the **latent heat of fusion**. For changes from the liquid state to the gas state and vice versa the heat energy is the **latent heat of vaporization**.

Heat added to ice at 0°C causes ice to melt and a change of phase from the solid state to the liquid state occurs. There is no change in temperature during the phase change. The heat added overcomes the potential energy associated with the attractive forces between the molecules in the solid (ice). There is no increase in kinetic energy of the molecules until all of the ice melts and there is no increase in temperature during the phase change.

The reverse process occurs when water returns to the solid phase from the liquid phase. Heat must be removed from the system even though there is no change of temperature.

Table 14-2 lists the latent heats of fusion and vaporization for a number of substances. For water, the heat of fusion is 79.7 kcal/kg or 333 kJ/kg and the heat of vaporization is 539 kcal/kg or 2260 kJ/kg. The heat required for a phase change is given by

$$Q \; = \; m \; \ell \qquad \text{where} \; \ell \; \text{is the heat of fusion } (\ell_F) \text{ or heat of vaporization } (\ell_V)$$

Heat Transfer: Conduction, Convection, and Radiation

There are three ways to transfer heat from one object or place to another. In a particular situation, one or more of the processes may be involved in the heat transfer.

Conduction

Conduction of heat is the result of collisions between molecules in a material. In a solid, the molecules are not free to move throughout the volume of the solid. The increase in kinetic energy of the molecules is in the form of vibrational kinetic energy.

If one end of the solid, e.g., a metal rod, is held in a flame, the molecules in that end will have a higher average vibrational kinetic energy than molecules further along the rod. The higher energy molecules transfer some of this energy to adjacent molecules via molecular collisions. As a result, heat energy is gradually transferred through the object.

The rate of heat flow through a substance due to conduction is given by the relation

$$\Delta Q/\Delta t = K \, A \, (T_1 - T_2)/\ell$$

$\Delta Q/\Delta t$ is the heat transferred per unit time, A is the cross-sectional area of the object, and ℓ is the object's thickness. $T_1 - T_2$ is the temperature difference between the ends of the object in °C, where T_1 is greater than T_2. K is the **thermal conductivity** of the material and the unit of thermal conductivity is kcal/(s m °C).

Convection

Convection is transfer of heat due to mass movement of warm material from one region to another. In the process of moving, the warm material displaces cold material. An example of this is a convection current present in a pan of water heated on a stove. The hot water rises from the bottom of the pan to the top while the cold water on the top sinks to the bottom. As the cold water sinks it is heated and the convection process continues until the boiling point is reached.

Radiation

Radiation is the transfer of energy due to electromagnetic waves. Electromagnetic waves, e.g., light from the sun, travel at the speed of light. They require no material medium to transport the waves and light from very distant objects, e.g. the stars, is able to travel through the vacuum of outer space to the Earth. The rate at which an object radiates electromagnetic energy is given by the Stefan-Boltzmann's equation $\Delta Q/\Delta t = e \, \sigma \, A \, T^4$.

$\Delta Q/\Delta t$ is the rate at which energy leaves the object, A is the object's surface area, and T is the object's temperature in K. e is the **emissivity** of the material and is a characteristic property of the material. A perfect absorber is also a perfect emitter and has an emissivity value of 1. A perfect absorber is known as a perfect black body. A perfect reflector has an emissivity value of 0. $\sigma = 5.67 \times 10^{-8}$ J/(s m^2 K^4) and is known as the Stefan-Boltzmann constant. The net rate of radiant energy flow between an object and its surroundings is given by

$$\Delta Q/\Delta t = e \; \sigma \; A \; (T_1{}^4 - T_2{}^4)$$

T_1 and T_2 are the temperatures in degrees Kelvin of the object and its surroundings.

Thermal equilibrium ($\Delta Q/\Delta t = 0$) between an object and its surroundings is reached when they reach the same temperature, i.e., $T_1 = T_2$.

ANSWERS TO SELECTED TEXT QUESTIONS

QUESTION 7. Explain why burns caused by steam on the skin are often so severe.

ANSWER: The heat of vaporization of water is 539 Kcal/kg and this heat is released when the steam strikes the skin and condenses into water.

If the heat present in 1 gram of steam as it condenses into water at 100°C is completely absorbed by the skin, the skin absorbs 539 cal of heat. If this same 1 gram of water now cools from 100°C to 37°C (body temperature) it releases only 73 cal of heat. The burns caused by steam at 100°C are more severe than an equal mass of water at 100°C.

QUESTION 9. Will potatoes cook faster if the water is boiling faster?

ANSWER: Under 1 atmosphere of pressure, water boils at approximately 100°C whether the water is boiling quickly or slowly. The time required for the potatoes to cook will be the same but energy will be saved if the water is set for a slow boil.

QUESTION 22. Why is the liner of a thermos bottle silvered and why does it have a vacuum between its two walls?

ANSWER: Silver reflects heat back into the liquid in the bottle and therefore heat loss due to radiation is kept to a minimum. The partial vacuum between the liner and the outer wall keeps heat loss due to convection to a minimum. The cork or styrofoam cap is a poor conductor of heat and if the seal is tight the rate of heat loss due to conduction is very low. Thus the hot liquid in the thermos will be kept hot for several hours.

The same arguments can be used to explain why cold liquids will remain cold for extended periods in a thermos.

PROBLEM SOLVING SKILLS

For problems involving calorimetry:

1. Complete a data table for information both given and implied in the problem. This information should list the initial and final temperature of each substance, the specific heat of each substance, and the mass of each substance. If a phase change occurs, the list should include the latent heat of fusion or vaporization of the substance involved in the phase change.
2. Apply the law of conservation of energy and solve the problem. Remember that the sum of the heat gain and loss equals zero.

For problems involving heat transfer by conduction:

1. Complete a data table listing information both given and implied as well as the unknown quantity. This information should include the temperature on either side of the material, the thickness of the material, the cross-sectional area through which the heat transfer occurs, and the thermal conductivity of the substance. Also, the heat transferred and the time required for the transfer to occur should be included in the list.
2. Apply the formula for the rate of heat transfer through a material and solve for the unknown quantity.

For problems involving heat transfer by radiation:

1. Complete a data table listing information both given and implied. This list should include the temperature of the object and its surroundings, the surface area of the object, the object's emissivity, and the Stefan-Boltzmann constant. Also, the heat transferred and the time required for the transfer should be included in the list.
2. Apply the formula for rate of heat transfer by radiation and solve for the unknown quantity.

PROGRAMMED PROBLEMS

PROBLEM 1. A 200 gram piece of aluminum at 90.0°C is placed in a 100 gram glass container which holds an 1860 grams of water at 20.0°C. Determine the equilibrium temperature of the system. Note: the specific heats of water, glass, and aluminum are 1.00 kcal/kg°C, 0.200 kcal/kg°C, and 0.220 kcal/kg°C, respectively.

---

a. 1

Determine the change in temperature for each substance in the system.

Solution: (Section 14-4)

glass and water: $\Delta T_g = \Delta T_w = T_f - 20.0°C$

aluminum: $\Delta T_{Al} = T_f - 90.0°C$

---

a. 2

Apply the law of conservation of energy to determine the amount of water in the container.

heat gained by glass and water + heat lost by aluminum = 0

heat gained by glass and water = - heat lost by aluminum

$m_g \, c_g \, \Delta T_g \; + \; m_w \, c_w \, \Delta T_w = \; - (m_{Al} \, c_{Al} \, \Delta T_{Al})$ but $\Delta T_g = \Delta T_w$

$[m_g \, c_g + m_w \, c_w] \, \Delta T_w = \; - (m_{Al} \, c_{Al} \, \Delta T_{Al})$

$[(0.100 \text{ kg})(0.200 \text{ kcal/kg°C}) + (1.860 \text{ kg})(1.00 \text{ kcal/kg°C})](T_f - 20.0°C)$

$= - (0.200 \text{ kg})(0.220 \text{ kcal/kg °C})(T_f - 90.0°C)$

$(1.88 \text{ kcal/°C})(T_f - 20.0°C) = -(0.0440 \text{ kcal/°C})(T_f - 90.0°C)$

$42.7(T_f - 20.0°C) = - T_f + 90.0°C$

$42.7 \, T_f - 855°C = - T_f + 90.0°C$

$43.7 \, T_f = 945°C$

$T_f = 21.6°C$

PROBLEM 2. A student uses a microwave oven to heat a 0.300 kg ceramic coffee cup which contains 0.100 kg of water and 0.0500 kg of ice from 0.00°C to 50.0°C. Determine the time required if the microwave oven is rated at 1350 watts. Assume all of the energy produced by the oven is absorbed by the water + ice + cup. Note: the heat of fusion of ice is 79.7 kcal/kg, the specific heat of water is 1.00 kcal/kg °C, and the specific heat of the ceramic cup is 0.220 kcal/kg °C.

a. 1	Solution: (Sections 14-4 and 14-5)
Determine the heat which must be added to melt the ice.	$Q_{ice} = m_{ice} \; \ell_F = (0.0500 \text{ kg})(79.7 \text{ kcal/kg})$ $Q_{ice} = 3.99 \text{ kcal}$
a. 2 Determine the change of temperature of the water + cup.	$\Delta T_{Al} = \Delta T_w = 50.0°C - 0.00°C = 50.0 \text{ C}°$ $\Delta T_{Al} = \Delta T_w = 50.0°C$
a. 3 Determine the total energy required to raise the temperature of the system to 50°C.	Note: after the ice is melted, the total amount of water to be heated is 0.100 kg + 0.0500 kg = 0.150 kg. $Q_{total} = m_{ice} \; \ell_F + m_{cup} \; c_{cup} \; \Delta T_{cup} + m_w \; c_w \; \Delta T_w$ $\quad = 3.99 \text{ kcal} + (0.300 \text{ kg})(0.220 \text{ kcal/kg}°C)(50.0 \text{ C}°)$ $\qquad + (0.150 \text{ kg})(1.00 \text{ kcal/kg}°C)(50.0 \text{ C}°)$ $Q_{total} = 3.99 \text{ kcal} + 3.30 \text{ kcal} + 7.50 \text{ kcal}$ $Q_{total} = 14.8 \text{ kcal}$
a. 4 The work done by the the microwave oven equals the heat gained by the system. Determine the time required to heat the system to 50.0°C.	The electrical energy converted to heat energy by the microwave oven is determined by using the equation work = power x time work = (power)(time) Note: 1350 watt = 1350 J/s. $(14.8 \text{ kcal})(1000 \text{ cal/kcal})(4.18 \text{ J/cal}) = (1350 \text{ J/s}) \text{ time}$ $61900 \text{ J} = (1350 \text{ J/s}) \text{ time}$ time = 45.8 s

PROBLEM 3. A 0.0100 kg lead bullet moving at 300 m/s imbeds itself in a large block of wood. All of the kinetic energy lost by the lead bullet is transferred into heat which is shared equally by the bullet and the block of wood. Determine the temperature change of the bullet. The specific heat of lead is 0.031 kcal/kg°C.

a. 1 Determine the amount of mechanical energy dissipated in the form of heat.	Solution: (Sections 14-4 and 14-5) $Q = - \Delta KE = - (KE_f - KE_i)$ $\quad = - (\tfrac{1}{2} m \; v_f^2 - \tfrac{1}{2} m \; v_i^2)$ $\quad = - [\tfrac{1}{2}(0.0100 \text{ kg})(0 \text{ m/s})^2 - \tfrac{1}{2}(0.0100 \text{ kg})(100 \text{ m/s})^2]$ $Q = 50.0 \text{ J}$

a. 2	The heat transferred to the bullet $Q = \frac{1}{2}(50.0 \text{ J}) = 25.0 \text{ J}$
Determine the change in temperature of the bullet.	$Q = m c \Delta T$
	$(25.0 \text{ J})(1.0 \text{ kcal}/4180 \text{ J}) = (0.0100 \text{ kg})(0.031 \text{ kcal/kg °C}) \Delta T$
	$\Delta T = 19.3 \text{ C}°$

PROBLEM 4. A 1.00 kg aluminum pot holds 2.00 kg of water at 20.0°C. Determine the amount of steam added at 100°C which is required to raise the temperature of the water and pot to 25.0°C. Note: the heat of vaporization of steam is 540 kcal/kg, the specific heat of water is 1.00 kcal/kg °C, and the specific heat of aluminum is 0.220 kcal/kg °C.

a. 1	Solution: (Sections 14-4 and 14-5)
Determine the temperature change for each substance in the system.	Steam at 100°C to water at 100°C: no temperature change occurs during the change of phase to hot water.
	Hot water at 100°C to water at 25.0°C:
	$\Delta T_{hw} = 25.0°C - 100°C = -75.0 \text{ C}°$
	Aluminum pot: $\Delta T_{Al} = 25.0°C - 20.0°C = 5.00 \text{ C}°$
	Cold water at 20.0°C to water at 25.0°C:
	$\Delta T_w = 25.0°C - 20.0°C = 5.00 \text{ C}°$
a. 2	The steam becomes hot water which then loses heat until its temperature reaches 25.0°C. Therefore, the mass of the steam equals the mass of the hot water, i.e., $m_{hw} = m_{steam}$
Derive a formula for the heat lost by the steam and the hot water.	heat lost by steam + hot water = $Q_{steam} + Q_{hw}$
	$Q_{steam} + Q_{hw} = m_{steam} \, \ell_v + m_{hw} \, c_w \, \Delta T_{hw}$
	$= m_{steam} (-540 \text{ kcal/kg}) + m_{hw} (1.0 \text{ kcal/kg°C})(-75.0°C)$
	but $m_{steam} = m_{hw}$
	$Q_{steam} + Q_{hw} = -[540 \text{ kcal/kg} + 75.0 \text{ kcal/kg}] \, m_{steam}$
	heat lost = $-(615 \text{ kcal/kg}) \, m_{steam}$

a. 3	heat gained by pot + cold water = $m_{Al} c_{Al} \Delta T_{Al} + m_{cw} c_w \Delta T_{cw}$
Determine the heat gained by the pot and cold water. water. Note: let the mass of the cold water be m_{cw}.	heat gained = (1.00 kg)(0.220 kcal/kg°C)(5.00°C) $\qquad\qquad$ + (2.0 kg)(1.0 kcal/kg°C)(5.00°C) heat gained = 1.10 kcal + 10.0 kcal = 11.1 kcal
a. 4 Apply the law of conservation of energy and solve for the mass of the steam.	The heat gained by the aluminum and cold water plus the heat lost by the steam and the hot water must equal zero. heat gained + heat lost = 0 11.1 kcal + -(615 kcal/kg)m_{steam} = 0 m_{steam} = (- 11.1 kcal)/(-615 kcal/kg) = 0.0180 kg or 18.0 grams

PROBLEM 5. 0.0500 kg of ice at 0.00°C is added to a 0.100 kg glass container which holds 0.150 kg of water at 30.0°C. Determine the final temperature of the mixture. The heat of fusion of ice is 79.7 kcal/kg, the specific heat of water is 1.00 kcal/kg °C, and the specific heat of glass is 0.200 kcal/kg °C.

a. 1	Solution: (Sections 14-5 and 14-6)
Determine the heat gained by the ice as it changes into cold water and then reaches its final temperature.	heat gained = $m_{ice} \ell_f + m_{cw} c_w \Delta T_{cw}$ $\qquad$ = (0.0500 kg)(79.7 kcal/kg) + $\qquad\qquad$ (0.0500 kg)(1.00 kcal/kg °C)(T_f - 0.0°C) Heat gained = 3.99 kcal + (0.0500 kg/°C)T_f
a. 2 Write an equation for the heat lost by the water and the glass.	heat lost = $m_g c_g (T_f - 30.0°C) + m_{cw} c_w (T_f - 30.0°C)$ $\qquad$ = (0.100 kg)(0.200 kcal/kg°C)(T_f - 30.0°C) $\qquad\qquad$ + (0.150 kg)(1.0 kcal/kg°C)(T_f - 30.0°C) $\qquad$ = [(0.0200 kcal/°C + 0.150 kcal/°C)](T_f - 30.0°C) $\qquad$ = (0.170 kcal/°C)(T_f - 30.0°C) heat lost = (0.170 kcal/°C)t_f - 5.10 kcal
a. 3 Apply the law of conservation of energy and solve for T_f.	heat gained + heat lost = 0 3.99 kcal + (0.0500 kg/°C)T_f + (0.170 kcal/°C)T_f - 5.10 kcal = 0 (0.220 kcal/°C)T_f = 1.11 kcal T_f = 5.05°C

a. 1	Solution: (Section 14-7)
Complete a data table and apply the equation for the conduction of heat through a substance.	ℓ = (5.0 mm)(1 m/1000 mm) = 5.0 x 10⁻³ m = 0.0050 m
	A = (370 cm²)(1 m²/10⁻⁴ cm²) = 3.7 x 10⁻² m² = 0.037 m²
	$T_1 - T_2$ = 80°C - 20°C = 60 C°
	$\Delta Q/\Delta t = k\,A\,(T_1 - T_2)/\ell$
	$\qquad$ = (5.5 x 10⁻⁶ kcal/s m°C)(0.037 m²)(60 C°)/(0.005 m)
	$\Delta Q/\Delta t$ = 2.4 x 10⁻³ kcal/s

b. 1	The time for the temperature of the coffee to change from 80°C to 70°C may be determined as follows:
Determine the time required for the temperature of the coffee to cool from 80°C to 70°C. Note: assume that the rate of heat loss over the time interval is constant.	heat loss = $(\Delta Q/\Delta t)$(time interval) but
	heat loss = Q = m c ΔT
	$(\Delta Q/\Delta t)$(time interval) = m c ΔT
	(2.4 x 10⁻³ kcal/s) t = - (0.50 kg)(1.0 kcal/kg°C)(70°C - 80°C)
	t = 2100 s or 35 minutes

Note: the answer is unrealistic due to a number of assumptions that would affect the time interval. For example, we have assumed that the cup is tightly sealed so that convective heat losses are negligible, rate of heat loss is constant, thickness of the cup is uniform. However, even with these assumptions we can see the advantage of using styrofoam cups for holding hot or cold liquids.

a. 1 Calculate the temperature of the bar in degrees Kelvin.	Solution: (Section 14-9) 273°C + 727°C = 1000 K
a. 2 Determine the surface area of the bar.	The lateral surface area of the bar = $2\pi rL$ and the area of each end = πr^2; therefore, $A = 2\pi rL + \pi r^2 + \pi r^2$ $A = 2\pi (1.00 \text{ cm})(10.0 \text{ cm}) + \pi (1.00 \text{ cm})^2 + \pi (1.00 \text{ cm})^2$ $A = 62.8 \text{ cm}^2 + 3.14 \text{ cm}^2 + 3.14 \text{ cm}^2 = 69.1 \text{ cm}^2$ $A = (69.1 \text{ cm}^2)(1.0 \text{ m}/100 \text{ cm})^2 = 6.91 \times 10^{-3} \text{ m}^2$
a. 3 Determine the rate at which energy is radiated.	The emissivity of the iron bar is 0.60. The rate at which energy is radiated can now be determined. $\Delta Q/\Delta t = e \, \sigma \, A \, T^4$ $\Delta Q/\Delta t = (0.60)(5.67 \times 10^{-8} \text{ J/s m}^2 \text{ K}^4)(6.91 \times 10^{-3} \text{ m}^2)(1000 \text{ K})^4$ $\Delta Q/\Delta t = 235 \text{ J/s} = 235 \text{ watts}$

PRACTICE PROBLEMS

PROBLEM 1. A 40.0 gram sample of water at 22.0°C is held in a 100 gram metal cup. 60.0 grams of water at 65.0°C is stirred into the cool water and the final temperature of the water is 47.0°C. Determine the specific heat of the metal of the cup.

ANS. 0.032 kcal/kg°C

PROBLEM 2. 1000 joules of heat is needed to melt 40 grams of a certain substance at its melting point. Determine the heat of fusion of the substance in kcal/kg.

ANS. 6.0 kcal/kg

PROBLEM 3. Determine the amount of heat required to change a 0.50 kg ice cube at -10°C into water at 20°C. Note: the specific heat of ice is 0.50 kcal/kg°C.

ANS. 53 kcal

PROBLEM 4. A copper rod of cross-sectional area 1.0 cm² and length 0.50 m has one end in hot water that is at 100°C and the other end in a very large container of cold water at 10°C. Determine the time required for 1.0 kcal of heat to be transferred to the cold water. Assume that all of the heat entering the rod is conducted to the cold water and the temperature rise of the cold water is negligible. Note: the thermal conductivity of copper is 9.2×10^{-2} kcal/s m°C.

ANS: 600 seconds

CHAPTER 15

THE FIRST AND SECOND LAWS OF THERMODYNAMICS

KEY TERMS AND PHRASES

thermodynamics is the study of energy transformations in natural processes and involves relations between heat, work, and energy.

system is any object or sets of objects which are under consideration; everything else in the universe is called the environment. In thermodynamics, a **closed** system is one where mass may not enter or leave. In an **open** system mass may be exchanged with the environment.

first law of thermodynamics is a statement of the law of conservation of energy. The first law states that the change in the internal energy (ΔU) of a closed system is due to heat added or removed from the system (Q) and/or work done on or by the system (W). $\Delta U = Q - W$.

pressure-volume diagram or PV diagram is used to determine the work done by a gas undergoing expansion or compression in a closed system.

isothermal process occurs whe the temperature of the gas remains constant.

isobaric process occurs when the pressure is constant, the work done on or by the gas can be determined by using $W = P \Delta V$.

isochoric process occurs at constant volume, $\Delta V = 0$; therefore, $W = P \Delta V = 0$ and $Q = \Delta U$.

adiabatic process occurs when no heat flows into or out of the system. An adiabatic process usually occurs when a gas is compressed or expands so rapidly that there is no time for the heat to flow in or out of the system.

heat engine is a device which is capable of changing thermal energy (Q_H), also known as the input heat or heat of combustion of the fuel, into useful work (W).

Carnot engine is an idealized engine where energy losses due to internal friction, turbulence present in the fuel after ignition, etc. are not considered. Carnot determined that the maximum efficiency that can be realized from a heat engine depends on the temperature of the input heat and the exhaust heat.

Refrigerators and air conditioners operate by removing heat from a low temperature (cold) reservoir and exhausting the heat to the higher temperature (hot) reservoir. In order to accomplish this task, work is done to cause heat to travel opposite from its normal direction.

entropy is a quantitative measure of the disorder in a system.

second law of thermodynamics can be stated in several equivalent ways, three of which are:
1) Heat energy flows spontaneously from a hot object to a cold object but not vice versa.
2) It is impossible to construct a heat engine which is 100% efficient. Thus a heat engine can convert some of the input heat into useful work, but the rest must be exhausted as waste heat.
3) The entropy of an isolated system never decreases. It can only stay the same or increase.

SUMMARY OF MATHEMATICAL FORMULAS

first law of thermodynamics	$\Delta U = Q - W$	The **first law of thermodynamics** is a statement of the law of conservation of energy. The first law states that the change in the internal energy (ΔU) of a closed system is due to heat added or removed from the system (Q) and/or work done on or by the system (W).
heat engine	$Q_H = W + Q_L$ or $W = Q_H - Q_L$	A heat engine is a device which is capable of changing thermal energy (Q_H). While part of the heat energy is converted to useful work (W), the remaining heat energy will be rejected to the environment as exhaust heat (Q_L).
maximum efficiency or Carnot efficiency of a heat engine	$e = W/Q_H,$ $e = (Q_H - Q_L)/Q_H$ $e = (1 - Q_L/Q_H)$ $e = (T_H - T_L)/T_H$ $e = (1 - T_L/T_H)$	The maximum efficiency or Carnot efficiency (e) of a heat engine is equal to the ratio of the useful work (W) to the input heat (Q_H). The Carnot efficiency can be written in terms of the input temperature (T_H) and the exhaust temperature (T_L).
coefficient of performance (CP) for refrigerators or air conditioners	$CP = Q_L/W$ or $CP = (Q_L)/(Q_H - Q_L)$ $CP_{ideal} = (T_L)/(T_H - T_L)$	The CP for a refrigerator is the ratio of the heat removed from the cold region (Q_L) to the work (W) performed to remove the heat. The CP for an "ideal" refrigerator in terms of the temperatures of the low temperature reservoir (T_L) and the high temperature reservoir(T_H).

Entropy	$\Delta S = Q/T$	Entropy (S) is a quantitative measure of the disorder in a system. The change in entropy (ΔS) for a reversible process is directly proportional to the heat (Q) added to the system and inversely related to the temperature of the system (T). The unit of entropy is kcal/K, where the heat is measured in kcal and T in degrees Kelvin.
Statistical Interpretation of Entropy	$S = k \ln W$ or $S = 2.3\ k \log W$	The entropy of the system (S) is proportional to the number of ways or microstates that can occur. $k = 1.38 \times 10^{-23}$ J/K (Boltzmann's constant), ln is the logarithm to the base e, $e = 2.718$, log is the logarithm to the base 10, and W is the number of microstates corresponding to to the given macrostate.

CONCEPT SUMMARY

Thermodynamics is the study of energy transformations in natural processes and involves relations between heat, work, and energy. In this chapter we shall study the first and second laws of thermodynamics, their significance, and application.

Physical Systems

A **system** is any object or sets of objects which are under consideration; everything else in the universe is called the environment. In thermodynamics, a **closed** system is one where mass may not enter or leave. In an **open** system mass may be exchanged with the environment.

First Law of Thermodynamics

The **first law of thermodynamics** is a statement of the law of conservation of energy. The first law states that the change in the internal energy (ΔU) of a closed system is due to heat added or removed from the system (Q) and/or work done on or by the system (W).

$\Delta U = Q - W$

As a sign convention, heat added to a closed system is positive (Q = +), while heat removed is negative (Q = -). If work is done on the closed system, the internal energy and temperature increase and W is negative (W = -). If an ideal gas is compressed in a cylinder with a moveable piston, the temperature and internal energy increase. If the system does work on the surroundings (environment), the ideal gas pushes the piston outward and the gas expands, the internal energy and temperature decrease, and W is positive (W = +).

PV Diagrams

The work done by a gas undergoing expansion or compression in a closed system can be determined through use of a **pressure-volume diagram** (PV diagram). The work done during an incremental volume change (ΔV) equals the area under the PV curve. This area may be determined using the process of graphical integration. The following diagrams represent typical PV processes on an ideal gas.

In an **isothermal** process, (AB) in fig. A, the temperature of the gas remains constant. $\Delta U = 0$, $Q = W$, and based on the general gas law, $PV = nRT$. n and T are constant; therefore, $PV = $ constant.

In an **isobaric** process, (DB) in fig. A, the pressure is constant, the work done on or by the gas can be determined by using $W = P \Delta V$. n and P are constant; therefore, $V/T = $ constant.

An **isochoric** process, (AD) in fig. A, occurs at constant volume, $\Delta V = 0$; therefore, $W = P \Delta V = 0$, and $Q = \Delta U$. For this process, $P/T = $ constant.

In an **adiabatic** process, (AC) in fig. B, no heat flows into or out of the system, $Q = 0$, and $W = -\Delta U$. An adiabatic process usually occurs when a gas is compressed or expands so rapidly that there is no time for the heat to flow into or out of the system. It should be noted that an adiabatic process is quite different from an isothermal process, although the PV diagram for each process appears similar.

The Second Law of Thermodynamics

The **second law of thermodynamics** can be stated in several equivalent ways, three of which are:

1) Heat energy flows spontaneously from a hot object to a cold object but not vice versa.
2) It is impossible to construct a **heat engine** which is 100% efficient. Thus a heat engine can convert some of the input heat into useful work, but the rest must be exhausted as waste heat.
3) The **entropy** of an isolated system never decreases. It can only stay the same or increase. If the system is not isolated, then the change in entropy of the system (S_s) plus the change in entropy of the environment (S_{env}) must be greater than or equal to zero. The total entropy of any system plus that of its environment increases as a result of any natural process: $\Delta S = \Delta S_s + \Delta S_{env} > 0$.

First Statement of the Second Law

The first statement of the second law is a statement from common experience. When two objects, one hot and the other cold, come into contact, heat energy will be transferred from the system at higher temperature to the system at lower temperature, but not vice versa. While the first law states that energy must be conserved, i.e., the sum of the energy lost and gained in any process must equal zero, it does not say that heat must flow from the hot object to the cold object. The first law would not be violated if the hot object became hotter while the cold object became colder. The second law states that the direction of the heat flow must be from hot to cold.

Second Statement of the Second Law: Heat Engines

A **heat engine** is a device which is capable of changing thermal energy (Q_H), also known as the input heat or heat of combustion of the fuel, into useful work (W). Heat engines cannot be made to be 100% efficient and while part of the heat energy is converted to useful work, the remaining heat energy will be rejected to the environment or surroundings as waste heat (Q_L), e.g., exhaust from a car engine. Therefore,

$$Q_H = W + Q_L$$

In an idealized engine, known as a Carnot engine, energy losses due to internal friction, turbulence present in the fuel after ignition, etc. are not considered. Carnot determined that the maximum efficiency (e) that could be realized from a heat engine depends on the temperature of the input heat (T_H) and the exhaust heat (T_L), where T_H and T_L are expressed in degrees Kelvin.

The maximum efficiency (e) or **Carnot efficiency** of a heat engine is determined by the following formulas:

$$e = W/Q_H, \quad \text{but since} \quad W = Q_H - Q_L, \quad \text{then}$$

$$e = (Q_H - Q_L)/Q_H \quad \text{and} \quad e = (1 - Q_L/Q_H)$$

Using the input and waste heat temperatures:

$$e_{ideal} = (T_H - T_L)/T_H = (1 - T_L/T_H)$$

In order for a Carnot engine to be 100% efficient, it would be necessary for the temperature of the exhaust heat to be at absolute zero (zero degrees Kelvin). This is a practical as well as a theoretical impossibility.

Refrigerators and Air Conditioners

Refrigerators and air conditioners operate by removing heat from a low temperature (cold) reservoir and exhausting the heat to the higher temperature (hot) reservoir. In order to accomplish this task, work is done to cause heat to travel opposite from its normal direction.

The effectiveness of a particular refrigerator or air conditioner in accomplishing the removal of heat from the low temperature reservoir is measured by the coefficient of perormance (CP).

The CP for a refrigerator is the ratio of the heat removed from the cold region (Q_L) to the work (W) performed to remove the heat, i. e.,

$$CP = Q_L /W \quad \text{but} \quad W = Q_H - Q_L \quad \text{so that} \quad CP = Q_L /(Q_H - Q_L)$$

The coefficient of performance for an "ideal" refrigerator can written in terms of the temperatures of the low temperature reservoir and the temperature of the high temperature reservoir as follows:

$$CP_{ideal} = T_L /(T_H - T_L)$$

Reversible and Irreversible Processes

In an ideal gas, a reversible process is one in which the values of P, V, T, and U will have the same values in the reverse order if the process is returned to its original state. In this type of process, the ideal gas can be returned to is original state with no change in the magnitude of the work done or the heat exchanged. To be reversible, the process must be done very slowly, with no loss of energy due to dissipative forces such as friction and no heat conduction due to a temperature difference. In reality, these conditions cannot be met and all real processes are irreversible.

Third Statement of the Second Law: Entropy

Entropy (S) is a quantitative measure of the disorder in a system. The change in entropy (ΔS) for a reversible process is directly proportional to the heat (Q) added to the system and inversely related to the temperature of the system (T).

$$\Delta S = Q/T$$

The unit of entropy is kcal/K, where the heat Q is measured in kcal and the temperature T in degrees Kelvin. The Kelvin temperature at which heat is added must remain constant for a process to be reversible.

As heat is added to a system, the average kinetic energy of the molecules increases and the motion becomes more disordered. At low temperatures we would expect to find a high degree of order while at high temperatures the system is likely to be very disordered. Thus energy added while the system is at low temperature would introduce considerably more disorder than the same amount of energy introduced when the system is at a high temperature. Therefore, the change in entropy is inversely proportional to the temperature at which the heat is added.

Statistical Interpretation of Entropy

An equivalent definition of entropy can be given from a detailed analysis of the position and velocity (i.e., microstate) of every molecule which makes up the system. The entropy of the system is proportional to the number of ways that the microstates can occur and is given by the following formula:

$$S = k \ln W = 2.3 \, k \log W$$

where k is Boltzmann's constant, k = 1.38 x 10^{-23} J/K, ln is the logarithm to the base e, e = 2.718, log is the logarithm to the base 10, and W is the number of microstates corresponding to the given macrostate.

The state of highest entropy is the state that can be achieved in the largest number of ways and is therefore the most probable.

ANSWERS TO SELECTED TEXT QUESTIONS

QUESTION 3. In an isothermal process, 3700 J of work is done by an ideal gas. Is this enough information to tell how much heat has been added to the system? If so, how much?

ANSWER: The first law of thermodynamics can be written as Q = ΔU + W. In an isothermal process the temperature remains constant. Since the temperature remains constant, the internal energy remains constant and ΔU = 0. Therefore, all of the heat added during the process went into the work done by the ideal gas (Q = W). The amount of heat added to the system is 3700 J.

QUESTION 16. What happens if you remove the lid of a bottle containing chlorine gas? Does the reverse process ever happen? Why or why not? Can you think of other examples of irreversibility?

ANSWER: Once the lid is removed, chlorine gas would gradually diffuse throughout the room. The gas molecules tend to move in random directions reaching a state of maximum disorder in agreement with the second law of thermodynamics. Chlorine gas will not return to the bottle. This process is irreversible, the gas returning to the bottle would violate the second law of thermodynamics.

Another example of irreversibility would be placing 100 coins heads-up in a box and shaking the box vigorously. After shaking, there is only one way (chance) in 1 x 10^{29} that the coins could remain heads-up. The same odds apply for subsequent shaking of the box. A third example would be when a new deck of cards is opened the cards are in order according to their suit, i.e., hearts, diamonds, clubs, or spades. Subsequent shuffling of the cards gives a random arrangement with high odds against ever finding the original arrangement occurring again.

QUESTION 18. Suppose you collect a lot of papers strewn all over the floor and put them in a neat stack; does this violate the second law of thermodynamics?

ANSWER: Contrary to what teenagers tell their parents, putting the papers in order does not violate the second law of thermodynamics. An outside agent (teenager) but probably parent does work in restoring order. There is a decrease of entropy for the papers but an overall increase in entropy due to the work done by the outside agent.

PROBLEM SOLVING SKILLS

For problems involving PV diagrams:

1. Take note whether the process was isobaric, isothermic, isochoric or adiabatic.
2. Use the appropriate equation(s) to complete a data table of P vs V.
3. Use the data table to construct the PV diagram.
4. Use the technique of graphical integration to determine the work done during the process.
5. Use the ideal gas equation, the equation for the internal energy, and the first law of thermodynamics to complete the solution of the problem.

For problems involving the Carnot engine:

1. Complete a data table listing the input and output heat, the input and output temperatures, the efficiency, and the useful work performed.
2. Use the equations involving Carnot efficiency and the first law of thermodynamics to solve the problem.

For problems related to change in entropy:

1. Determine the energy added or removed from the system.
2. Determine the temperature in degrees Kelvin at which the energy is added or removed.
3. Use $\Delta S = Q/T$ to determine the change in entropy.

For problems involving the statistical interpretation of entropy:

1. Determine the number of microstates corresponding to the given macrostate.
2. Use $S = k \ln W = 2.3 \, k \log W$ to determine the entropy of the system.

PROGRAMMED PROBLEMS

PROBLEM 1. a) How much energy must be added to a 0.010 kg ice cube at 0°C in order to change it to water at 0°C. Determine the b) work done on the system during the change and c) change in the internal energy of the system.

a. 1	Solution: (Sections 15-1 and 15-2)
Determine the energy required to melt the ice cube.	The ice cube is undergoing a change of phase. Use the methods of section 14-6 to solve for the energy required to melt the ice cube.
	$Q = m \, \ell_F$
	$= (0.010 \text{ kg})(80 \text{ kcal/kg})(4180 \text{ J/kcal})$
	$Q = 3340 \text{ J}$
b. 1	The change in volume of an ice cube when it melts is negligible. The external air pressure can be considered to be constant. Therefore,
Determine the work done on the system during the change.	$W = P \, \Delta V$
	$W = P \, (0 \text{ liters}) = 0 \text{ liter atm } = 0 \text{ joules}$
c. 1	The work done on the system is zero joules, therefore, all of the heat energy added increases the internal energy of the system. However, the temperature of the system does not change during the melting process and this indicates that the increase in the internal energy cannot be in the form of kinetic energy. The increase in the internal energy is in the form of potential energy as the molecules overcome the attractive forces which hold them in the solid phase. The increase in the internal energy may be determined as follows:
Determine the change in the internal energy of the system.	
	$Q = \Delta U + W$
	$3340 \text{ J} = \Delta U + 0 \text{ J}$
	$\Delta U = 3340 \text{ J}$

PROBLEM 2. One mole of an ideal gas is allowed to expand isothermally at -29.3°C from a volume of 4.0 liters and pressure 5.0 atm to a volume of 10.0 liters and pressure 2.0 atm. a) Draw a P-V diagram for the process and b) determine the work done by the gas during the expansion.

a. 1

The process is isothermal, i.e., PV = constant. Complete a data table for P vs V with enough data points so that a reasonably accurate graph may be drawn.

Solution: (Section 15-2)

sample calculation

$P_1 V_1 = P_2 V_2$

$(5.0 \text{ atm})(4.0 \text{ liters}) = (3.0 \text{ atm}) V_2$

$V_2 = 6.7$ liters

data table

P (atm)	V (liters)
5.0	4.0
4.0	5.0
3.0	6.7
2.0	10.0

a. 2

Draw the PV diagram.

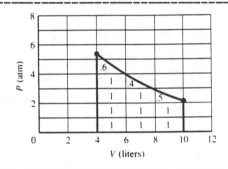

b. 1

Determine the work done by the gas during the expansion.

The work done by the gas during the expansion is represented by the area under the curve between points A and B. The work done can be estimated by graphical integration, i.e., counting the complete complete and partial blocks under the curve and multiplying by the work represented by the area of 1 block. This method was previously used in chapter 2.

1.0 atm ▭
2.0 liter

work represented by one block = (1.0 atm)(2.0 liter) = 2.0 liter atm

sum of complete blocks = 9.0

sum of partial blocks = 0.4 + 0.6 + 0.5 + 0.1 = 1.6

total number of blocks under the curve = 9.0 + 1.6 = 10.6

work done = (10.6 blocks)(2.0 liter atm/block) = 21.2 liter atm

Because of significant figures, the answer is 21 liter atm = 2100 J.

PROBLEM 3. 1.00 mole of an ideal gas, initially at 273 K, is heated at a constant pressure of 2.0 atm until the volume occupied by the gas is double the initial volume. a) Determine the initial volume and final volume occupied by the gas. b) Draw a P-V diagram for the process. Determine the c) work done during the expansion, d) change in the internal energy of the gas; and e) heat added to the gas during the change.

a. 1	Solution: (Section 15-2)
Use the ideal gas equation to determine the initial volume occupied by the gas.	$P_1 V_1 = n R T_1$ $(2.0 \text{ atm}) V_1 = (1.00 \text{ mole})(0.0821 \text{ l atm/mole K})(273 \text{ K})$ $V_1 = 11.2$ liters
a. 2 Determine the final volume.	$V_2 = 2 V_1$, then $V_2 = 22.4$ liters
b. 1 Draw a PV diagram for this process.	The process is isobaric because it occurs at constant pressure. The P-V diagram is:
c. 1 Determine the work done during the expansion.	The work done could be determined by graphical analysis. However, the pressure is constant, the process is isobaric, and the work done can be calculated as follows: $W = P \Delta V = (2.00 \text{ atm})(22.4 \text{ liters} - 11.2 \text{ liters})$ $W = 22.4$ liter atm $= 2240$ J
d. 1 Determine the final temperature of the gas.	The initial temperature is known but the final temperature must be determined. The process is isobaric; V/T = constant. $V_i / T_i = V_f / T_f$ $(11.2 \text{ liters})/(273 \text{ K}) = (22.4 \text{ liters})/T_f$ $T_f = 546$ K

d. 2 Determine the change in the internal energy of the gas.	From chapter 14, the internal energy of a gas is given by $U = 3/2 \, n \, R \, T$ $\Delta U = U_f - U_i = 3/2 \, n \, R \, T_f - 3/2 \, n \, R \, T_i$ $\Delta U = 3/2 \, n \, R \, (T_f - T_i)$ but $T_f - T_i = 546 \, K - 273 \, K = 273 \, K$ $\Delta U = 3/2 \, (1.0 \text{ mole})(0.0821 \text{ liter atm/mole K})(273 \, K)$ $\Delta U = 33.6$ liter atm $= 3360$ J
e. 1 Determine the heat added to the gas.	The heat added to the system (Q) equals the sum of the increase in internal energy and the work done by the gas during the expansion. $Q = \Delta U + W = 3360 \text{ J} + 2240 \text{ J} = 5600$ J

PROBLEM 4. A Carnot engine operates at an input temperature of 1000 K while producing 1000 J of useful work per hour. The engine is 30% efficient. Determine the a) input heat produced each hour, b) exhaust heat produced each hour, and c) exhaust temperature.

a. 1 Complete a data table based on information both given and implied in the problem.	Solution: (Section 15-5) $Q_H = ?$ $\qquad\qquad$ $T_L = ?$ $Q_L = ?$ $\qquad\qquad$ $e = 30\%$ $T_H = 1000 \, K$ $\qquad$ $W = 1000$ J
a. 2 Determine the heat input.	$e = W/Q_H$ $0.30 = 1000 \text{J}/Q_H$ $Q_H = 1000 \text{ J}/0.30 = 3300$ J
b. 1 Solve for the exhaust heat.	$Q_H = W + Q_L$ $3300 \text{ J} = 1000 \text{ J} + Q_L$ $Q_L = 2300$ J
c. 1 Solve for the temperature of the exhaust heat.	The efficiency and temperature of the input heat are known; therefore, $\quad e = (T_H - T_L)/T_H$ $\qquad 0.30 = (1000 \, K - T_L)/1000 \, K$ $\qquad T_L = 700 \, K$

PROBLEM 5. The input temperature of a Carnot engine is 527°C and the exhaust temperature is 327°C. The engine produces 1500 J of exhaust heat each hour. Determine the a) maximum efficiency of the engine, b) input heat, and c) useful work performed by the engine during one hour.

a. 1 Complete a data table using the information both given and implied in the problem.	Solution: (Section 15-5) Q_H = ? T_L = 327°C + 273°C = 600 K Q_L = 1500 J e = ? T_H = 527°C + 273°C = 800 K
a. 2 Determine the Carnot efficiency of the engine.	Both the input heat temperature and waste heat temperature are known; therefore, e = $(T_H - T_L)/T_H$ = (800 K - 600 K)/800 K e = 0.25 or 25 %
b. 1 Calculate the input heat.	e = $(Q_H - Q_L)/Q_H$ 0.25 = $(Q_H - 1500 J)/Q_H$ $0.25Q_H = Q_H - 1500$ J $-0.75Q_H = -1500$ J Q_H = 2000 J
c. 1 Calculate the useful work.	$Q_H = W + Q_L$ 2000 J = W + 1500 J W = 500 J

PROBLEM 6. An ideal gas is enclosed in a cylinder which has a tight fitting but moveable piston. The gas expands slowly and the temperature of the gas remains "just" slightly greater than the external air which is at 293 K. During the expansion the volume occupied by the gas increases by 10.0 m³ while the difference between the internal pressure and external pressure exerted by the piston and outside air remains constant at 20.0 N/m². Determine the change in entropy for the gas.

a. 1	Solution: (Section 15-7)
Determine the work done and the heat added by the gas during the expansion.	The temperature of the gas remains constant, the process is isothermal, and $\Delta U = 0$. Determine the work done by the gas during the expansion and then use the appropriate equation to determine the change in entropy.
	$Q = \Delta U + W = 0 + P \Delta V$
	$= (20.0 \text{ N/m}^2)(10.0 \text{ m}^3)$
	$Q = + 200 \text{ J}$
a. 2	$\Delta S = \Delta Q / T$
Determine the entropy change for the gas.	$= (200 \text{ J})/(293 \text{ K})$
	$\Delta S = + 0.68 \text{ J/K}$

PROBLEM 7. Two identical compartments contain an equal mass of the same ideal gas. The temperature of the hot gas is 67°C while the temperature of the cold gas is 27°C. Three joules of heat transfer through the partition which separates the gases without significantly affecting the temperature of either gas. Determine the a) entropy change of each gas and b) entropy change of the universe.

a. 1	Solution: (Section 15-7)
Convert the centigrade temperatures to kelvin.	67°C + 273°C = 340 K
	27°C + 273°C = 300 K
a. 2	The hot gas loses 3.0 J in the process; therefore, $Q = - 3.0$ J for the hot gas. The cold gas gains 3.0 J of heat; therefore, $Q = + 3.0$ J for the cold gas.
Determine the change in entropy for each gas.	$\Delta S_h = Q/T_h = - 3.0 \text{ J}/(340 \text{ K})$
	$\Delta S_h = - 8.8 \times 10^{-3} \text{ J/K}$
	$\Delta S_c = Q/T_c = + 3.0 \text{ J}/(300 \text{ K})$
	$\Delta S_c = + 1.0 \times 10^{-2} \text{ J/K}$
	The hot gas lost heat energy with the result that the molecular motion became slightly more ordered and its entropy decreased. The cold gas gained heat energy with the result that the motion of its molecules became more disordered and its entropy increased.

b. 1 Determine the change in the entropy of the universe.	The change in entropy of the universe is equal to the sum of the changes in entropy for the two gases. $\Delta S_{universe} = \Delta S_h + \Delta S_c$ $= -8.8 \times 10^{-3}$ J/K $+ 1.0 \times 10^{-2}$ J/K $\Delta S_{universe} = +1.2 \times 10^{-3}$ J/K The change in entropy is positive; therefore, the entropy of the universe increased during the process.

PROBLEM 8. 100 coins are placed heads up in a box. After the box is shaken vigorously, only 50 coins are found to be heads up. Determine the change in entropy of the system.

a. 1 Determine the entropy for each macrostate.	Solution: (Section 15-11) Based on Table 15-2 in the text, there is only 1.0 microstate that corresponds to the macrostate where 100 heads are found. This is because in order to have 100 heads, each coin must come up heads. However, there are 1.0×10^{29} microstates which correspond to the macrostate where 50 coins come up heads. The entropy for each macrostate can be determined from the formula $S = 2.3$ k log W. 1 head $S = 2.3 (1.38 \times 10^{-23}$ J/K) (log 1.0) but log 1.0 = 0, therefore S = 0. 50 heads $S = 2.3 (1.38 \times 10^{-23}$ J/k)(log 1.0×10^{29}) but log $1.0 \times 10^{29} = 29$ Therefore, $S = 2.3 (1.38 \times 10^{-23}$ J/K)(29) and $S = 9.2 \times 10^{-22}$ J/K
a. 2 Determine the change in entropy of the universe for this process.	$\Delta S = S_{50\ heads} - S_{100\ heads}$ $= 9.22 \times 10^{-22}$ J/K $- 0$ J/K $\Delta S = +9.22 \times 10^{-22}$ J/K The change in entropy is positive, the entropy of the universe has increased in the process.

PRACTICE PROBLEMS

PROBLEM 1. One mole of an ideal gas occupies a volume of 0.0100 m³ at 244 K and is held in a cylinder with a frictionless piston under 2.02 x 10⁵ N/m² of pressure. Heat is added to the gas and the gas expands in an isobaric process to a volume of 0.0300 m³. The temperature of the gas at this point is 731 K. a) Draw a P-V diagram of the process. Determine the b) work done by the gas and c) heat added to the gas during the expansion.

ANS. b) 4040 J,   c) 1.01 x 10⁴ J

PROBLEM 2. A Carnot engine takes in 3000 calories of input heat and rejects 2000 calories as waste heat. The temperature of the waste heat is 600°C. Determine the a) efficiency of the engine, b) useful work done by the engine, and c) temperature of the input heat.

ANS. a) 0.333 or 33.3%,   b) 1000 cal,   c) 1310 K

PROBLEM 3. The actual work done by a heat engine in 1 hour is 2.7 x 10⁶ J while frictional losses within the engine are 6.0 x 10⁵ J. The engine operates between temperatures of 500 K and 1000 K. Determine the a) Carnot efficiency of the engine, b) number of joules of input heat taken in per hour by the engine, and c) overall efficiency of the engine.

ANS. a) 0.50 or 50%, b) 6.6 x 10⁶ J, c) 0.41 or 41 %

PROBLEM 4. Four gas molecules are held in a container which has a partition which divides the container into two equal halves. The partition has a small opening through which the molecules can pass from one side of the container to the other. Determine the entropy of the system if a) all of the molecules are on one side of the partition and b) two of the molecules are on one side and two are on the other side. Hint: first determine the number of microstates for each event.

ANS. a) zero, b) 2.47 x 10⁻²³ J/K

CHAPTER 16

ELECTRIC CHARGE AND ELECTRIC FIELD

KEY TERMS AND PHRASES

electrostatics is the study of interaction between electric charges which are not moving.

electric charge is a fundamental property of matter. Electric charge appears as two kinds, arbitrarily designated **positive** and **negative**. The negative charges are carried by particles called **electrons** while the positive charge carriers are known as **protons**.

law of conservation of electric charge states that the net amount of electric charge produced in any process is zero. Another way of saying this is that in any process electric charge cannot be created or destroyed, however, it can be transferred from one object to another.

force of attraction or repulsion between charged particles can be summarized as unlike charges attract and like charges repel. A negative charge and a positive charge will attract one another while two negative charges or two positive charges repel .

coulomb (C) is the SI unit of charge where 1 C = 6.25 x 10^{18} electrons or protons. The charge carried by the electron is represented by the symbol -e, and the charge carried by the proton is +e. 1 e = 1.6 x 10^{-19} coulomb, $m_{electron}$ = 9.11 x 10^{-31} kg and m_{proton} = 1.672 x 10^{-27} kg. The neutron carries no electrical charge. The neutron has a mass of 1.675 x 10^{-27} kg.

insulators are materials in which the electrons are tightly held by the nucleus and are not free to move through the material. There is no such thing as a perfect insulator, however, examples of good insulators include substances such as glass, rubber, plastic, and dry wood.

conductors are materials through which electrons are free to move. Just as in the case of the insulators, there is no such thing as a perfect conductor. Examples of good conductors include metals, such as silver, copper, gold, and mercury.

semiconductors are materials where there are a few free electrons and the material is a poor conductor of electricity. As the temperature rises, electrons break free and move through the material. Examples of elements which are semiconductors are silicon, germanium, and carbon.

charging by contact occurs when electric charge is transferred from a charged object to an uncharged object.

charging by induction occurs when an electrically charged object is brought near but does not touch an uncharged second object. The negative and positive charges on the second object separate. Overall, the second conductor is still electrically neutral and if the first object is removed a re-distribution of the negative charge occurs.

Coulomb's law states that two point charges exert a force (F) on one another that is directly proportional to the product of the magnitudes of the charges (Q) and inversely proportional to the square of the distance (r) between their centers.

electric fields exist in the space surrounding a charged particle or object. Electric fields are represented by lines of force that start on a positive charge and end on a negative charge.

SUMMARY OF MATHEMATICAL FORMULAS

electric charge	$q = n e$ where $e = 1.6 \times 10^{-19}$ C	Relates the total charge on an object (q) to the fundamental unit of charge (e) and the total number of charges on the object (n).
Coulomb's law	$F = k \, Q_1 \, Q_2 \, /r^2$ where $k = 9 \times 10^9$ N m^2/C^2	Coulomb's law states that two point charges exert a force (F) on one another that is directly proportional to the product of the magnitudes of the charges (Q) and inversely proportional to the square of the distance (r) between their centers.
electric field (E)	$\mathbf{E = F/q}$ or $\mathbf{F = q \, E}$	The magnitude of the electric field (E) at any point in space can be determined by the ratio of the force (F) exerted on a test charge placed at the point to the magnitude of the charge on the test particle (q).
electric field due to a point charge	$\mathbf{E = k \, Q/r^2}$	The magnitude of the electric field (E) a distance (r) from a single point charge (q) is related to the magnitude of the charge (q) and is inversely proportional to the square of the distance (r) from the charge.

CONCEPT SUMMARY

Electric Charge

There are two types of **electric charge**, arbitrarily called **positive** and **negative**. Rubbing certain electrically neutral objects together (e.g., a glass rod and a silk cloth) tends to cause the electric charges to separate. In the case of the glass and silk, the glass rod loses negative charge and becomes positively charged while the silk cloth gains negative charge and therefore becomes negatively charged. After separation, the negative charges and positive charges are found to attract one another.

If the glass rod is suspended from a string and a second positively charged glass rod is brought near, a force of electrical repulsion results. Negatively charged objects also exert a repulsive force on one another. These results can be summarized as follows: **unlike charges attract and like charges repel**.

Conservation of Electric Charge

In the process of rubbing two solid objects together, electrical charges are **not** created. Instead, both objects contain both positive and negative charges. During the rubbing process, the negative charge is transferred from one object to the other and this leaves one object with an excess of positive charge and the other with an excess of negative charge. The quantity of excess charge on each object is exactly the same. This is summarized by the **law of conservation of electric charge**: the net amount of electric charge produced in any process is zero. Another way of saying this is that in any process electric charge **cannot** be created or destroyed, however, it can be transferred from one object to another.

During the past century, the negative charges have been shown to be carried by particles which are now called **electrons** while the positive charge carriers are known as **protons**. Both particles, as well as many others, are found in the atoms which make up a substance. Note: in a later chapter, the wave properties of electrons, protons, and other particles will be discussed. However, for ease of visualization and discussion, the term particle will be applied for the present.

The SI unit of charge is the coulomb (C). The amount of charge transferred when objects like glass or silk are rubbed together is in the order of microcoulombs (μC).

$1 \text{ C} = 6.25 \times 10^{18}$ electrons or protons and $1 \text{ } \mu\text{C} = 10^{-6} \text{ C}$

The charge carried by the electron is represented by the symbol -e, and the charge carried by the proton is +e.

$e = 1.6 \times 10^{-19}$ coulomb   and $m_{electron} = 9.11 \times 10^{-31}$ kg   while $m_{proton} = 1.672 \times 10^{-27}$ kg

A third particle, which carries no electrical charge, is the **neutron**. The neutron has a mass of 1.675×10^{-27} kg.

Experiments performed early in this century have led to the conclusion that protons and neutrons are confined to the nucleus of the atom while the electrons exist outside of the nucleus. When solids are rubbed together, it is the electrons which are transferred from one object to the other. The positive charges, which are located in the nucleus, do not move.

Insulators, Semiconductors, and Conductors

An **insulator** is a material in which the electrons are tightly held by the nucleus and are not free to move through the material. There is no such thing as a perfect insulator, however, but examples of good insulators include substances such as glass, rubber, plastic, and dry wood.

A **conductor** is a material through which electrons are free to move. Just as in the case of the insulators, there is no such thing as a perfect conductor. Examples of good conductors include metals such as silver, copper, gold, and mercury.

A few materials, such as silicon, germanium, and carbon, are called **semiconductors.** At ordinary temperatures, there are a few free electrons and the material is a poor conductor of electricity. As the temperature rises, electrons break free and move through the material. As a result, the ability of a semiconducting material to conduct improves with temperature.

Charging by Induction and Charging by Contact

If a negatively charged rod is brought near an uncharged electrical conductor the negative charges in the conductor travel to the far end of the conductor (see diagram).

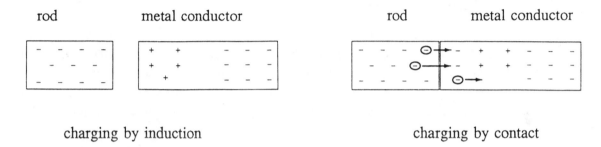

charging by induction   charging by contact

The positive charges are not free to move through a solid object and a charge is temporarily **induced** at the two ends of the conductor. Overall, the conductor is still electrically neutral and if the rod is removed a redistribution of the negative charge would occur.

If the metal conductor is touched by a person's finger or a wire connected to ground, it is said to be "grounded." The negative charges would flow from the conductor to ground. If the ground is removed and then the rod is removed, a permanent positive charge would be left on the conductor. The electrons would move until the excess positive charge was uniformly distributed over the conductor.

If the rod touches the metal conductor, some of the negative charges on the rod transfer to the metal. This charge distributes uniformly over the metal. The metal has been charged by **contact** and a permanent charge remains when the rod is removed.

Coulomb's Law

Coulomb's law states that two point charges exert a force (F) on one another that is directly proportional to the product of the magnitudes of the charges (Q) and inversely proportional to the square of the distance (r) between their centers. The formula relating the force to the charges and the distance is

charging by induction occurs when an electrically charged object is brought near but does not touch an uncharged second object. The negative and positive charges on the second object separate. Overall, the second conductor is still electrically neutral and if the first object is removed a re-distribution of the negative charge occurs.

Coulomb's law states that two point charges exert a force (F) on one another that is directly proportional to the product of the magnitudes of the charges (Q) and inversely proportional to the square of the distance (r) between their centers.

electric fields exist in the space surrounding a charged particle or object. Electric fields are represented by lines of force that start on a positive charge and end on a negative charge.

SUMMARY OF MATHEMATICAL FORMULAS

electric charge	$q = n\,e$ where $e = 1.6 \times 10^{-19}$ C	Relates the total charge on an object (q) to the fundamental unit of charge (e) and the total number of charges on the object (n).
Coulomb's law	$F = k\,Q_1\,Q_2\,/r^2$ where $k = 9 \times 10^9$ N m^2/C^2	Coulomb's law states that two point charges exert a force (F) on one another that is directly proportional to the product of the magnitudes of the charges (Q) and inversely proportional to the square of the distance (r) between their centers.
electric field (E)	$E = F/q$ or $F = q\,E$	The magnitude of the electric field (E) at any point in space can be determined by the ratio of the force (F) exerted on a test charge placed at the point to the magnitude of the charge on the test particle (q).
electric field due to a point charge	$E = k\,Q/r^2$	The magnitude of the electric field (E) a distance (r) from a single point charge (q) is related to the magnitude of the charge (q) and is inversely proportional to the square of the distance (r) from the charge.

CONCEPT SUMMARY

Electric Charge

There are two types of **electric charge**, arbitrarily called **positive** and **negative**. Rubbing certain electrically neutral objects together (e.g., a glass rod and a silk cloth) tends to cause the electric charges to separate. In the case of the glass and silk, the glass rod loses negative charge and becomes positively charged while the silk cloth gains negative charge and therefore becomes negatively charged. After separation, the negative charges and positive charges are found to

a) positive
   point charge

b) negative point
   charge

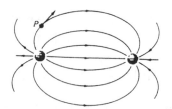

c) positive and negative
   point charges

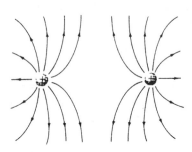

d) two positive
   point charges

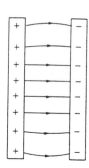

e) oppositiely charged
   parallel plates

The electric field is strongest in regions where the lines are close together and weak where the lines are farther apart. Thus in diagrams (a) and (b), the field is strongest close to the point charges. In diagram (c) the field weakens as the lines diverge as they leave the positive charge and strengthens as the lines converge on the negative charge. In diagram (e), two parallel plates of opposite charge produce an electric field where the lines are parallel near the center of the plates. In this region the electric field is uniform and E is constant in magnitude and direction.

Coulomb's law can be used to predict that the electric field inside a closed conductor is zero. An example of a closed conductor is a hollow metal sphere which contains an excess of static electric charge. The charge on the conductor tends to reside on its outer surface. Inside the conductor, the electric field is zero. Outside the conductor, the electric field is not zero and the electric field lines are drawn perpendicular to the surface.

Electric charges tend to distribute throughout the volume of a non-conductor which contains an excess of static charge. It can be shown that a charged non-conductor has an electric field inside as well as outside the non-conductor.

SELECTED TEXT QUESTIONS WITH ANSWERS

QUESTION 4. A positively charged rod is brought close to a neutral piece of paper, which it attracts. Draw a diagram showing the separation of charge and explain why attraction occurs.

ANSWER: Paper is a poor conductor, and only a slight displacement of the electric charge occurs when the positively charged rod is brought near the neutral piece of paper. The negative charges in the paper are attracted toward the rod while the positive charges are located in the atomic nucleus and cannot move. As shown in the diagram below, the piece of paper remains neutral overall, but a temporary separation of electric charge is induced in the paper. Because the negative charges in the paper are closer to the rod than the positive charges, the force of attraction is

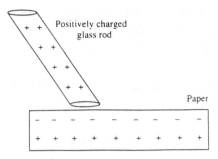

greater than the force of repulsion between the rod and the positive charges in the paper. Thus the paper is attracted to the rod.

QUESTION 13. When a charged ruler attracts small pieces of paper, sometimes a piece jumps quickly away after touching the ruler. Explain.

ANSWER: Let us assume that the rod in question 4 represents a plastic ruler. Plastic is a very poor conductor of electricity and paper is a poor conductor. However, if the paper touches the plastic it is possible for some of the negative charges in the paper to transfer to the plastic. If this should happen then both the plastic ruler and the paper would contain an excess of positive charge. The paper would then be repelled from the ruler.

QUESTION 19. Why can lines of force never cross?

ANSWER: A line of force indicates the direction of the force on a positive test charge and the path the test charge would follow. At the point where two lines cross, the force on the test charge would be in two different directions at the same time. Since the test charge cannot move in two directions at the same time, the lines cannot cross.

PROBLEM SOLVING SKILLS

For problems involving Coulomb's law:

1. Complete a data table listing the charge on each object and the distance between the objects. If more than two charged objects are given, draw a diagram showing the position of each object.
2. If the objects touch then charge transfer occurs and the law of conservation of charge must be applied to determine the charge on each object.
3. If more than two charges are given it may be necessary to use the methods of vector algebra discussed in Chapter 3 to solve the problem.
4. Apply Coulomb's law and solve the problem.

For problems involving the motion of a charged particle in an electric field.

1. Draw an accurate diagram showing the motion of the particle in the field.
2. Complete a data table based on information both given and implied in the problem.
3. Determine the magnitude and direction of the electric force acting on the particle. Use Newton's second law to determine the rate of acceleration.
4. Use the kinematics equations of Chapter 2 to solve the problem.

For problems involving the magnitude and direction of the resultant electric field due to two or more point charges:

1. Draw an accurate diagram locating the position of each charge.
2. Determine the magnitude and direction of the electric field at the point in question due to each charge.
3. Use the methods of vector algebra discussed in Chapter 3 to solve for the resultant electric field.

PROGRAMMED PROBLEMS

PROBLEM 1. $Q_1 = +4.0$ μC and $Q_2 = +1.0$ μC are located on the x-axis at $x = 0$ and $x = 6.0$ m, respectively. Determine the point on the x-axis where a third charge $Q_3 = +1.0$ μC can be placed and experience no net force.

a. 1	Solution: (Sections 16-5 and 16-6)
Determine the point where the net force equals zero.	At the point in question, the vector sum of the two forces equals zero. The force exerted by Q_1 on Q_3 must be equal in magnitude but opposite in direction from the force exerted by Q_2 on Q_3. The only location that would satisfy this condition would be a location on the x-axis between Q_1 and Q_2.
a. 2	Let r represent the distance from Q_1 to Q_3 and 6 - r represent the distance from Q_2 to Q_3.
Draw a diagram locating each charge.	Q_1 $\qquad$ Q_3 $\qquad$ Q_2 ⊕ $\qquad$ $F_{2\ on\ 3}$⇐ ⊕ ⇒$F_{1\ on\ 3}$ $\qquad$ ⊕ ⊢← $\qquad$ r $\qquad$ →⊢← $\quad$ 6 - r $\quad$ →⊣
a. 3	$F_{1\ on\ 3} = k\ Q_1\ Q_3/r^2$ and $F_{2\ on\ 3} = k\ Q_2\ Q_3/(6 - r)^2$
Use Coulomb's law to write an equation for the force that 1 exerts on 3 and 2 exerts on 3.	$F_{1\ on\ 3} = F_{2\ on\ 3}$ $k\ Q_1\ Q_3/r^2 = k\ Q_2\ Q_3/(6 - r)^2$ both k and Q_3 cancel algebraically, and rearranging $r^2/(6 - r)^2 = Q_1/Q_2$ but $Q_1/Q_2 = +4.0$ μC$/+1.0$ μC $= 4$ Solving algebraically, $r^2/(6 - r)^2 = 4$ and $r/(6 - r) = 2$ $r = 2(6 - r)$ and $r = 12 - 2r$ and $3r = 12$ $r = 4.0$ m. Therefore, Q_3 is located on the x-axis 4.0 m to the right of Q_1.

PROBLEM 2. Solve problem 1 if charge Q_2 is replaced with a - 1.0 μC located at $x = 6.0$ m.

a. 1	Solution: (Sections 16-5 and 16-6)
Determine the point where the net force equals zero.	At the point in question, the vector sum of the two forces equals zero. The force exerted by Q_1 on Q_3 must be equal in magnitude but opposite in direction from the force exerted by Q_2 on Q_3. Since the magnitude of the charge on Q_1 is much greater than the charge on Q_1, the only location that would satisfy this condition would be a location on the x-axis to the right of charge Q_2.

a. 2	Let r represent the distance from Q_2 to Q_3 and 6 m $+$ r represent the distance from Q_1 to Q_3.
Draw a diagram locating each charge.	Q_1 $\qquad\qquad$ Q_2 $\qquad$ Q_3 $\oplus$ $\qquad\qquad\qquad$ $\ominus$ $\ F_{2 \text{ on } 3} \Leftarrow \oplus \Rightarrow F_{1 \text{ on } 3}$ $\mid\leftarrow$ $\quad$ 6.0 m $\quad$ $\rightarrow\mid\leftarrow$ $\ $ r $\ $ $\rightarrow\mid$

a. 3	$F_{1 \text{ on } 3} = k\,Q_1\,Q_3/(6 \text{ m} + r)^2$ and $F_{2 \text{ on } 3} = k\,Q_2\,Q_3/r^2$
Use Coulomb's law to write an equation for the force that 1 exerts on 3 and 2 exerts on 3.	$F_{1 \text{ on } 3} = -F_{2 \text{ on } 3}$ $k\,Q_1\,Q_3/(6 \text{ m} + r)^2 = -\,k\,Q_2\,Q_3/r^2$ Both k and Q_3 cancel algebraically, and rearranging
Note: the magnitude of forces are equal; however, $F_{2 \text{ on } 3}$ has a negative value because $Q_2 = -1.0$ μC.	$(6 + r)^2/r^2 = -\,Q_1/Q_2$ but $Q_1/Q_2 = -(+4.0 \ \mu C/-1.0 \ \mu C) = 4$ $(6 + r)^2/r^2 = 4$ and $(6 + r)/r = 2$ $6 + r = 2r$ and $r = 6$ m Therefore, Q_3 is located on the x-axis 6.0 m to the right of charge Q_2 at 12 m on the x-axis .

PROBLEM 3. Two small, identical metal spheres contain excess charges of –10.0 μC and +6.0 μC, respectively. The spheres are mounted on insulated stands and placed 0.40 m apart. a) Determine the magnitude and direction of the force between the spheres. b) The spheres are touched together and then returned to their original 0.40 m separation. Determine the magnitude and direction of the force between the spheres.

a. 1	Solution: (Sections 16-5 and 16-6)
Use Coulomb's law to determine the magnitude of the force between the charges.	$Q_1\,Q_2 = (-10.0 \times 10^{-6} \text{ C})(+6.0 \times 10^{-6} \text{ C})$ $Q_1\,Q_2 = -6.0 \times 10^{-11} \text{ C}$ $F = k\,Q_1\,Q_2/r^2$

$$= (9.0 \times 10^9 \text{N m}^2/\text{C}^2)(-6.0 \times 10^{-11}\text{C}^2)/(0.40 \text{ m})^2$$

$F = -3.4 \text{ N}$ The negative value indicates that the force between the charges is one of attraction.

b. 1 Determine the magnitude of the charge on each sphere after contact.	When the spheres touch, electrons travel from the negatively charged sphere to the positively charged sphere. The magnitude of the negative charge is greater than that of the positive charge; therefore, the negative charge completely neutralizes the positive charge. The excess negative charge will distribute uniformly over the two spheres until they both contain equal amounts of negative charge. The magnitude of the charge on each sphere after separation can be determined as follows: the overall excess charge on the spheres is $- 10.0 \ \mu\text{C} + 6.0 \ \mu\text{C} = - 4.0 \ \mu\text{C}$. This excess charge is shared equally between the two identical metal spheres. $Q_1' = Q_2' = - 4.0 \ \mu\text{C}/2 = - 2.0 \ \mu\text{C}$
b. 2 Apply Coulomb's law to determine the magnitude of the force between the spheres.	$Q_1 \, Q_2 = (-2.0 \times 10^{-6} \text{ C})(-2.0 \times 10^{-6} \text{ C})$ $Q_1 \, Q_2 = +4.0 \times 10^{-12} \text{ C}^2$ $F = k \, Q_1 \, Q_2/r^2$ $= (9.0 \times 10^9 \text{ N m}^2/\text{C}^2)(4.0 \times 10^{-12}\text{C}^2)/(0.40 \text{ m})^2$ $F = + 0.23 \text{ N}$ Note: the positive value indicates that the force between the charges is one of repulsion.

PROBLEM 4. a) Determine the electric field 0.10 m from a point charge Q_1 which carries a charge of 0.20 μC. b) Determine the magnitude and direction of the electric force exerted on a second charge Q_2 = 0.80 μC which is placed 0.10 m from Q_1. c) Q_2 is replaced by a third charge Q_3 = -4.0 μC. Determine the magnitude and direction of the electric force exerted on a charge Q_3.

a. 1 Use the formula for the electric field due to a point charge.	Solution: (Section 16-7) $E_1 = k \, Q_1/r^2$ $= (9.0 \times 10^9 \text{ N m}^2/\text{C}^2)(2.0 \times 10^{-7} \text{ C})/(0.10 \text{ m})^2$ $E_1 = 1.8 \times 10^5 \text{ N/C away from } Q_1$
b. 1 Calculate the magnitude and direction of the force.	$F_{\text{on } 2} = Q_2 \, E_1 = (0.80 \times 10^{-6} \text{ C})(1.8 \times 10^5 \text{ N/C})$ $F_{\text{on } 2} = 0.144 \text{ N away from } Q_1$

c. 1

Calculate the magnitude and direction of the force.

$F_{on\ 3} = Q_3\ E_1$

$= (-4.0\ x\ 10^{-6}\ C)(1.8\ x\ 10^5\ N/C)$

$F_{on\ 3} = -0.72\ N$ toward Q_1

PROBLEM 5. A particle of mass $1.0\ x\ 10^{-3}$ kg has an excess charge of $+1.0\ \mu C$. The particle is located in a region between two oppositely charged parallel plates where the electric field is uniform and has a magnitude of $1000\ N/C$. Determine the a) magnitude of the force acting on the particle and b) rate of acceleration of the particle. c) The distance between the plates is 0.010 m and the particle is initially at rest at a point close to the positive plate. Determine the velocity of the particle "just" before it strikes the negative plate.

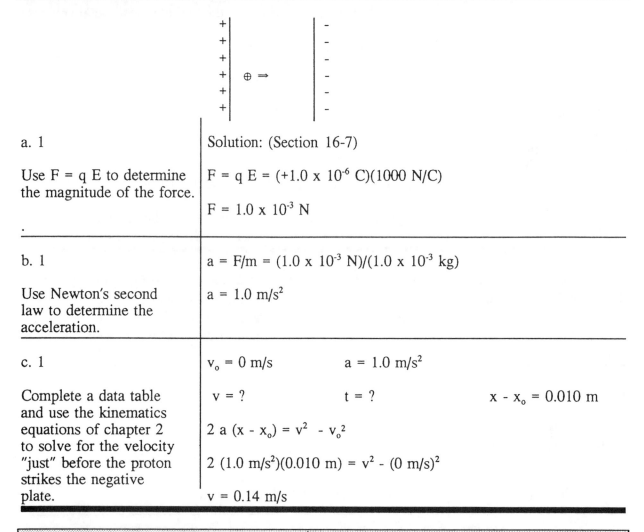

a. 1

Use F = q E to determine the magnitude of the force.

Solution: (Section 16-7)

$F = q\ E = (+1.0\ x\ 10^{-6}\ C)(1000\ N/C)$

$F = 1.0\ x\ 10^{-3}\ N$

---

b. 1

Use Newton's second law to determine the acceleration.

$a = F/m = (1.0\ x\ 10^{-3}\ N)/(1.0\ x\ 10^{-3}\ kg)$

$a = 1.0\ m/s^2$

---

c. 1

Complete a data table and use the kinematics equations of chapter 2 to solve for the velocity "just" before the proton strikes the negative plate.

$v_o = 0\ m/s$ $a = 1.0\ m/s^2$

$v = ?$ $t = ?$ $x - x_o = 0.010\ m$

$2\ a\ (x - x_o) = v^2 - v_o^2$

$2\ (1.0\ m/s^2)(0.010\ m) = v^2 - (0\ m/s)^2$

$v = 0.14\ m/s$

PROBLEM 6. A proton traveling at $3.0\ x\ 10^6$ m/s enters a region where the electric field has a magnitude of $3.0\ x\ 10^5$ N/C. The electric field is uniform and retards the proton's motion. Determine the a) distance the proton will travel before coming to a momentary halt and b) time required for the proton to travel this distance.

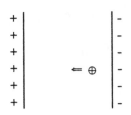

a. 1 Use the work-energy theorem to solve for the distance traveled.	Solution: (Section 16-7) $W = \Delta KE$ $F\,d = \frac{1}{2}\,m\,v^2 - \frac{1}{2}\,m\,v_o^2$ but $v_o = 3.0 \times 10^6$ m/s The force on the proton is negative since the electric field retards the motion ($F = -q\,E$). $-(+1.6 \times 10^{-19}\ C)(3.0 \times 10^5\ N/C)\,d =$ $\qquad\qquad\qquad 0 - \frac{1}{2}(1.67 \times 10^{-27}\ kg)(3.0 \times 10^6\ m/s)^2$ $d = 0.16$ m
b. 1 Use Newton's second law to solve for the acceleration.	$F = m\,a$ but $F = q\,E = -\,(+e)\,E$ $-(1.6 \times 10^{-19}\ C)(3.0 \times 10^5\ N/C) = (1.67 \times 10^{-27}\ kg)\,a$ $a = -\,2.9 \times 10^{13}\ m/s^2$
b. 2 Use the kinematics equations to solve for for the time.	$v = v_o + a\,t$ $0 = 3.0 \times 10^6\ m/s + (-\,2.9 \times 10^{13}\ m/s^2)\,t$ $t = 1.0 \times 10^{-7}$ s

---

PROBLEM 7. Two point charges, $Q_1 = +\,5.0\ \mu C$ and $Q_2 = -\,5.0\ \mu C$, are located on the y-axis at $y = +\,3.0$ m and $y = -\,3.0$ m, respectively. Determine the magnitude and direction of the electric field on the x-axis at $x = 4.0$ m.

---

a. 1 Draw a diagram locating the position of each charge.	Solution: (Section 16-7) 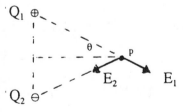

---

a. 2	$r = [(3.0 \text{ m})^2 + (4.0 \text{ m})^2]^{1/2} = 5.0 \text{ m}$

Determine the distance r and the angle θ shown in the diagram.

$\theta = \tan^{-1} (3.0 \text{ m})/(4.0 \text{ m}) = 37°$

---

a. 3.

Determine the magnitude and direction of the electric field produced by each point charge at point p.

$E_1 = k \, Q_1 / r^2$

$= (9.0 \times 10^9 \text{ N m}^2/\text{C}^2)(5.0 \times 10^{-6} \text{ C})/(5.0 \text{ m})^2$

$E_1 = 1.8 \times 10^3$ N/C $\angle 37°$ below the horizontal and away from Q_1

$E_2 = k \, Q_2 / r^2$

$= (9.0 \times 10^9 \text{ N m}^2/\text{C}^2)(-5.0 \times 10^{-6} \text{ C})/(5.0 \text{ m})^2$

$E_1 = -1.8 \times 10^3$ N/C $\angle 37°$ below the horizontal and toward Q_2

---

a. 4

Use the trigonometric component method to determine the magnitude and direction of the resultant vector.

Note: use the sign convention adopted in chapter 3 in designating the direction of each component.

$E_{1x} = E_1 \cos 37° = (1.8 \times 10^3 \text{ N/C}) \cos 37°$

$E_{1x} = + 1.4 \times 10^3$ N/C

$E_{1y} = E_1 \sin 37° = (1.8 \times 10^3 \text{ N/C}) \sin 37°$

$E_{1y} = - 1.1 \times 10^3$ N/C

$E_2 = E_2 \cos 37° = (1.8 \times 10^3 \text{ N/C}) \cos 37°$

$E_{2x} = - 1.4 \times 10^3$ N/C

$E_{2y} = E_2 \sin 37° = (1.8 \times 10^3 \text{ N/C}) \sin 37°$

$E_{2y} = - 1.1 \times 10^3$ N/C

$\Sigma X = E_{1x} + E_{2x}$

$\Sigma X = + 1.4 \times 10^3 \text{ N/C} + - 1.4 \times 10^3 \text{ N/C} = 0$

$\Sigma Y = E_{1y} + E_{2y} = -1.1 \times 10^3 \text{ N/C} + -1.1 \times 10^3 \text{ N/C}$

$\Sigma Y = - 2.2 \times 10^3$ N/C $= -2200$ N/C

The resultant electric field has a magnitude of 2200 N/C and is directed in the negative y direction at point p.

PRACTICE PROBLEMS

PROBLEM 1. In the Bohr model of the hydrogen atom, the electron orbits the nucleus at a distance of 0.53×10^{-10} m. The nucleus consists of a single proton. Calculate the a) gravitational force between the proton and electron and b) electric force between the proton and electron. c) Which force is greater and by what factor.

Note: $m_{electron} = 9.1 \times 10^{-31}$ kg, $m_{proton} = 1.67 \times 10^{-27}$ kg, $q_{proton} = q_{electron} = 1.6 \times 10^{-19}$ C.

ANS. a) 3.6×10^{-47} N, b) 8.2×10^{-8} N, c) The electric force is much greater than the gravitational force; $F_{electric}/F_{gravitational} = 2.3 \times 10^{39}$.

PROBLEM 2. Two identical pith balls are suspended from the same point by light strings 1.0 meters long. The balls each have a mass of 0.0020 kg and carry equal amounts of positive charge with the result that the balls are separated by 0.050 meters. Determine the magnitude of the electric charge on each ball.

ANS. 1.2×10^{-8} C

PROBLEM 3. Given that Q_1 = 5.00 μC and Q_2 = 1.00 μC, determine the magnitude and direction of the resultant electric field at point p.

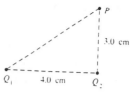

ANS. 2.53 x 10^7 N/C ∠55.2° above the horizontal and directed toward the right.

PROBLEM 4. Two point charges, Q_1 = 10 μC and Q_2 = - 5.0 μC are arranged at x = 0 m and x = 2.0 m, respectively. Determine the magnitude and direction of the resultant electric field at each of the following positions on the x-axis: a) x = 1.0 m, b) x = 3.0 m, and c) x = -1.0 m.

ANS. a) 1.4 x 10^5 N/C in the positive x direction, b) 3.5 x 10^4 N/C in the negative x direction, and c) 8.5 x 10^3 in the negative x direction.

CHAPTER 17

ELECTRIC POTENTIAL AND POTENTIAL ENERGY; CAPACITANCE

KEY TERMS AND PHRASES

electric potential at point a (V_a) equals the electric potential energy (PE_a) per unit charge (q) placed at that point.

electric potential difference between two points (V_{ab}) is measured by the work required to move a unit of electric charge from point b to point a.

equipotential lines are lines along which each point is at the same potential. On an equipotential surface, each point on the surface is at the same potential.

voltage is a common term used for potential difference. The SI unit of electric potential and potential difference is the volt (V), where 1 V = 1 J/C.

electron volt (eV) is the unit used for the energy gained by a charged particle which is accelerated through a potential difference. 1 eV = 1.6×10^{-19} J.

electric dipole is two equal point charges (q), of opposite sign, separated by a distance ℓ. The SI unit of the dipole moment (p) is the **debye**, where 1 debye = 3.33×10^{-33} C m. In polar molecules, such as water, the molecule is electrically neutral but there is a separation of charge in the molecule. Such molecules have a net dipole moment.

capacitor stores electric charge and consists of two conductors separated by an insulator known as a dielectric. The ability of a capacitor to store electric charge is referred to as **capacitance** (C).

SUMMARY OF MATHEMATICAL FORMULAS

electric potential	$V = PE_a /q$	The electric potential at point a (V_a) equals the electric potential energy (PE_a) per unit charge (q) placed at that point.

electric potential due to a point charge	$V = kq/r$	The electric potential (V) due to a point charge is related to the magnitude of the charge and the distance from the charge (r) to the point in question. If more than one point charge is present, the potential at a particular point is equal to the arithmetic sum of the potential due to each charge at the point in question.
potential difference	$V_{ab} = V_a - V_b = W_{ab}/q$	The electric potential difference between two points (V_{ab}) is measured by the work required to move a unit of electric charge from point b to point a.
	$\Delta PE_{ab} = PE_a - PE_b = q\,V_{ab}$ $V_{ab} = \Delta PE/q$	The potential difference can also be discussed in terms of the change in potential energy of a charge q when it is moved between points a and b.
	$V_{ab} = E\,d\,\cos\theta$	If the charged particle is in an electric field which is uniform, i.e., constant in magnitude and direction, then the potential difference (V_{ab}) equals the product of the electric field strength (E), the distance (d) between points a and b, and the cosine of the angle ($\cos\theta$) between the electric field vector and the displacement vector.
potential gradient	$E_x = -\,\Delta V/\Delta x$	If the electric field is non-uniform, i.e., varies in magnitude and direction between points a and b, then the electric field strength can only be properly defined over an incremental distance. If we consider the x component of the electric field (E_x), then $E_x = -\,\Delta V/\Delta x$, where ΔV is the change in potential over a very short distance Δx. The minus sign indicates that E points in the direction of decreasing V.

electric potential due to an electric dipole	$V = k\, q\, \ell \cos \theta / r^2$ or $V = k\, p \cos \theta / r^2$	The potential due to two equal point charges (q), of opposite sign, separated by a distance ℓ, depends on the angle θ and the distance r. $q\,\ell$ is the dipole moment (p).
capacitance	$C = Q/V$	Capacitance (C) is ratio of the charge stored (Q) to the potential difference (V) between the conducting surfaces.
	$C = K\, \epsilon_o\, A/d$	For a parallel plate capacitor, the capacitance (C) depends on the dielectric constant (K) of the insulating material between the plates, the permittivity of free space(ϵ_o), the surface area (A) of one side of one plate which is opposed by an equal area of the other plate, and the distance between the plates (d).
energy stored in a charged capacitor	energy = ½ Q V energy = ½ C V^2 energy = ½ Q^2/C energy = ½ $\epsilon_o\, E^2\, A\, d$	The electric energy stored in charging a capacitor depends on the charge stored (Q) and the potential difference (V) across the conducting surfaces. The energy stored by a parallel plate capacitor is related to the electric field existing between the plates. The product of the area (A) and the distance between the plates (d) equals the volume between the conducting surfaces.
energy density	energy density = ½ $\epsilon_o\, E^2$	The **energy density** is the energy stored per unit volume.

CONCEPT SUMMARY

Electric Potential and Potential Difference

The **electric potential** at point a (V_a) equals the electric potential energy (PE_a) per unit charge (q) placed at that point.

$V_a = PE_a/q$

The **electric potential difference** between two points (V_{ab}) is measured by the work required to move a unit of electric charge from point b to point a.

$$V_{ab} = V_a - V_b = W_{ab}/q$$

The potential difference can also be discussed in terms of the change in potential energy of a charge (q) when it is moved between points a and b.

$$\Delta PE = PE_a - PE_b = q\ V_{ab} \quad \text{and therefore} \quad V_{ab} = \Delta PE/q$$

Potential difference is often referred to as **voltage**. Both potential and potential difference are scalar quantities which have dimensions of joules/coulomb. The SI unit of electric potential and potential difference is the **volt** (V), where $1\ V = 1\ J/C$.

If the charged particle is in an electric field which is uniform, i.e., constant in magnitude and direction, then the potential difference is related to the electric field as follows:

$$V_{ab} = E\ d\ \cos\theta$$

E is the electric field strength in N/C, d is the distance between points a and b in meters and θ is the angle between the electric field vector and the displacement vector.

If the electric field is non-uniform, i.e., varies in magnitude and direction between points a and b, then the electric field strength can only be properly defined over an incremental distance. If we consider the x component of the electric field (E_x), then $E_x = -\Delta V/\Delta x$, where ΔV is the change in potential over a very short distance Δx. The minus sign indicates that E points in the direction of decreasing V.

Equipotential Lines

Equipotential lines are lines along which each point is at the same potential. On an equipotential surface, each point on the surface is at the same potential. The equipotential line or surface is perpendicular to the direction of the electric field lines at every point. Thus, if the electric field pattern is known, it is possible to determine the pattern of equipotential lines or surfaces and vice versa.

In the following diagrams, the dashed lines represent equipotential lines and the solid lines the electric field lines.

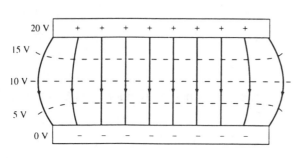

parallel plate conductors

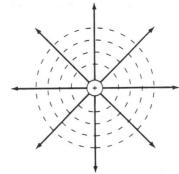

single positive charge

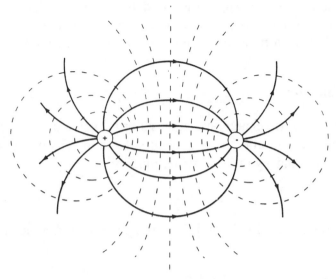

two point charges of opposite sign

Electron Volt

The energy gained by a charged particle which is accelerated through a potential difference can be expressed in **electron volts** (eV) as well as joules. Higher amounts of energy can be measured in KeV or MeV. 1 eV = 1.6 x 10^{-19} J, 1 KeV = 10^3 eV, and 1 MeV = 10^6 eV.

Electric Potential Due to a Point Charge

The electric potential due to a point charge (q) at a distance r from the charge is given by

V = kq/r

Note that the zero of potential is arbitrarily taken to be at infinity (r = ∞). For a negative charge, the potential at distance r from the charge is less than zero. As r increases, the potential increases toward zero, reaching zero at r = ∞. If more than one point charge is present, the potential at a particular point is equal to the arithmetic sum of the potential due to each charge at the point in question.

Electric Dipole

Two equal point charges (q), of opposite sign, separated by a distance ℓ, are called an **electric dipole**. If r >> ℓ, then the potential at point P due to the dipole is given by

V = k q ℓ cos θ/r²

The product qℓ is the **dipole moment** (p) and the potential can be written as

V = k p cos θ/r²

The SI unit of the dipole moment (p) is the **debye**, where 1 debye = 3.33 x 10^{-33} C m. In polar molecules, such as water, the molecule is electrically neutral but there is a separation of charge in the molecule. Such molecules have a net dipole moment.

Capacitance and Dielectrics

A **capacitor** stores electric charge and consists of two conductors separated by an insulator known as a dielectric. The ability of a capacitor to store electric charge is referred to as **capacitance** (C) and is found by the following equation:

C = Q/V

Q is the charge stored in coulombs and V is the potential difference between the conducting surfaces in volts.

The SI unit of capacitance is the farad (F), where 1 farad = 1 coulomb/volt (1 F = 1 C/V). Typical capacitors have values which range from 1 picofarad (1 pF) to 1 microfarad (1 μF) where

1 pF = 1 x 10^{-12} F   and   1 μF = 1 x 10^{-6} F

The capacitance of a capacitor depends on the physical characteristics of the capacitor as well as the insulating material which separates the conducting surfaces which store the electric charge. For a parallel plate capacitor, the capacitance is given by

C = K ϵ_o A/d   or   C = ϵA/d   where ϵ = K ϵ_o and ϵ_o = 8.85 x 10^{-12} C^2/N^{2}m^2

K is the dielectric constant of the insulating material between the plates. The constant is dimensionless and depends on the material. For dry air at 20°C, the constant is 1.0006. ϵ_o is the permittivity of free space. A is the surface area of one side of one plate which is opposed by an equal area of the other plate, and d is the distance between the plates. ϵ is the permittivity of the material between the plates.

Energy Stored in a Capacitor

A charged capacitor stores energy. The electric energy stored in charging a capacitor from an uncharged condition to a charge of Q and a potential difference V is given by

energy = ½ Q V = ½ C V^2 = ½ Q^2/C

The energy stored by a parallel plate capacitor is related to the electric field existing between the plates:

C = ϵ_o A/d   and   V = E d,   then   energy = ½ C V^2 = ½ ϵ_o E^2 A d

The product of the area (A) and the distance between the plates (d) equals the volume between the conducting surfaces. The **energy density** is the energy stored per unit volume:

energy density = ½ ϵ_o E^2 If a dielectric is present, then ϵ_o is replaced by ϵ.

SELECTED TEXT QUESTIONS WITH ANSWERS

QUESTION 3. Draw a few equipotential lines in Fig. 16-29b.

ANSWER: The equipotential lines are drawn so that they are perpendicular to the electric field lines at the points where the equipotential lines cross the electric field lines.

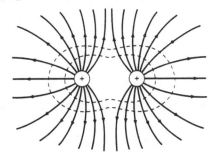

QUESTION 4. Is there a point between two equal positive charges where the electric field is zero? Where the electric potential is zero? Explain.

ANSWER: Electric field is a vector quantity. At the midpoint between the two charges the two vectors would be equal in magnitude but oppositely directed. The magnitude of the resultant electric field is zero.

$$Q_1 + \quad \longleftarrow\!\!\cdot\!\!\longrightarrow \quad + Q_2$$
$$ E_2 \qquad\qquad E_1$$

Electric potential is a scalar quantity. It has magnitude but no direction. At the midpoint between the charges the electric potential due to each charge is positive and equal in magnitude. The resultant electric potential is the arithmetic sum of the two potentials and is greater than zero. The charges are both positive, therefore, the arithmetic sum of the potential of each charge would be greater than zero at every point.

QUESTION 8. If the potential at a point is zero, must the electric field also be zero? Give an example.

ANSWER: At a point where the potential equals zero, the electric field need NOT equal zero. In question 5, it was shown that the electric potential need not be zero at a point where the electric field equals zero. The reverse is also true. The reason is that electric field is a vector while electric potential is a scalar.

For example, consider the possibility that the two point charges are equal in magnitude but one is positive and the other negative. As shown in the diagram, the electric field due to the positive charge is directed away from the positive charge and toward the negative. The electric field due to the negative charge is directed toward the negative charge. The resultant electric field equals the sum of the two vectors and is directed toward the negative charge.

$$Q_1 + \quad \overset{\longrightarrow\, E_1}{\underset{\longrightarrow\, E_2}{}} \; - Q_2$$

Electric potential is a scalar quantity. At the midpoint between the two charges the magnitude of the potential due to the positive charge equals the potential due to the negative charge. However, one potential is positive and the other is negative and the arithmetic sum of the two potentials equals zero.

PROBLEM SOLVING SKILLS

For the problems involving a charged particle accelerated through a potential difference:

1. Draw an accurate diagram showing the motion of the particle.
2. Complete a data table.
3. If necessary, review the work-energy theorem, the concept of kinetic energy, Newton's laws of motion, and the kinematics equations.
4. Use the concepts of electric field, electric force, and electric potential along with the concepts of step 3 to solve the problem.

For problems involving the electric potential due to a point charge(s):

1. Remember that potential is a scalar quantity.
2. Solve for the potential due to each charge. The potential due to a negative charge is negative and for a positive charge the potential is positive.
3. The total potential equals the arithmetic sum of the potentials due to the individual point charges.

For problems involving a parallel plate capacitor based on the physical characteristics of the capacitor and the dielectric constant of the material between the plates:

1. Note the value of the dielectric constant.
2. Determine the area of one plate and the distance between the plates.
3. Apply the formula for the capacitance of a parallel plate capacitor.

For problems involving capacitance when a dielectric inserted after the initial conditions are described:

1. If the battery is disconnected before the dieletric is inserted, the charge stored cannot increase. However, the magnitude of both the electric field and potential difference decrease.
2. If the battery remains connected, the charge stored increases. The magnitude of both the final electric field and the potential difference between the plates equals the value before the dielectric was inserted.
3. In each case, the final capacitance is greater than the original.

PROGRAMMED PROBLEMS

PROBLEM 1. The two horizontal parallel plates of a capacitor are 0.0300 m apart and the electric potential difference between the plates is 100 V. A deuterium nucleus is initially at rest near the positive plate. Determine the a) magnitude of the electric force acting on the particle, b) work done on the particle as it passes between the plates, and c) particle's kinetic energy and speed just before it strikes the negative plate. Note: the mass of the nucleus is 3.34×10^{-27} kg and the charge is 1.60×10^{-19} C.

```
+  +  +  +  +  +  +
_____

          ⊕
          ⇓

_____
 -  -  -  -  -  -  -
```

a. 1	Solution: (Section 17-1)
Determine the work done on the particle as it passes between the plates.	$W = q\,V = (1.60 \times 10^{-19}$ C$)(100$ V$)$
	$W = 1.60 \times 10^{-17}$ J
b. 1	work $= \Delta KE = KE_f - KE_i$
Use the work-energy theorem to determine the particle's final kinetic energy.	1.60×10^{-17} J $= KE_f - 0$ J
	$KE_f = 1.60 \times 10^{-17}$ J
Assume that the particle was initially at rest and determine the particle's final velocity.	$KE_f = \frac{1}{2}\,m\,v^2$
	1.60×10^{-17} J $= \frac{1}{2}(3.34 \times 10^{-27}$ kg$)\,v_f^{\,2}$
	$v_f = 9.79 \times 10^4$ m/s

PROBLEM 2. The two horizontal parallel plates of a capacitor are 0.0400 m apart and the potential difference between the plates is 100 V. A particle which has a excess charge of +2e, i. e., $+3.20 \times 10^{-19}$ C, is in static equilibrium between the plates. Calculate the a) magnitude and direction of the electric field between the plates and b) mass of the particle.

```
 -  -  -  -  -  -  -  -  -  -  -
_____
        ↑ F_electric
      ⊕
        ↓ F_gravitational

_____
 +  +  +  +  +  +  +  +  +  +  +
```

a. 1 Determine the magnitude and direction of the electric field.	Solution: (Sections 17-1 and 17-2) The electric field is uniform in the region between the two parallel plates and is directed from the positive plate toward the negative plate. The magnitude of the electric field is given by $V = E\,d\,\cos\theta$ $100\ V = E\ (0.0400\ m)\cos 0°$ $E = (100\ V)/(0.0400\ m) = 2500\ V/m = 2500\ N/C$
b. 1 Determine the mass of the particle.	Since the particle is in static equilibrium, the electric force must be equal but oppositely directed from the gravitational force. $F_{electric} = F_{gravitational}$ $q\,E = m\,g$ $(3.20 \times 10^{-19}\ C)(2500\ N/C) = m\ (9.80\ m/s^2)$ $m = (3.20 \times 10^{-19}\ C)(2500\ N/C)/(9.80\ m/s^2)$ $m = 8.16 \times 10^{-17}\ kg$

PROBLEM 3. a) Determine the potential at point p due to a point charge of magnitude 3.00 μC if point p is 0.400 m from the charge. b) Determine the work which must be done in order to move a 0.100 μC charge from infinity to point p.

a. 1 Use the formula for the electric potential a distance r from a single point charge.	Solution: (Section 17-5) Assume that the potential at infinity is zero. $V = k\,q/r = (9.0 \times 10^9\ N\ m^2/C^2)(3.00 \times 10^{-6}\ C)/(0.400\ m)$ $V = 6.75 \times 10^3\ volts$
b. 1 Determine the work required to move a 0.100 μC charge from infinity to point p.	At infinity the potential due to the 3.00 μC charge is arbitrarily defined as being equal to zero. The work done in moving the 0.100 μC charge from infinity to point p can be determined by the product of the magnitude of the 0.100 μC and the potential difference between the potential due to the 3.00 μC charge at point p as compared to the 3.00 μC charge's potential at infinity. $W = q\,V = (0.100 \times 10^{-6}\ C)(6.75 \times 10^3\ V)$ $W = 6.75 \times 10^{-4}\ J$

PROBLEM 4. Four point charges are placed at the four corners of a square 0.0500 m on a side. Determine the potential at the center of the square if a) each charge has a magnitude of 3.00 µC and b) two of the charges have a magnitude of 3.00 µC while the other two have a magnitude of -3.00 µC.

a. 1 Draw a diagram locating the four charges. Determine the distance from each charge to the center of the square.	Solution: (17-5) $r = [(0.0500 \text{ m})^2 + (0.0500 \text{ m})^2]^{\frac{1}{2}}$ $r = 0.0707 \text{ m}$
a. 2 Determine the magnitude of the potential due to one of the charges.	Each charge has the same magnitude and the distance from p is the same. The potential due to one of the charges is given by $V = kq/r$ $V_1 = (9.0 \times 10^9 \text{ Nm}^2/\text{C}^2)(3.00 \times 10^{-6} \text{ C})/(0.0707 \text{ m})$ $V_1 = 3.82 \times 10^5 \text{ volts}$
a. 3 Determine the total potential of the four charges.	Potential is a scalar quantity. The total potential equals the arithmetic sum of the potential due to the individual charges. $V_{total} = V_1 + V_2 + V_3 + V_4$ $V_{total} = (4) \ k \ q/r = 4 \ V_1$ $V_{total} = 4 \ V_1 = 1.53 \times 10^6 \text{ volts}$
b. 1 Assume that charges 3 and 4 are negative. Determine V_{total}.	$V_1 = V_2 = 3.82 \times 10^5 \text{ volts}$ $V_3 = V_4 = - \ 3.82 \times 10^5 \text{ volts}$ $V_{total} = V_1 + V_2 + V_3 + V_4 = 0$

Note: Although the potential at point p is zero, the electric field strength does not equal zero. It is left to the student to show that if charges 1 and 4 are positive and 2 and 3 are negative, then both the potential and the electric field strength equal zero.

PROBLEM 5. A parallel plate capacitor is rated at 100 pF. The potential difference between the plates is 200 volts. Determine the a) charge on each plate and b) energy stored.

a. 1	Solution: (Sections 17-7 and 17-9)
Determine the charge on each plate. Note: 1 pf = 10^{-12} F	$C = Q/V$ and $Q = CV$ $Q = (100 \times 10^{-12} \text{ F})(200 \text{ volts}) = 2.00 \times 10^{-8} \text{ C}$
b. 1 Determine the energy stored on the capacitor.	energy = ½ C V^2 = ½ $(100 \times 10^{-12} \text{ F})(200 \text{ V})^2$ energy = 2.00×10^{-6} J

PROBLEM 6. A 12.0 volt battery is connected to a parallel plate capacitor which is rated at 1.00 μF. The plates of the capacitor are separated by 0.00600 m of dry air. Determine the a) charge on each plate and, b) magnitude of the electric field between the plates. c) After the battery is disconnected, a sheet of paraffin 0.00600 m thick and dielectric constant 2.20 is slipped between the plates. Determine the electric field strength and the magnitude of the new capacitance.

a. 1	Solution: (Sections 17-7 and 17-8)
Determine the charge stored on each plate.	$Q = C V = (1.00 \times 10^{-6} \text{ F})(12.0 \text{ V}) = 1.20 \times 10^{-5}$ C
b. 1 Determine the electric field between the plates.	$E = V/d = (12.0 \text{ volts})/(0.00600 \text{ m})$ $E = 2.00 \times 10^3$ V/m or 2.00×10^3 N/C
c. 1 Remember that the battery has been disconnected. Determine the new poten-potential and then determine the electric field.	Since the battery is disconnected before the dielectric is inserted, the quantity of charge on the plates remains the same. The electric field decreases by a factor equal to the dielectric constant of the paraffin. Since the electric field strength decreases, the potential difference between the plates also decreases. $V = V_o/K = (12.0 \text{ volts})/(2.20) = 5.45 \text{volts}$ $E = V/d = (5.45 \text{ volts})/(0.00600 \text{ m}) = 909$ V/m
c. 2 Determine the magnitude of the new capacitance.	The capacitance increases as a result of the insertion of the dielectric. $C = Q/V = (1.20 \times 10^{-5} \text{ C})/(5.45 \text{ V}) = 2.20 \text{ μF}$ or $C = K C_o = (2.20)(1.00 \text{ μF}) = 2.20 \text{ μF}$

PROBLEM 7. A parallel plate capacitor consists of two metal plates separated by 0.00600 m of dry air. The capacitor is connected to a 100 volt source. The area of each plate is 0.0400 m². Determine the a) capacitance, b) charge on each plate, c) energy stored, and d) electric field between the plates. e) A sheet of mica 0.00600 m thick is inserted between the plates. The dielectric constant of mica is 7.00. Determine the capacitance with the mica inserted, the total energy stored and the increase in the charge stored on the plates.

a. 1 Determine the initial capacitance (C_i).	Solution: (Sections 17-7, 17-8, and 17-9) The dielectric constant of dry air to three significant figures is 1.00. $C_i = K \epsilon_o A/d$ $= (1.00)(8.85 \times 10^{-12} \ C^2/N \ m^2)(0.0400 \ m^2)/(0.00600 \ m)$ $C_i = 5.90 \times 10^{-11} \ F$ or $59.0 \ pF$
b. 1 Determine the charge stored.	$Q_i = C \ V = (5.90 \times 10^{-11} \ F)(100 \ V)$ $Q_i = 5.90 \times 10^{-9} \ C$
c. 1 Determine the energy stored.	energy $= \frac{1}{2} \ Q \ V = \frac{1}{2} \ (5.90 \times 10^{-9} \ C)(100 \ V)$ energy $= 2.95 \times 10^{-7} \ J$
d. 1 Determine the initial electric field (E_i).	$V = E \ d$ and $E = V/d = (100 \ V)/(0.0060 \ m)$ $E_i = 1.67 \times 10^4 \ V/m = 16,700 \ N/C$
e. 1 Determine the capacitance with the dielectric inserted (C_f).	Since the battery remains connected to the capacitor, the amount of charge stored on the plates will increase. The electric field will decrease immediately after the dielectric is inserted but will return to 16,700 N/C when the charge stops forming on the plates. The final voltage will be 100 volts. $C_f = K \epsilon_o A/d$ but $\epsilon_o A/d = 59.0 \ pF$ $C_f = (7.00)(59.0 \ pF) = 413 \ pF$
e. 2 Determine the energy stored.	energy $= \frac{1}{2} \ C \ V^2$ energy $= \frac{1}{2}(413 \times 10^{-12} \ F)(100 \ V)^2$ energy $= 2.07 \times 10^{-6} \ J$
e. 3 Determine the final charge stored on the plates and the increase in the charge stored.	$Q_f = C \ V = (413 \times 10^{-12} \ F)(100 \ V)$ $Q_f = 4.13 \times 10^{-8} \ C$ $\Delta Q = Q_f - Q_i = 4.13 \times 10^{-8} \ C - 5.90 \times 10^{-9} \ C = 3.54 \times 10^{-8} \ C$

PRACTICE PROBLEMS

PROBLEM 1. A battery rated at 1.50 volts is connected to the plates of a parallel plate capacitor. The plates are separated by 0.00300 m and a proton is released from rest near the positive plate. Determine the a) magnitude of the electric field between the plates, b) work done on the proton as it passes between the plates and, c) speed of the proton as it is just about to strike the negative plate.

ANS. a) 500 V/m, b) 1.50 eV or 2.40 x 10^{-19} J, c) 1.70 x 10^4 m/s

PROBLEM 2. Two point charges, q_1 = 10.0 μC and q_2 = -2.00 μC are arranged along the x axis at x = 0 m and x = 4.00 m, respectively. Determine the a) positions along the x-axis where V = 0 and b) magnitude of the electric field strength (E) at the point(s) where V = 0.

ANS. a) x = 3.30 m and x = 5.00 m, b) 4.50 x 10^4 N/C toward right at x = 3.30 m,
1.40 x 10^4 N/C toward left at x = 5.00 m

PROBLEM 3. A potential difference of 200 volts is applied to a 3.00 μF parallel plate capacitor. Determine the a) charge stored on each plate and b) energy stored by the capacitor.

ANS. a) 6.00×10^{-4} C, b) 0.0600 J

PROBLEM 4. A parallel plate capacitor has circular plates 0.100 m in diameter and separated by a piece of mica 0.00200 m thick. The dielectric constant of mica is 7.00. Determine the capacitance of the capacitor.

ANS. 243 pF

CHAPTER 18

ELECTRIC CURRENTS

KEY TERMS AND PHRASES

source of emf is a device which transforms one form of energy, chemical, mechanical, etc. to electric energy. Examples of sources of emf are a chemical battery and an electric generator.

emf refers to electromotive force. Emf is a measure of the potential difference across the source of voltage.

electric current is the rate of flow of electric charge. The magnitude of the current is measured in **ampere**s (I), where 1 ampere = 1 coulomb/second.

conventional current is the direction of positive charge flow. In gases and liquids both positive and negative ions move. Only negative charges, i.e., electrons, move through solids and this is referred to as **electron current**. For historical reasons, conventional current is used in referring to the direction of electric charge flow.

electrical resistance refers to the opposition offered by a substance to the flow of electrical current. The resistance of a metal conductor is a property which depends on its dimensions, material and temperature. The unit of resistance is the ohm (Ω).

Ohm's law states that magnitude of the electric current that flows through a closed circuit depends directly on the voltage between the terminals of the source of emf and inversely to the electrical resistance.

electric power is the rate at which work is done to maintain an electric current in a circuit. The SI unit of power is the watt (W), where 1 W = 1 J/s. The kilowatt is a commonly used unit where 1 kilowatt = 1000 watts.

direct current (dc) refers to a current which flows in one direction only.

alternating current (ac) refers to a current where the direction of current flow through the circuit changes usually at a particular frequency (f). The frequency used in the United States is 60 cycles per second or 60 Hz.

root mean square current (I_{rms}) is the square root of the mean of the square of the current flowing through a circuit.

SUMMARY OF MATHEMATICAL FORMULAS

electric current	$I = Q/t$	Electric current is measured in amperes (I), where 1 ampere = 1 coulomb/second and $I = Q/t$.
electrical resistance	$R = \rho\, L/A$	At a specific temperature, the resistance (R) of a metal wire is related to the length (L), cross-sectional area (A), and a constant of proportionality called the **resistivity** (ρ).
	$\rho_T = \rho_o(1 + \alpha\, T)$ $R_T = R_o(1 + \alpha\, T)$	As the temperature of a conductor changes, both the resistivity and resistance of a conductor change. ρ_T and R_T are the values of the resistivity and resistance at temperature T, while ρ_o and R_o are the values of the resistivity and resistance at 20°C. α is the temperature coefficient of resistance which remains constant over a certain range of temperature. The value of α depends on the material and its units are $(C°)^{-1}$.
Ohm's law	$I = V/R$ or $V = I\,R$	The magnitude of the electric current (I) that flows through a closed circuit depends directly on the voltage (V) between the battery terminals and inversely to the circuit resistance (R). The current (I) is measured in amperes, the voltage (V) in volts and the resistance (R) in ohms.
electric power in a dc circuit	$P = I\,V$	Electric power equals the product of the current I and the potential difference V.
	$P = I^2\,R$	In a circuit of resistance R, the rate at which electrical energy (P) is converted to heat energy is referred to as joule heating. Joule heating is related to the product of the square of the current and the resistance.
	$P = V^2/R$	Since $I = V/R$, an alternate formula for power can be written in terms of the voltage and the resistance.

alternating current electricity	$V = V_o \sin 2\pi f t$	The emf (V) produced by an ac electric generator is sinusoidal.
	$I = I_o \sin 2\pi f t$	The current produced in a closed circuit connected to an ac generator is sinusoidal. V_o and I_o represent the peak or maximum voltage and current respectively, and t represents the time in seconds.
	$P = I_o^2 R \sin^2 2\pi f t$	The power delivered to a resistance R at any instant (t).
	$\bar{P} = \frac{1}{2} I_o^2 R$	The average power $(\bar{P})$ delivered to the resistance where I_{rms} is the root mean square current.
	$\bar{P} = (I_{rms})^2 R$	
	$\bar{P} = I_{rms} V_{rms}$	where $I_{rms} = 0.707 I_o$ and $V_{rms} = 0.707 V_o$

CONCEPT SUMMARY

The Electric Battery

A **battery** is a source of electric energy. A simple battery contains two dissimilar metals, called **electrodes,** and a solution called the **electrolyte,** in which the electrodes are partially immersed. An example of a simple battery would be one in which zinc and carbon are used as the electrodes, while a dilute acid, such as sulfuric acid (dilute), acts as the electrolyte. The acid dissolves the zinc and causes zinc ions to leave the electrode. Each zinc ion which enters the electrolyte leaves two electrons on the zinc plate. The carbon electrode also dissolves but at a slower rate. The result is a difference in potential between the two electrodes.

The potential difference between the terminals of a battery in which no internal energy losses occur is referred to as the **electromotive force** (emf) and is measured in volts. The symbol electromotive force is ξ. The schematic symbol for a battery is $\underline{\quad -\!|\!|^+\quad}$.

Electric Current

An electric **current** exists whenever electric charge flows through a region, e.g., a simple light bulb circuit. The magnitude of the current is measured in **ampere**s (I), where 1 ampere = 1 coulomb/second and $I = Q/t$. The direction of **conventional current** is in the direction in which positive charge flows. In gases and liquids both positive and negative ions move. Only negative charges, i.e., electrons, move through solids and this is referred to as **electron current.** For historical reasons, conventional current is used in referring to the direction of electric charge flow.

Resistivity

When electric charge flows through a circuit it encounters electrical **resistance**. The resistance of a metal conductor is a property which depends on its dimensions, material and temperature. A resistor is represented by a jagged line, e.g., ———⋀⋀⋀———.

At a specific temperature, the resistance (R) of a metal wire of length L and cross-sectional area A is given by

$$R = \rho \ L/A$$

ρ is a constant of proportionality called the **resistivity**. The unit of resistance is the ohm (Ω) and the unit of resistivity is ohm m. Within a certain range of temperature, the resistivity of a conductor changes according to the following equation:

$$\rho_T = \rho_o(1 + \alpha \ T)$$

while the resistance changes according to the equation

$$R_T = R_o(1 + \alpha \ T)$$

ρ_o and R_o are the values of the resistivity and resistance at 20°C while ρ_T and R_T are the values of the resistivity and resistance at temperature T, where T is measured in °C. α is the **temperature coefficient of resistance**. The value of α depends on the material and its units are $(C°)^{-1}$.

Ohm's Law

The magnitude of the electric current that flows through a closed circuit depends directly on the voltage between the battery terminals and inversely to the circuit resistance. The relationship that connects current, voltage, and resistance is known as **Ohm's law** and is written as follows:

$$I = V/R \quad \text{or} \quad V = IR$$

The current (I) is measured in amperes, the voltage (V) in volts, and the resistance (R) in ohms.

The following is the schematic for a simple direct current circuit, e.g., a flashlight circuit.

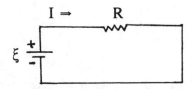

where ξ represents the battery voltage and R represents the electrical resistance of the filament of the light bulb. I is the current. The arrow represents the direction of conventional current is directed away from the positive side of the battery.

Electric Power

Work is required to transfer charge through an electric circuit. The work required depends on the amount of charge transferred through the circuit and the potential difference between the terminals of the battery: $W = QV$. The rate at which work is done to maintain an electric current in a circuit is termed **electric power**. Electric power equals the product of the current I and the potential difference V, i.e., $P = IV$. The SI unit of power is the watt (W), where $1 \ W = 1 \ J/s$. The kilowatt is a commonly used unit where $1 \ kilowatt = 1000 \ watts$.

The electric energy produced by the source of emf is dissipated in the circuit in the form of heat. The kilowatt hour (KWH) is commonly used to represent electric energy production and consumption where $1 \ KWH = 3.6 \times 10^6 \ J$.

In a circuit of resistance R, the rate at which electrical energy is converted to heat energy is given by

$P = IV$ but $V = IR$,   then $P = I(IR) = I^2 R$ $I^2 R$ is known as **joule heating**.

Since $I = V/R$, an alternate formula for power can be written; then

$P = IV = (V/R)V = V^2/R$

Alternating Current

In a **direct current** (dc) circuit the current flows in one direction only. In an **alternating current** (ac) circuit the direction of current flow through the circuit changes at a particular frequency (f). The frequency used in the United States is 60 cycles per second or 60 Hz.

The emf produced by an ac **electric generator** is **sinusoidal**. The current produced in a closed circuit connected to the generator is also sinusoidal. The equations for the voltage and current are as follows:

$V = V_o \sin 2 \pi f t$ and $I = I_o \sin 2 \pi f t$

V_o and I_o represent the peak or maximum voltage and current respectively, and t represents the time in seconds.

The power delivered to a resistance R at any instant is

$P = I^2 R = I_o^2 R \sin^2 2 \pi f t$

The average power delivered to the resistance is

$\overline{P} = \frac{1}{2} I_o^2 R = (I_{rms})^2 R$

I_{rms} is the **root mean square** current. This current is the square root of the mean of the square of the current. It can be shown that

$I_{rms} = 0.707 \ I_o$ and $V_{rms} = 0.707 \ V_o$

A direct current whose values of I and V equal the rms values of I and V of an alternating current produce the same amount of power. In ac circuits, it is usually the rms value that is specified. For example, ordinary ac line voltage is 120 volts. The 120 volts is V_{rms}, while the peak voltage V_o is 170 volts.

SELECTED TEXT QUESTIONS WITH ANSWERS

QUESTION 1. Car batteries are often rated in ampere-hours (A h). What does this rating mean?

ANSWER: An ampere-hour is a way of expressing the total charge the battery can supply at its rated voltage. For example, if the battery is rated at 55 A h, then $Q = I t = (55 C/s)(3600 s)$ and $Q = 198,000 C$.

QUESTION 5. Can a copper wire and an aluminum wire of the same length have the same resistance? Explain.

ANSWER: Resistance of a wire depends on the resistivity, length, cross-sectional area, and temperature. The resistivity of aluminum is greater than that of copper. Therefore, if the length, area and temperature are the same for both, the aluminum wire would have greater resistance. However, an aluminum wire of larger cross-sectional area could be made so that the resistance is the same. Also, it would be possible to keep the aluminum wire at room temperature but raise the temperature of the copper wire until the resistance is the same as that of the aluminum wire.

QUESTION 12. Electric power is transferred over large distances at very high voltages. Explain how the high voltage reduces power loss in the transmission lines.

ANSWER: Power losses in transmission lines is the result of Joule heating $P = I^2 R$. Because of joule heating, it is necessary to have a low current (I) in the line. Since $P = I V$, this can be accomplished by stepping up the voltage at the power plant. The high voltage is sent at low current through the transmission lines. When it reaches an electrical sub-station in your neighborhood, a step down transformer reduces the voltage and the current increases.

PROBLEM SOLVING SKILLS

For problems involving resistivity and temperature coefficient of resistance:

1. Complete a data table listing information both given and implied. For example, the resistivity, length, cross-sectional area, temperature coefficient of resistance and the change in temperature.
2. Solve for the resistance and the resistance at some higher temperature.

For problems involving electric power, electric energy and the cost of electric energy:

1. Complete a data table listing information both given and implied. For example, the current, voltage and resistance and time the device is used.
2. Determine the power dissipated in watts and kilowatts.
3. Determine the number of KWH of energy used and multiply the number of KWH by the cost per KWH.

For problems involving alternating current:

1. Complete a data table listing information both given and implied. For example, include the rms current and voltage, peak current and voltage, and the resistance.
2. Use the data table to solve for the average and instantaneous power dissipated in the resistor.

PROGRAMMED PROBLEMS

PROBLEM 1. The starter motor on a particular car draws a current of 50 amperes through a copper cable. Determine the number of a) coulombs of charge which pass a particular point in the cable in 15 s, and b) electrons which pass a particular point in the wire in 15 seconds.

a. 1	Solution: (Section 18-2)
Determine the charge that passes a point in 15 seconds.	$I = Q/t$, then $Q = I t$ $Q = I t = (50 A)[(1 C/s)/(1 A)](15 s) = 750 C$
b. 1 Determine the number of electrons in 750 C.	$1 C = 6.25 \times 10^{18}$ electrons, then $Q = (750 C)(6.25 \times 10^{18}$ electrons/1 C) $Q = 4.7 \times 10^{21}$ electrons

PROBLEM 2. a) Determine the electrical resistance of a 20.0 m length of tungsten wire of radius 0.200 mm. b) If the temperature of the wire does not change, determine the resistance of the same wire if it is stretched to a length of 60.0 m. The resistivity of tungsten is $5.60 \times 10^{-8} \Omega$ m.

a. 1	Solution: (Section 18-4)
Complete a data table.	$A = \pi r^2 = \pi (0.200$ mm$)^2(1.0$ m/1000 mm$)^2 = 1.26 \times 10^{-7}$ m^2 $L = 20.0$ m, $\quad \rho = 5.60 \times 10^{-8} \Omega$ m
a. 2 Determine the resistance of 20.0 m of wire.	$R = \rho L/A$ $= (5.60 \times 10^{-8} \Omega$m$)(20.0$ m$)/(1.26 \times 10^{-7}$ m$^2)$ $R = 8.89 \Omega$
b. 1 Determine the cross-sectional area of the wire when it is stretched to 60.0 m.	Stretching the wire will affect the wire's length and cross-sectional area. However, the volume (V) of the wire, which equals the product of the length and cross-sectional area (V = A L), must remain the same. $V_f = V_o$ $A_o L_o = A_f L_f$ $(1.26 \times 10^{-7}$ m$^2)(20.0$ m$) = A_f (60.0$ m$)$ $A_f = 4.20 \times 10^{-8}$ m^2

b. 2	$R_f = \rho\, L/A$
Determine the resistance of 60.0 m of wire.	$= (5.60 \times 10^{-8}\ \Omega\text{m})(60.0\ \text{m})/(4.20 \times 10^{-8}\ \text{m}^2)$
	$R_f = 80.0\ \Omega$

PROBLEM 3. a) The resistance of the tungsten wire in the previous problem is 80.0 Ω at 0.00°C. Determine the resistance of the same wire if the temperature is raised to 50.0°C. b) The resistance of a carbon wire is 80.0 Ω at 0.00°C. Determine the resistance of the wire at 50.0°C. Explain why the resistance of each substance changes as it does with the increasing temperature. Note: the temperature coefficient of resistance is 0.0045/C° for tungsten and -0.00050/C° for carbon.

a. 1	Solution: (Section 18-4)
Determine the resistance of tungsten at 50.0°C.	$R = R_0\,(1 + \alpha\,T)$
	$= 80.0\ \Omega\ [1 + (0.0045/\text{C}°)(50.0°\text{C} - 0.00°\text{C})]$
	$R = 80.0\ \Omega\ (1 + 0.225) = 80.0\ \Omega\ (1.23) = 98.0\ \Omega$

a. 2	Tungsten is a metal and its resistance increases with temperature. At room temperature the outer electron is free to move throughout the metal. As the temperature increases, the atoms are vibrating more rapidly. The electric field of the individual atoms has a greater probability of interfering with the electrons as they move through the metal, therefore, the resistance increases.
Explain why the resistance of tungsten increases with temperature.	

b. 1	$R = R_0\,(1 + \alpha\,T)$
Determine the resistance of carbon at 50.0°C.	$= 80.0\ \Omega\ [1 + -0.00050/\text{C}°(50.0°\text{C} - 0.00°\text{C})]$
	$= 80.0\ \Omega\ (1 + -0.025) = 80.0\ \Omega\ (0.975)$
	$R = 78.0\ \Omega$

b. 2	Carbon is a semiconductor and the increase in temperature causes more electrons to break free and become part of the electron current. Because of the increased number of free electrons, the resistance of carbon wire to the flow of current decreases.
Explain why the resistance of carbon decreases as the temperature increases.	

PROBLEM 4. It is estimated that the typical American child watches 31 hours of television per week. If the average television draws 1.00 ampere of current from a 120 volt line, determine the a) power rating of the television in watts, b) number of kilowatt-hours of electrical energy consumed in one year during the time that the child watches the TV and c) yearly cost to the parents if electricity costs 8.00 cents per kilowatt-hour.

a. 1 Determine the power rating of the TV.	Solution: (Section 18-6) $P = I \, V = (1.00 \text{ ampere})(120 \text{ volt}) = 120 \text{ watts}$
b. 1 Determine the total number of KWH used in one year.	$W = P \, t$ $\quad = (120 \text{ watts})(1.0 \text{ KW}/1000 \text{ watts})(31 \text{ h}/1 \text{ week})(52 \text{ weeks}/1 \text{ y})$ $W = 193 \text{ KWH}$
c. 1 Calculate the yearly cost.	$\text{cost} = (193 \text{ KWH})(8.00 \text{ cents}/1.0 \text{ KWH})$ $\text{cost} = 1550 \text{ cents}$ or $\$15.50$

PROBLEM 5. An electric immersion heater rated at 250 watts is inserted in a 100 gram aluminum cup which contains 200 grams of water. The initial temperature of the cup and water is 20.0°C. Determine the time required for the temperature of the cup + water to rise to 90.0°C. Note: the specific heat of aluminum is 0.220 cal/g C° and that for water is 1.00 cal/g C°.

a. 1 Determine the amount of heat required to raise the temperature of the system from 20.0°C to 90.0°C.	Solution: (Section 18-6) $Q = m_w \, c_w \, T_w + m_{Al} \, c_{Al} \, T_{Al}$ But $T = 90.0°C - 20.0°C = 70.0 \text{ C}°$. $Q = (200 \text{ g})(1.0 \text{ cal/g°C})(70.0 \text{ C}°) + (100 \text{ g})(0.22 \text{ cal/g°C})(70.0 \text{ C}°)$ $Q = 1.40 \times 10^4 \text{ cal} + 1.54 \times 10^3 \text{ cal} = 1.55 \times 10^4 \text{ cal}$
a. 2 Convert the heat energy from calories to joules.	Heat energy is measured in calories while electrical energy is measured in joules, where 4.18 J = 1.00 calories. $(1.55 \times 10^4 \text{ cal})(4.18 \text{ J/cal}) = 6.5 \times 10^4 \text{ J}$
a. 3 Determine the time required for the temperature change. Note: 1 W = 1 J/s	$\text{work} = (\text{power})(\text{time})$ $6.5 \times 10^4 \text{ J} = (250 \text{ watts}) [(1 \text{ J/s})/(1 \text{ watt})] \text{ (time)}$ $6.5 \times 10^4 \text{ J} = (250 \text{ J/s}) \text{ time}$

time = $(6.5 \times 10^4 \text{ J})/(250 \text{ J/s}) = 260 \text{ s}$ or   4.3 min

Note: the answer assumes that all of the electrical energy which has been converted into heat energy has remained in the cup + water. Since some of the heat will be dissipated into the surroundings, the actual length of time required to raise the temperature from 20°C to 90°C will be greater than 4.3 min.

PROBLEM 6. A 10.0 Ω resistor is connected to a 120 volt ac line. Determine the a) average power dissipated in the resistor, b) rms current through the resistor, c) maximum instantaneous current through the resistor, and d) maximum instantaneous power dissipated in the resistor.

a. 1	Solution: (Section 18-8)
Determine the average power dissipated in the resistor.	$\overline{P} = I\,V$ where I and V refer to the rms values of current and voltage. Since I = V/R,
	$\overline{P} = V^2/R$
	$\overline{P} = (120 \text{ volts})^2/(10.0 \text{ Ω}) = 1400 \text{ watts}$
b. 1	$I_{rms} = V_{rms}/R = (120 \text{ volts})/(10.0 \text{ Ω})$
Determine I_{rms}.	$I_{rms} = 12.0 \text{ amp}$
c. 1	$I_{rms} = 0.707\,I_o$, where I_o is the maximum instantaneous (peak) current that flows through the resistor, therefore,
Determine the maximum instantaneous current that flows through the resistor.	$I_o = I_{rms}/0.707$
	$I_o = (12.0 \text{ amp})/(0.707) = 17.0 \text{ amp}$
d. 1	$P = I_o^2\,R\,\sin^2 2\pi\,f\,t$
Determine the maximum instantaneous power dissipated in the resistor.	But the maximum value of $\sin^2 2\pi\,f\,t = 1.0$
	$P_{max} = I_o^2\,R$
	$P_{max} = (17.0 \text{ amp})^2(10.0 \text{ Ω}) = 2890 \text{ watts}$

PRACTICE PROBLEMS

PROBLEM 1. A certain type of copper wire is rated at 160 Ω per 300 m at 0.00°C. a) Determine the diameter of the wire. b) Determine the resistance of 300 m of wire if the temperature of the wire is raised to 100°C. The resistivity of copper is 1.68×10^{-8} Ω m and the temperature coefficient of resistance is 0.0068/C°.

ANS. a) 2.0×10^{-4} m or 0.20 mm, b) 270 Ω

PROBLEM 2. According to the electrical code, the maximum allowable current for 14 gauge copper wire is 15.0 A. The resistivity of copper at 20.0°C is 1.68×10^{-8} Ω m and the diameter of the wire is 1.628 mm. Calculate the a) electrical resistance of 20.0 m of gauge 14 wire, and b) potential difference between the ends of the wire when 15.0 A flows through it.

ANS. a) 0.161 Ω,   b) 2.42 volts

PROBLEM 3. A certain tungsten filament light bulb has 2.00 Ω of electrical resistance when cold but 120 Ω when operated at a temperature of 1800°C. Determine the a) average temperature coefficient of resistance for the filament. b) Table 18-1 in your text lists the value of the temperature coefficient of resistance as 0.0045/°C at 20°C. Based on the answer to the first part of this problem, is the temperature coefficient of resistance constant over all temperatures?

ANS. a) 0.033/C°,   b) no

PROBLEM 4. An electric toaster draws 10.0 amperes of electric current and has an electrical resistance of 12.0 Ω. a) Determine the power rating of the toaster. b) The toaster operates an average of 5.00 minutes each day. Determine the yearly cost of operation if electrical energy costs 8.00 cents per KWH.

ANS. a) 1.20 KW, b) $2.92

CHAPTER 19

DC CIRCUITS AND INSTRUMENTS

KEY TERMS AND PHRASES

series circuit is an electric circuit with only a single path for electric current to travel. The current through each circuit element is the same.

parallel circuit is an electric circuit with more than one path for electric current to travel. The current is divided among the branches of the circuit. The voltage drop is the same across each branch.

equivalent resistance is the resistance of a single resistor which is equivalent to the total resistance of a network of resistors.

emf refers to electromotive force. emf is a measure of the potential difference across the source of voltage, i.e., a battery or generator.

internal resistance refers to the resistance to electric current inside the voltage source. The internal resistance of the source of emf is always considered to be in a series with the external resistance present in the electric circuit.

terminal voltage is the potential difference available to the circuit outside the source of emf. The terminal voltage equals the difference between the emf and the voltage drop across the internal resistance.

Kirchhoff's first rule or **junction rule** states that the sum of all currents entering any junction point equals the sum of all currents leaving the junction point. This rule is based on the law of conservation of electric charge.

Kirchhoff's second rule or **loop rule** states that the algebraic sum of all the gains and losses of potential around any closed path must equal zero. This law is based on the law of conservation of energy.

RC circuit consists of a resistor and a capacitor connected in series to a dc power source.

time constant (τ) of an RC circuit equals the product of the resistance and the capacitance and is measured in seconds.

galvanometer consists of a moving coil placed in a magnetic field. When an electrical current flows through the galvanometer the interaction of the current in the coil and the magnetic field causes the coil to deflect. The galvanometer is used to detect and measure very low currents, usually in the range of 1 milliampere or less.

ammeter measures the amount of electric current passing a particular point in a circuit. The ammeter is placed in series in the circuit and consists of a galvanometer and a resistor of very low value, called the **shunt resistor**, placed in parallel with the galvanometer.

voltmeter measures the potential difference across a circuit element, e.g., a resistor. A voltmeter consists of a galvanometer of internal resistance r and a resistor of high resistance R placed in series with the galvanometer.

potentiometer circuit is used to give precise measurements of the voltage of the source of emf.

Wheatstone bridge is a device used to give precise measurements of electrical resistance.

SUMMARY OF MATHEMATICAL FORMULAS

resistors arranged in series	$I = I_1 = I_2 = ... = I_n$	The current (I) at every point in a series circuit equals the current leaving the battery.
	$V = V_1 + V_2 + ... + V_n$	The potential difference between the terminals of the battery (V) equals the sum of the potential differences across the resistors.
	$R = R_1 + R_2 + ... + R_n$	The equivalent electrical resistance (R) for a series combination equals the sum of the individual resistors.
resistors arranged in parallel	$I = I_1 + I_2 + ... + I_n$	The battery current (I) equals the sum of the currents in the branches.
	$V = V_1 = V_2 = ... = V_n$	The potential difference across each resistor in the arrangement is the same.
	$1/R = 1/R_1 + 1/R_2 + .. + 1/R_n$	The equivalent resistance (R) of a parallel combination is always less than the smallest of the individual resistors.

terminal voltage of a source of emf	$V = \xi - Ir$	The terminal voltage (V) equals the difference between the emf of the source (ξ) and the drop in potential due to internal resistance (I r).
Kirchhoff's first rule or junction rule	$\Sigma I = 0$	Kirchhoff's first rule or junction rule states that the sum of all currents entering any junction point equals the sum of all currents leaving the junction point. This rule is based on the law of conservation of electric charge.
Kirchhoff's second rule or loop rule	$\Sigma V = 0$	Kirchhoff's second rule or loop rule: The algebraic sum of all the gains and losses of potential around any closed path must equal zero. This law is based on the law of conservation of energy.
capacitors arranged in parallel	$V = V_1 = V_2 = ... = V_n$	The potential difference across each capacitor equals the potential difference (V) of the source of emf.
	$Q = Q_1 + Q_2 + ... + Q_n$	The total charge stored on the capacitor plates (Q) equals the sum of the charges stored on the individual capacitors.
	$C = C_1 + C_2 + ... + C_n$	The equivalent capacitance equals the sum of the individual capacitors.
capacitors arranged in series	$Q = Q_1 = Q_2 = ... = Q_n$	the charge (Q) that leaves the source of emf equals the charge that forms on each capacitor.
	$V = V_1 + V_2 + ... + V_n$	The potential difference across the source of emf (V) equals the sum of the potential differences across the capacitors.
	$1/C = 1/C_1 + 1/C_2 + .. + 1/C_n$	The equivalent capacitance (C) of a parallel combination is always less than the smallest of the individual capacitors.

charging an RC circuit	$I = I_o e^{-t/RC}$	The current I in the circuit at time t after the switch is closed depends on the initial current (I_o), the resistance (R) and the capacitance (C).
	$V = \xi(1 - e^{-t/RC})$	The potential difference (V) across the capacitor as it is being charged.
	$Q_t = Q (1 - e^{-t/RC})$	The amount of charge (Q_t) accumulated on the capacitor as it is being charged.
discharging an RC circuit	$I = I_o e^{-t/R'C}$	the current in an RC discharge circuit as a function of time
	$V = V_o e^{-t/R'C}$	the voltage across the resistance (R')in an RC discharge circuit as a function of time
	$Q_t = Q e^{-t/R'C}$	the charge (Q_t) on a capacitor in an RC discharge circuit as a function of time

CONCEPT SUMMARY

Resistors in Series and Parallel

A simple dc circuit may contain resistors arranged in series or in parallel or in a series-parallel combination. The symbol for a battery is ⊣|⊢ where the long line represents the anode or positive terminal and the short line the cathode or negative terminal. The symbol for a resistor is ⌇⌇⌇.

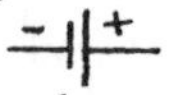

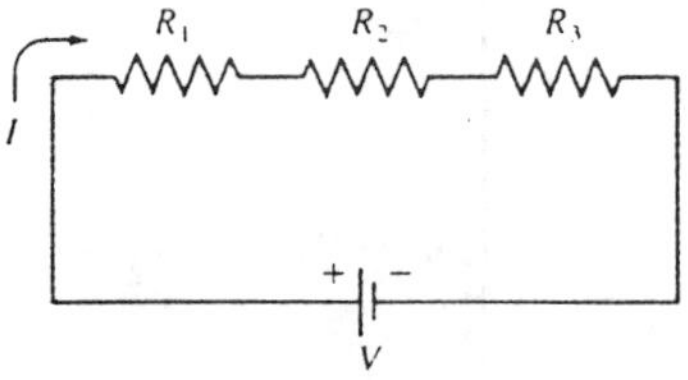

A simple **series circuit** is shown in the diagram. The current (I) at every point in a series circuit equals the current leaving the battery.

$I = I_1 = I_2 = I_3 = ... = I_n$

Assuming that the connecting wires offer no resistance to current flow, the potential difference between the terminals of the battery (V) equals the sum of the potential differences across the resistors, i.e.,

$V = V_1 + V_2 + V_3 + ... + V_n$

The equivalent electrical resistance (R) for this combination is equal to the sum of the individual resistors, i.e.,

$$R = R_1 + R_2 + R_3 + ... + R_n$$

In a simple **parallel circuit,** the current leaving the battery divides at junction point A in the diagram shown below and recombines at point B. The battery current (I) equals the sum of the currents in the branches. In general,

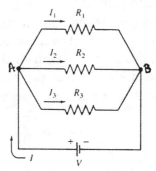

$$I = I_1 + I_2 + I_3 + ... + I_n$$

The potential difference across each resistor in the arrangement is the same, i.e., $V = V_1 = V_2 = V_3 = ... = V_n$. If no other resistance is present, the potential difference across each resistor equals the potential difference across the terminals of the battery.

The equivalent resistance (R) of a parallel combination is always less than the smallest of the individual resistors. The formula for the equivalent resistance is as follows:

$$1/R = 1/R_1 + 1/R_2 + 1/R_3 + ... + 1/R_n$$

Emf and Terminal Voltage

All sources of emf have what is known as **internal resistance** (r) to the flow of electric current. The internal resistance of a fresh battery is usually small but increases with use. Thus the voltage across the terminals of a battery is less than the emf of the battery. The **terminal voltage** (V) is given by the equation $V = \xi - Ir$, where ξ represents the emf of the source of potential in volts, I the current leaving the source of emf in amperes and r the internal resistance in ohms. The internal resistance of the source of emf is always considered to be in a series with the external resistance present in the electric circuit.

Kirchhoff's Rules

Kirchhoff's rules are used in conjunction with Ohm's law in solving problems involving complex circuits:

Kirchhoff's first rule or **junction rule**: the sum of all currents entering any junction point equals the sum of all currents leaving the junction point. This rule is based on the law of conservation of electric charge.

Kirchhoff's second rule or **loop rule**: the algebraic sum of all the gains and losses of potential around any closed path must equal zero. This law is based on the law of conservation of energy.

Suggestions for Using Kirchhoff's Laws

1. Assign a direction to the current in each independent branch of the circuit. Place a positive (+) sign on the side of each resistor where the current enters and a negative sign on the side

where the current exits, e.g., $I \Rightarrow \pm \text{\scriptsize WWW}$. This indicates that a drop in potential occurs as the current passes through the resistor. Place a (+) sign next to the long line of the battery symbol and a (-) sign next to the short line. If you choose the wrong direction for the flow of current in a particular branch, your final answer for the current in that branch will be negative. The negative answer indicates that the current actually flows in the opposite direction. The diagram shown below is an example of a complex circuit which can be solved using Kirchhoff's rules:

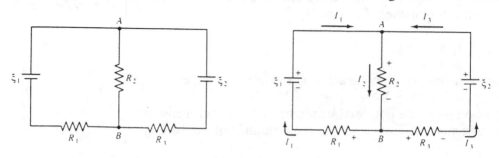

2. Select a **junction point** and apply the junction rule, e.g., at point A in the diagram:

$$I_1 + I_3 = I_2$$

The junction rule may be applied at more than one junction point. In general, apply the junction rule to enough junctions so that each branch current appears in at least one equation.

3. Apply Kirchhoff's loop rule by first taking note whether there is a gain or loss of potential at each resistor and source of emf as you trace the closed loop. Remember that the sum of the gains and losses of potential must add to zero. For example, for the left loop of the sample circuit above, start at point B and travel clockwise around the loop. Because the direction chosen for the loop is also the direction assigned for the current, there is a gain in potential across the battery (- to +), but a loss of potential across each resistor (+ to -). Following the path of the current shown in the diagram (below left) and using the loop rule, the following equation can be written:

$$- I_1 R_1 + \xi_1 - I_2 R_2 = 0$$

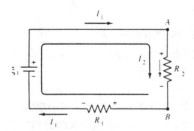

clockwise around loop

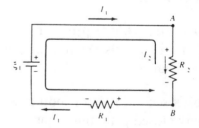

counterclockwise around loop

The direction taken around the loop is arbitrary. Tracing a counterclockwise path around the circuit starting at B, as shown in the above right diagram, there is gain in potential across each resistor (- to +) and a drop in potential across the battery (+ to -). The loop equation would then be

$$+ I_2 R_2 - \xi_1 + I_1 R_1 = 0$$

Multiplying both sides of the above equation by -1 and algebraically rearranging, it can be shown that the two equations are equivalent.

Be sure to apply the loop rule to enough closed loops so that each branch current appears in at least one loop equation. Solve for each branch current using standard algebraic methods.

Capacitors in Series and Parallel

A circuit with **capacitors in parallel** is shown below. According to Kirchhoff's loop rule, the potential difference (V) of the source of emf: $V = V_1 = V_2 = V_3 = ... = V_n$. The total charge stored on the capacitor plates (Q) equals the amount of charge which left the source of emf: $Q = Q_1 + Q_2 + ... + Q_n$ and since $Q = CV$ then

$$C V = C_1 V_1 + C_2 V_2 + ... + C_n V_n \quad \text{and} \quad C = C_1 + C_2 + ... + C_n$$

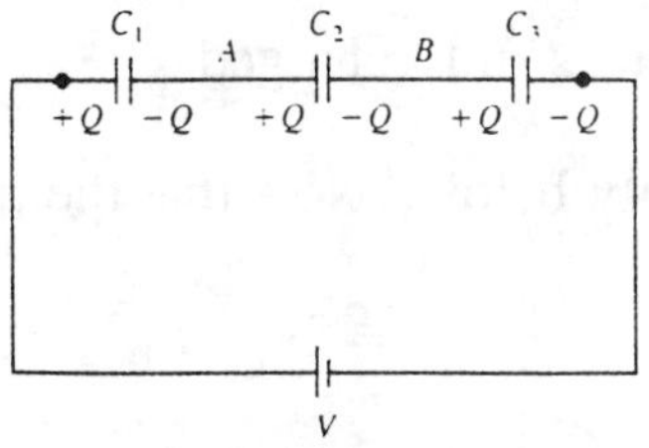

capacitors in series

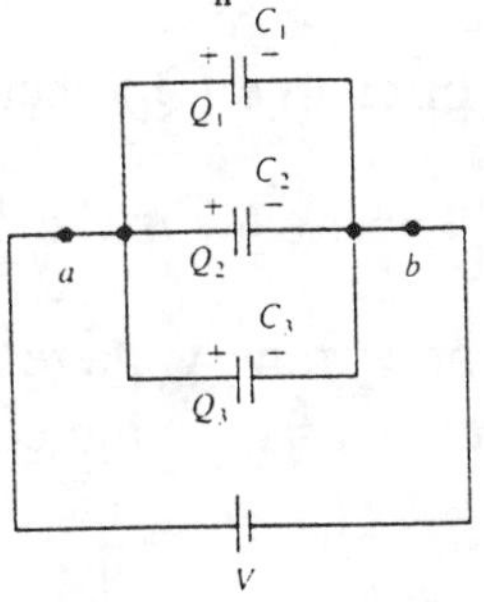

capacitors in parallel

For **capacitors in series,** the amount of charge (Q) that leaves the source of emf equals the amount of charge that forms on each capacitor: $Q = Q_1 = Q_2 = Q_3 = ... = Q_n$.

From Kirchhoff's loop rule, the potential difference across the source of emf (V) equals the sum of the potential differences across the individual capacitors:

$$V = V_1 + V_2 + V_3 + ... + V_n \quad \text{and since} \quad V = Q/C \quad \text{then}$$

$$Q/C = Q_1/C_1 + Q_2/C_2 + ... + Q_n/C_n \quad \text{and} \quad 1/C = 1/C_1 + 1/C_2 + ... + 1/C_n$$

Circuits Containing Resistor and Capacitor

An **RC circuit** consists of a resistor and a capacitor connected in series to a dc power source. When switch 1 (S_1), shown in the diagram below, is closed, the current will begin to flow from the source of emf and charge will begin to accumulate on the capacitor. Using Kirchhoff's loop rule it can be shown that

$\xi - I R - Q/C = 0$

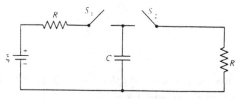

where I R refers to the drop in potential across the resistor and Q/C refers to the drop in potential across the capacitor.

As charge accumulates on the capacitor, the potential difference will increase until it equals ξ, and at that point the current ceases to flow.

By using the methods of calculus, it can be shown that the current I through the circuit at time t after the S_1 is closed is given by

$I = I_o \, e^{-t/RC}$

I_o is the initial value of the current just after switch 1 is closed. RC is the product of the resistance and the capacitance. RC is referred to as the time constant (τ) of the circuit and is measured in seconds. The potential difference across the capacitor as it is being charged is

$V = \xi(1 - e^{-t/RC})$

while the amount of charge (Q_t) accumulated on the capacitor as it is being charged is

$Q_t = Q \, (1 - e^{-t/RC})$ where   Q is the charge on the capacitor when it is fully charged.

The variation of current, voltage, and charge is shown graphically below. Note that the time is marked off in terms of the time constant (τ).

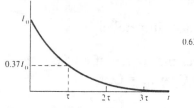 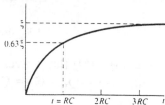 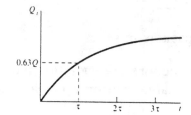

If switch 1 is opened and switch 2 (S_2) is closed then the capacitor will **discharge** through resistor R'. The charge on the capacitor (Q_t) decreases exponentially with time as follows:

$Q_t = Q \, e^{-t/R'C}$ where   Q is the charge on the capacitor when it is fully charged.

The following diagram shows the variation of charge with time as the capacitor discharges. The time is marked off in time constants.

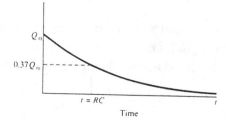

Time

The current in a discharge circuit as a function of time is given by $I = I_o e^{-t/R'C}$, where I_o represents the initial current as the discharge begins.

The voltage across the resistance (R') in a discharge circuit as a function of time is given by $V = V_o e^{-t/R'C}$ where V_o represents the initial voltage across the capacitor as the discharge begins.

Galvanometer, Ammeter, and Voltmeter

A **galvanometer** is a device used to detect and measure very low currents, usually in the range of 1 milliampere or less. It usually consists of a coil of wire suspended by a fine wire and arranged parallel to the magnetic field produced by a permanent magnet. When a current passes through the coil, a torque is produced. This torque causes the coil to twist until a restoring torque provided by the coil suspension balances the torque produced by the interaction of the current in the coil and the magnetic field of the permanent magnet. The torque on the coil and the angle through which the coil deflects is proportional to the current. As a result the galvanometer can be calibrated to determine the amount of current passing through the coil.

Because the coil of a galvanometer consists of wires, the galvanometer presents resistance to the flow of current through it. This internal resistance is usually low.

An **ammeter** is a device designed to measure the amount of electric current passing a particular point in a circuit. The ammeter is placed in series in the circuit and consists of 1) a galvanometer and 2) a resistor of very low value, called the **shunt resistor**, placed in parallel with the galvanometer. Since a potential drop occurs across it, an ammeter is designed to have low overall resistance. Therefore, if it is properly designed, the ammeter will have minimal effect on the external circuit.

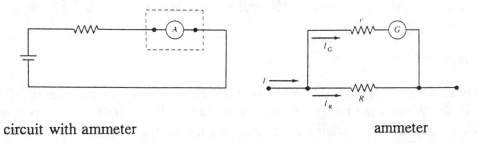

circuit with ammeter   ammeter

A **voltmeter** measures the potential difference across a circuit element, e.g., a resistor. The voltmeter is placed in parallel with the circuit element and therefore the potential difference across the voltmeter is the same as across the circuit element. A voltmeter consists of 1) a galvanometer of internal resistance r and 2) a resistor of high resistance R placed in series with the galvanometer. The resistance of the voltmeter is very large compared to the circuit element and as a result very little current flows through the voltmeter and it has a minimal effect on the circuit.

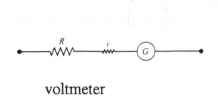

voltmeter

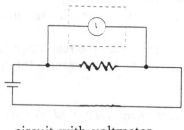

circuit with voltmeter

The Potentiometer

The **slide wire potentiometer** is used to give precise measurements of the source of emf. A typical potentiometer circuit is shown at right. A battery V is connected to a resistor R, the resistance of which can be varied so that a convenient potential difference occurs across the slide wire (AB).

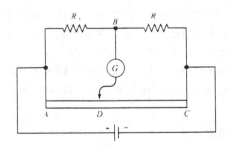

The resistance of the slide wire is proportional to the wire's length, $R = \rho L/A$. Since $V = IR$, then the potential difference between point A and point C along the length of the wire is proportional to the distance between A and C (L_{AC}).

A circuit containing a standard cell of known emf (ξ_s) is then placed in parallel with the slide wire. The position of the slide is then adjusted along the length of the wire until no current flows through the galvanometer (G). The length of the wire between A and C is measured. At this point the potential difference along the wire is equal to ξ_s and therefore $\xi_s \propto L_{AC}$.

The standard cell is then replaced with the unknown and the slide is again adjusted until no current flows. The length $L_{AC'}$ is noted. The emf of the unknown is ξ_x and $\xi_x \propto L_{AC'}$. The emf of the unknown is given by $\xi_x/\xi_s = L_{AC'}/L_{AC}$.

The Wheatstone Bridge

The **Wheatstone bridge** is a device used to give precise measurements of electrical resistance. In the Wheatstone bridge circuit, shown below, R_1, R_2 and R_3 are known while R_x has unknown resistance. The variable resistor R_3 is adjusted until no current flows through the galvanometer when the switch (S) is closed.

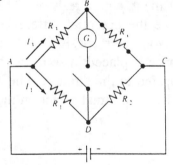

Wheatstone bridge circuit

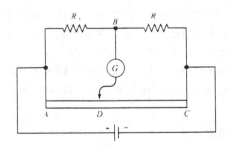

circuit with slide wire

At the point where the current through the galvanometer (G) is zero, the potential at points B and D is exactly the same, therefore

$V_{AB} = V_{AD}$ and $I_3 R_3 = I_1 R_1$

also $V_{BC} = V_{DC}$ and $I_3 R_x = I_1 R_2$. Therefore,

$R_x /R_3 = R_2 /R_1$ and R_x can be determined.

In practice, the slide wire used in the potentiometer circuit is often used since the wire has uniform resistance along its length. The slide is moved until the galvanometer current reaches zero. At this point $R_x /R_3 = L_{CD} /L_{AD}$ and R_x can be determined.

SELECTED TEXT QUESTIONS WITH ANSWERS

QUESTION 3. If all you have is a 120 V line, would it be possible to light several 6 V lamps without burning them out? How?

ANSWER: If the lamps are connected in series, the sum of the potential differences across the lamps would be 120 V. If the maximum voltage across each lamp is to be 6 V, then it would be necessary to connect 120 V/6 V = 20 lamps in series.

QUESTION 17. Suppose that three identical capacitors are connected to a battery. Will they have more energy if they are connected in series or in parallel?

ANSWER: The capacitors store more energy when they are connected in parallel. For example, let each capacitor be rated at 6.0 F and they are connected to a 6 volt battery. If the capacitors are connected in SERIES:

$1/C = 1/6.0$ F $+ 1/6.0$ F $+ 1/6.0$ F, $1/C = 3/6.0$ F and $C = 2.0$ F

The charge across each can now be determined since the charge on each equals the charge (Q) produced by the battery.

$Q = C V = (2.0$ F$)(6.0$ V$) = 12.0$ C

The energy stored on each is the same and can now be determined.

energy $= \frac{1}{2} Q_1^2/C_1 = \frac{1}{2}(12.0$ C$)^2/(6.0$ F$) = 12.0$ J

The total energy stored by the combination is

energy $= \frac{1}{2} Q^2/C = \frac{1}{2}(12.0$ C$)^2/(2.0$ F$) = 36.0$ J

The capacitors are identical and if they are connected in **parallel** equal amounts of charge form on each capacitor.

$Q_1 = C_1 V = (6.0 \text{ F})(6.0 \text{ V}) = 36.0 \text{ C}$

The energy stored on each is the same and can now be determined.

energy $= \frac{1}{2} Q^2/C_1 = \frac{1}{2} (36.0 \text{ C})^2/(6.0 \text{ F}) = 108 \text{ J}$

The total energy stored by the combination can now be determined.

$Q = Q_1 + Q_2 + Q_3 = 108 \text{ C}$ and $C = C_1 + C_2 + C_3 = 18.0 \text{ F}$

energy $= \frac{1}{2} Q^2/C = \frac{1}{2} (108 \text{ C})^2/(18.0 \text{ F}) = 324 \text{ J}$

QUESTION 23. Explain why an ideal ammeter would have zero resistance and an ideal voltmeter infinite resistance.

ANSWER: An ammeter is connected in series in the circuit. The total resistance to current flow is the sum of the ammeter resistance and the resistance of the circuit as a whole. If the ammeter has zero resistance then there is no voltage drop across it and it has no affect on the flow of current in the circuit.

The potential difference across resistors connected in parallel is the same. The voltmeter is connected in parallel with the circuit element and the voltage drop across each is the same. If the resistance of the voltmeter is much greater than that of the circuit, the equivalent resistance would equal the resistance of the circuit element. The voltmeter would draw a very small current and would have little affect on the circuit. However, it would record the voltage drop across the circuit element. The ideal voltmeter would have infinite resistance and would not alter the circuit.

PROBLEM SOLVING SKILLS

For problems involving equivalent resistance:

1. Determine whether the resistors are arranged in series or in parallel.
2. If necessary, simplify the circuit step by step until it is reduced to a simple series or parallel combination. Use the appropriate formula to determine the equivalent resistance.
3. Use Ohm's law and Kirchhoff's laws to solve for the current through each resistor and the potential difference across each resistor.

For problems involving a complex circuit:

1. Assign a direction to the current in each branch of the circuit. Place a + sign on the side of each resistor where the current enters and a - sign where the current exits.
2. Place a + sign at the positive terminal of each battery and a - sign at the negative terminal.
3. Select a junction point and use Kirchhoff's junction rule to write an equation for the currents entering the point.

4. Apply Kirchhoff's loop rule and write equations based on the gains and losses of potential around selected loops.
5. Use an algebraic technique to solve for the unknown currents. Recall that it is necessary to have as many equations as there are unknowns in order to solve the problem.

For problems involving equivalent capacitance:

1. Determine whether the capacitors are arranged in series or in parallel.
2. If necessary, simplify the circuit step by step until it is reduced to a simple series or parallel combination. Use the appropriate formula to determine the equivalent capacitance.
3. Use $C = Q/V$ and apply Kirchhoff's rules to solve for the charge stored on each capacitor and the potential difference across each capacitor.

For problems involving the construction of an ammeter or voltmeter from a galvanometer and an added resistor.

1. Recall that for an ammeter the shunt resistor is added in parallel while for a voltmeter the additional resistor is added in series.
2. Apply Ohm's law and Kirchhoff's rules and solve for the magnitude of the resistance necessary for the conversion.

PROGRAMMED PROBLEMS

PROBLEM 1. Three resistors rated at 2.0 Ω, 4.0 Ω, and 6.0 Ω are connected in a) series and b) parallel. Draw a diagram and determine the equivalent resistance for each arrangement.

a. 1 Draw a diagram for the series arrangement.	Solution: (Section 19-1) 2.0Ω ——ww—— 4.0Ω ——ww—— 6.0Ω ——ww——
a. 2 Determine the equivalent resistance.	$R = R_1 + R_2 + R_3$ $= 2.0\ \Omega + 4.0\ \Omega + 6.0\ \Omega$ $R = 12\ \Omega$
b. 1 Draw a diagram for the parallel arrangement.	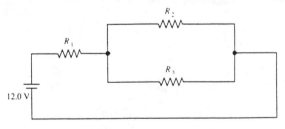
b. 2 Determine the equivalent resistance.	$1/R = 1/R_1 + 1/R_2 + 1/R_3$ $1/R = 1/2.0\Omega + 1/4.0\Omega + 1/6.0\Omega$ $1/R = (6.0 + 3.0 + 2.0)/12\Omega = 11/12\Omega$ $R = 12\Omega/11 = 1.1\ \Omega$

PROBLEM 2. Given $R_1 = 2.0\ \Omega$, $R_2 = 6.0\ \Omega$, and $R_3 = 12.0\ \Omega$. Determine the a) equivalent resistance of the following arrangement, b) total current leaving the battery, c) potential drop across each resistor, and d) current through each resistor.

a. 1 Determine the equivalent resistance of the two resistors in parallel.	Solution: (Section 19-1) $1/R_\| = 1/6.0\ \Omega + 1/12.0\ \Omega = (2 + 1)/12.0\ \Omega$ $R_\| = 12.0\Omega/3$ $R_\| = 4.0\ \Omega$

a. 2 Determine the total resistance.	The parallel combination is in series with the 2.0 Ω resistor. Re-draw the circuit showing the 2.0 Ω and $R_\parallel$ and solve for R_T. $R_T = 2.0\ \Omega + 4.0\ \Omega = 6.0\ \Omega$ The dotted line drawn about the 4.0 Ω resistor indicates that this resistor is equivalent to the parallel combination. 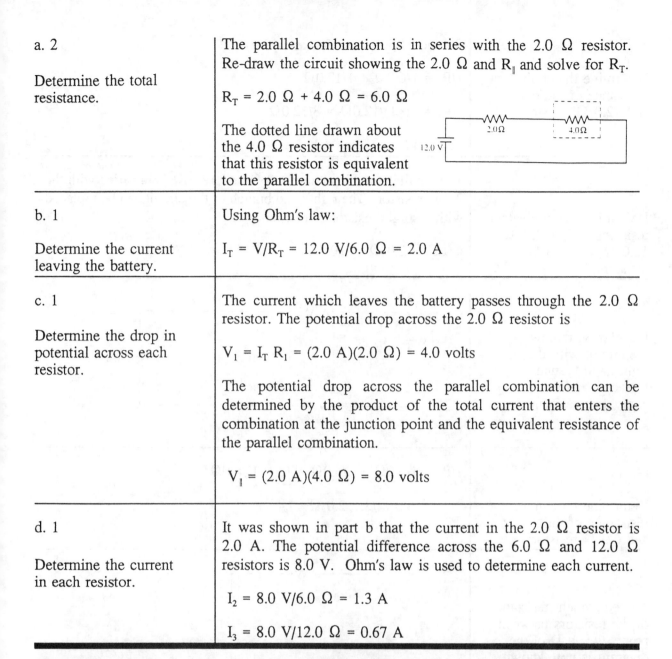
b. 1 Determine the current leaving the battery.	Using Ohm's law: $I_T = V/R_T = 12.0\ \text{V}/6.0\ \Omega = 2.0\ \text{A}$
c. 1 Determine the drop in potential across each resistor.	The current which leaves the battery passes through the 2.0 Ω resistor. The potential drop across the 2.0 Ω resistor is $V_1 = I_T R_1 = (2.0\ \text{A})(2.0\ \Omega) = 4.0\ \text{volts}$ The potential drop across the parallel combination can be determined by the product of the total current that enters the combination at the junction point and the equivalent resistance of the parallel combination. $V_\parallel = (2.0\ \text{A})(4.0\ \Omega) = 8.0\ \text{volts}$
d. 1 Determine the current in each resistor.	It was shown in part b that the current in the 2.0 Ω resistor is 2.0 A. The potential difference across the 6.0 Ω and 12.0 Ω resistors is 8.0 V. Ohm's law is used to determine each current. $I_2 = 8.0\ \text{V}/6.0\ \Omega = 1.3\ \text{A}$ $I_3 = 8.0\ \text{V}/12.0\ \Omega = 0.67\ \text{A}$

PROBLEM 3. A battery of emf 8.0 V and internal resistance r = 1.0 Ω is connected to an external circuit as shown in the diagram shown below. The internal resistance is represented by the resistor r. Determine the a) equivalent resistance of the circuit, b) total current leaving the battery, and c) potential difference between the terminals (A and B) of the battery.

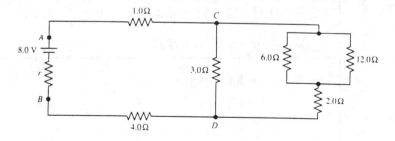

a. 1	Solution: (Sections 19-1 and 19-2)
Determine the equivalent resistance of the 6.0 Ω and 12.0 Ω resistors.	$1/R_\parallel = 1/6.0\Omega + 1/12.0\Omega$ $1/R_\parallel = (2+1)/12.0\Omega = 3/12.0\Omega$ $R_\parallel = 4.0\ \Omega$

---

a. 2

.Determine the equivalent resistance of the 0.6 Ω, 12.0 Ω, and 2.0 Ω resistors.

The parallel combination of 6.0 Ω and 12.0 Ω is in series with the 2.0 Ω resistor. The right hand branch of the circuit can be replaced with a single resistor of resistance

$4.0\ \Omega + 2.0\ \Omega = 6.0\ \Omega$.

---

a. 3

For clarity, redraw the circuit with the right hand branch replaced by the equivalent resistance.

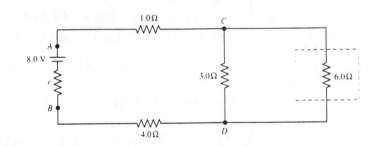

---

a. 4

Determine the equivalent resistance of the 3.0 Ω and 6.0 Ω resistors.

$1/R_\parallel = 1/3.0\Omega + 1/6.0\Omega = (2 + 1)/6.0\Omega$

$R_\parallel = 6.0\ \Omega/3 = 2.0\ \Omega$

---

a. 5

The equivalent resistance of the resistors between points C and D of the diagram is now known to be 2.0 Ω. Redraw the circuit and determine the total resistance in the circuit.

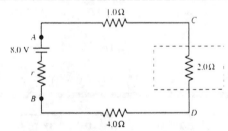

It can be seen that the 2.0 Ω resistor is in series with the rest of the circuit. The equivalent resistance of the entire circuit can now be determined.

$R_{eq} = 1.0\ \Omega + 2.0\ \Omega + 4.0\ \Omega + r$ but $r = 1.0\ \Omega$

therefore $R_{eq} = R_T = 8.0\ \Omega$

---

b. 1

Determine the current leaving the battery.

$I_T = V_T/R_{eq} = 8.0\ V/8.0\ \Omega$

$I_T = 1.0\ A$

c. 1	$V_{AB} = \xi - I_T r$
Determine the potential difference between points A and B.	$= 8.0\ V - (1.0\ A)(1.0\ \Omega) = 8.0\ V - 1.0\ V$
	$V_{AB} = 7.0\ V$

PROBLEM 4. The circuit shown below was used to clarify the use of Kirchhoff's rules in the chapter summary. Review the suggestions made for solving this type of complex circuit and then determine the current in each branch if

$\xi_1 = 4.0\ V,\ \xi_2 = 2.0\ V,\ R_1 = 1.0\ \Omega,\ R_2 = 2.0\ \Omega,\ and\ R_3 = 3.0\ \Omega$

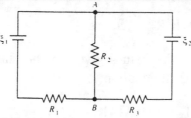

a. 1	Solution: (Sections 19-3 and 19-4)
Apply the junction rule at point A and write an equation for the current.	at point A: $I_1 + I_3 = I_2$ 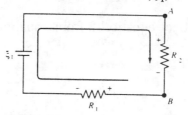
	rearranging gives
	$I_1 - I_2 + I_3 = 0$ (eqn. 1)

a. 2	Note that at each resistor there will be a potential drop (+ to -) while at the source of emf (ξ_1) there will be a gain.
Note the direction of the current through each resistor. Place + and - signs on the appropriate sides of each resistor. Apply the loop rule. Start at point B and proceed clockwise around the left loop.	
	$- I_1 R_1 + \xi_1 - I_2 R_2 = 0$ rearranging and simplifying gives
	$+ I_1 R_1 + I_2 R_2 = \xi_1$
	Substitute the values given in the statement of the problem and arrange the resulting equation in the form of equation 1.
	$(1.0\ \Omega)I_1 + (2.0\ \Omega)I_2 = 4.0\ V$
	$1.0\ I_1 + 2.0\ I_2 = 4.0\ A$ (equation 2)

a. 3

Again apply the loop rule. Start at point B and proceed clockwise around the right loop.

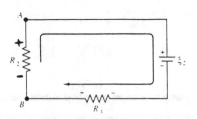

By proceeding clockwise there will be a potential gain (- to +) at each resistor while at the source of emf (ξ_2) there will be a potential drop (+ to -).

$+ I_2 R_2 - \xi_2 + I_3 R_3 = 0$ and rearranging gives

$+ I_2 R_2 + I_3 R_3 = \xi_2$

Substitute the values given in the statement of the problem and arrange as in equation 1.

$(2.0\ \Omega)\ I_2 + (3.0\ \Omega)\ I_3 = 2.0\ V$

$2.0\ I_2 + 3.0\ I_3 = 2.0\ A$ (equation 3)

---

a. 4

Solve the three equations using algebra.

The following method is known as substitution:

$I_1 - I_2 + I_3 \qquad = 0$ (equation 1)

$1.0\ I_1 + 2.0\ I_2 \qquad = 4.0$ (equation 2)

$\qquad 2.0\ I_2 + 3.0\ I_3 = 2.0$ (equation 3)

Subtract equation 2 from equation 1 in order to eliminate I_1.

$I_1 - \qquad I_2 + I_3 \qquad = 0$ (equation 1)

$1.0\ I_1 + \qquad 2.0\ I_2 \qquad = 4.0$ (equation 2)

$\overline{\qquad -3.0\ I_2 + 1.0\ I_3 \qquad = -4.0}$ (equation 4)

Multiply both sides of equation 4 by 3 and then subtract equation 3 from equation 4.

$-9.0\ I_2 + 3.0\ I_3 \qquad = -12.0$ (equation 4)

$\ 2.0\ I_2 + 3.0\ I_3 \qquad = \ 2.0$ (equation 3)

$\overline{-11.0\ I_2 \qquad\qquad = -14.0}$

$I_2 = (-14.0)/(-11.0) = 1.3\ A$

Substitute $I_2 = 1.3\ A$ into eqn. 3 and determine I_3.

$2.0\ I_2 + 3.0\ I_3 = 2.0$ (eqn. 3)

$(2.0)(1.3) + 3.0\ I_3 = 2.0$

$I_3 = -0.20$ A

The negative sign indicates that the actual direction of the current is opposite from the direction initially assumed. In this instance, the direction of current I_3 is away from point A.

Substitute the values of I_2 and I_3 into equation 1 to solve for I_1.

$I_1 - I_2 + I_3 = 0$ (equation 1)

$I_1 - 1.3$ A $+ -0.20$ A $= 0$

$I_1 = 1.5$ A

PROBLEM 5. Three capacitors are rated at 10 µF, 20 µF, and 30 µF, respectively. Determine the equivalent capacitance of the capacitors, the charge stored on each capacitor, and the potential difference across each capacitor if they are arranged in a) series and b) parallel and each arrangement is attached to a 120 volt source.

a. 1

Draw a diagram for the series arrangement.

Solution: (Section 19-6)

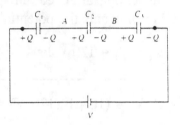

---

a. 2

Determine the equivalent capacitance of a series arrangement.

$1/C = 1/C_1 + 1/C_2 + 1/C_3$

$1/C = 1/10\mu F + 1/20\mu F + 1/30\mu F$

$1/C = (6 + 3 + 2)/60\ \mu F = 11/60\ \mu F$

$C = 60/11\ \mu F$

$C = 5.5\ \mu F$

---

a. 3

Determine the charge on each capacitor.

In a series arrangement, the charge across each capacitor equals the total charge which leaves the battery. The total charge can be determined as follows:

$Q = C\ V = (5.5\ \mu F)(120\ V) = 660\ \mu C = 6.6 \times 10^{-4}$ coul

a. 4	$V_1 = Q_1/C_1 = 660\ \mu C/10\ \mu F = 66$ V
Determine the potential difference across each capacitor.	$V_2 = Q_2/C_2 = 660\ \mu C/20\ \mu F = 33$ V
	$V_3 = Q_3/C_3 = 660\ \mu C/30\ \mu F = 22$ V

b. 1

Draw a diagram for a parallel arrangement.

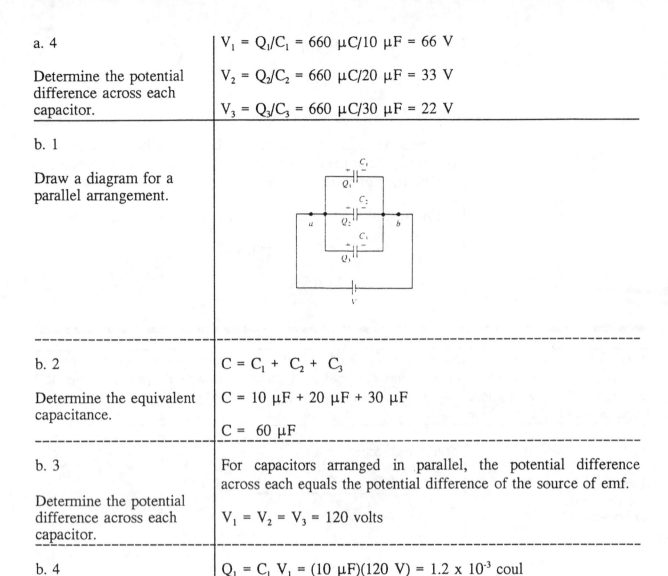

b. 2

Determine the equivalent capacitance.

$$C = C_1 + C_2 + C_3$$

$$C = 10\ \mu F + 20\ \mu F + 30\ \mu F$$

$$C = 60\ \mu F$$

b. 3

Determine the potential difference across each capacitor.

For capacitors arranged in parallel, the potential difference across each equals the potential difference of the source of emf.

$$V_1 = V_2 = V_3 = 120 \text{ volts}$$

b. 4

Determine the charge on each capacitor.

$$Q_1 = C_1 V_1 = (10\ \mu F)(120\ V) = 1.2 \times 10^{-3} \text{ coul}$$

$$Q_2 = C_2 V_2 = (20\ \mu F)(120\ V) = 2.4 \times 10^{-3} \text{ coul}$$

$$Q_3 = C_3 V_3 = (30\ \mu F)(120\ V) = 3.6 \times 10^{-3} \text{ coul}$$

PROBLEM 6. Three capacitors are arranged as shown in the diagram. Determine the a) equivalent capacitance of the combination, b) potential difference across each capacitor, and c) charge on each capacitor.

$C_1 = 10\ \mu F$

$C_2 = 20\ \mu F$

$C_3 = 30\ \mu F$

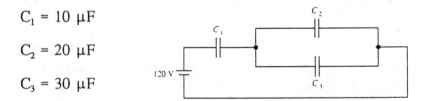

a. 1 Determine the equivalent capacitance of the two capacitors arranged in parallel.	Solution: (Section 19-6) $C_{\parallel} = C_2 + C_3$ $C_{\parallel} = 20\ \mu F + 30\ \mu F = 50\ \mu F$
a. 2 Redraw the diagram and solve for the total capacitance.	C_1 and $C_{\parallel}$ in series; therefore $1/C = 1/C_1 + 1/C_{\parallel}$ $1/C = 1/10\mu F + 1/50\mu F$ $1/C = (5 + 1)/50\mu F$ $C = 8.3\ \mu F$
b. 1 Determine the total charge that leaves the source of emf and the charge that forms on C_1.	$Q = C\ V = (8.3\ \mu F)(120\ V) = 1000\ \mu C$ $Q = 1.0 \times 10^{-3}\ C$ C_1 is in series with the source of emf. The charge that forms on C_1 equals the total charge that left the source of emf, $Q_1 = 1000\ \mu C$.
b. 2 Determine the potential difference across C_1.	$V_1 = Q_1/C_1$ $V_1 = 1000\ \mu C/10\ \mu F = 100\ V$
b. 3 Determine the potential difference across the two capacitors arranged in parallel.	The potential difference across the capacitors in parallel can be determined by applying Kirchhoff's loop rule. Starting at the source of emf and following the loop through C_1 and C_2: $120\ V - V_1 - V_2 = 0$ but $V_1 = 100$ volts. Therefore, $V_2 = 20\ V$ The potential difference is the same across capacitors in parallel; therefore, $V_3 = 20\ V$.
c. 1 Determine the charge stored on each of the capacitors in the parallel combination.	The charge stored on C_1 was determined to be 1000 C. The charge stored on C_2 and C_3 can be determined as follows: $Q_2 = C_2\ V_2 = (20\ \mu F)(20\ V) = 400\ \mu C$ $Q_3 = C_3\ V_3 = (30\ \mu F)(20\ V) = 600\ \mu C$ Note that the total charge stored on the parallel combination equals 1000 μC and is equal to the amount stored on C_1.

PROBLEM 7. A 5.00 ma current causes a galvanometer of internal resistance 10.0 ohms to deflect full scale. Determine the resistance that must be added in order to convert the galvanometer into a) a voltmeter which measures potential differences from 0 to 50.0 volts and b) an ammeter which measures currents from 0 to 1.00 A.

a. 1 Draw a diagram indicating the location of the resistance which must be added to convert the galvanometer into a voltmeter.	Solution: (Section 19-10) In order to convert the galvanometer into a voltmeter, a resistance R must be placed in series with the galvanometer.

In order to convert the galvanometer into a voltmeter, a resistance R must be placed in series with the galvanometer.

a. 2

Determine the magnitude of the resistance which must be added in order to convert the galvanometer into a voltmeter.

The voltage drop across the circuit must be 50.0 volts when a current of 5.00 ma, i.e., 0.00500 A, passes through the galvanometer.

$$V = I (r + R)$$

$$50.0 \ V = (5.00 \times 10^{-3} \ A)(10.0 \ \Omega + R)$$

$$10.0 \ \Omega + R = (50.0 \ V)/(5.00 \times 10^{-3} \ A)$$

$$10.0 \ \Omega + R = 10,000 \ \Omega$$

$$R = 10,000 \ \Omega - 10.0 \ \Omega$$

$$R = 9,990 \ \Omega$$

b. 1

Draw a diagram showing the resistance which must be added to convert the galvanometer into an ammeter.

In order to convert the galvanometer into an ammeter which reads from 0 to 1.00 A, it is necessary to add a shunt resistor (R_s) in parallel with the galvanometer.

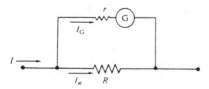

b. 2

Determine the magnitude of the current which passes through the shunt.

Only 5.00 ma may pass through the galvanometer when it reads full scale deflection. Therefore, if the ammeter is to measure 1.00 A when it is fully deflected, the shunt current is

$$I_{shunt} = 1.00 \ A - 0.0050 \ A = 0.995 \ A$$

b. 3	The galvanometer and shunt resistance are in parallel. The potential difference across each will be equal, i.e., $V_{galv} = V_{shunt}$.
Determine the voltage drop across the galvanometer and across the shunt.	$V_{galv} = I_{galv}\ r = (0.0050\ A)(10.0\ \Omega)$
	$V_{galv} = 0.050\ V$
	but $V_{shunt} = V_{galv} = 0.050\ V$

b. 4	$V_{shunt} = I_{shunt}\ R_{shunt}$
Determine the shunt resistance.	$0.050\ V = (0.995\ A)\ R_{shunt}$
	$R_{shunt} = 0.050\ \Omega$

PROBLEM 8. The emf of the standard cell used in the potentiometer circuit shown on page 19-9 is 1.50 V. Determine the emf of the unknown if the galvanometer current is zero and L_{AC} is 50 cm when the standard cell is in the circuit and 30 cm when the unknown is in the circuit.

a. 1	Solution: (Section 19-10)
Determine the emf of the unknown cell.	The emf of the unknown can be determined by using the following equation:
	$\xi_x/\xi_s = L_{AC}/L_{AC}$ and substituting values gives
	$\xi_x/1.50\ V = 30\ cm/50\ cm$
	$\xi_x = 0.90\ V$

PROBLEM 9. The Wheatstone bridge circuit shown on page 19-10 is used to measure the resistance of an unknown resistor. If $R_1 = 10\ \Omega$, $R_2 = 20\ \Omega$, and the galvanometer current is zero when R_3 is adjusted to 30 ohms, determine the value of R_x.

a. 1	Solution: (Section 19-10)
Determine the resistance of the unknown resistor.	The resistance of the unknown can be determined by using the following equation:
	$R_x/R_3 = R_2/R_1$; therefore, $R_x/30\Omega = 20\Omega/10\Omega$.
	$R_x = 60\ \Omega$

PROBLEM 10. The slide wire Wheatstone bridge circuit shown on page 19-10 is used to determine the resistance of an unknown resistor. The galvanometer current is zero when R_3 = 30 Ω, L_{DC} = 40 cm, and L_{AD} = 60 cm. Determine the value of R_x.

a. 1

Determine the magnitude of the unknown resistance.

Solution: (Section 19-10)

The resistance of the unknown can be determined as follows:

$R_x/R_3 = L_{DC}/L_{AD}$ and substituting,

$R_x/30\ \Omega$ = 40 cm/60 cm

R_x = 20 Ω

PRACTICE PROBLEMS

PROBLEM 1. A 2.0 amp current flows through the 6.0 Ω resistor shown in the diagram below. Determine the a) current in the 2.0 Ω and 12.0 Ω resistors and b) potential difference across the battery.

ANS. a) 3.0 A and 1.0 A,   b) 18 V

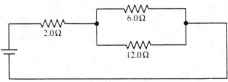

PROBLEM 2. Determine the equivalent resistance of the following arrangement of resistors.

ANS. 4.0 Ω

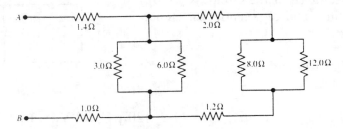

PROBLEM 3. Solve for the current in each branch of the circuit shown below. Note: $\xi_1 = 10$ V, $R_1 = 10$ Ω, $\xi_2 = 20$ V, $\xi_3 = 30$ V, $R_2 = 20$ Ω, and $R_3 = 30$ Ω.

Ans. $I_1 = -0.64$ A

   $I_2 = 0.18$ A

   $I_3 = 0.45$ A

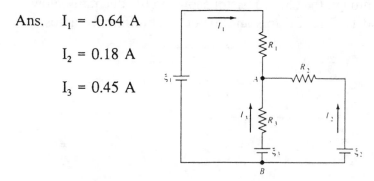

PROBLEM 4. For the circuit shown at right, determine the a) equivalent capacitance of the circuit (C_{eq}), b) charge on each capacitor, and c) potential difference across each capacitor if

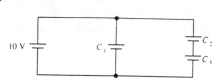

$C_1 = 8.0$ μF, $C_2 = 20.0$ μF, and $C_3 = 30.0$ μF.

ANS. a) $C_{eq} = 20$ μF,   b) $Q_1 = 80$ μC, $Q_2 = Q_3 = 120$ μC,   c) $V_1 = 10$ V, $V_2 = 6$ V, $V_3 = 4$ V

CHAPTER 20

MAGNETISM

KEY TERMS AND PHRASES

north pole or "north seeking pole" of a bar magnet tends to align with the Earth's magnetic field and point toward magnetic north. The south pole of a bar magnet tends to point toward magnetic south.

magnetic field surrounds every magnet and is also produced by a charged particles in motion relative to some reference point. The direction of the field at any point is indicated by the north pole of a compass needle placed at that point. The SI unit magnetic field strength is the tesla (T).

right-hand rule is used to predict the direction of the magnetic field produced by a current-carrying wire. The thumb of the right hand points in the direction of the conventional current in the wire. The fingers encircle the wire in the direction of the magnetic field.

magnetic force acts on a charged particle traveling through a magnetic field. The magnetic force always acts perpendicular to the direction of the magnetic field and the velocity vector. A second right-hand rule is used to predict the direction of the force on the wire.

velocity selector allows only charged particles which have a particular velocity to pass undeflected. The velocity is the same regardless of the magnitude of the charge or the mass of the particle.

mass spectrograph uses charged particles traveling through magnetic fields to determine the relative mass of the particle.

galvanometer movement is the basis of most meters, i.e., ammeters, voltmeters, and ohmeters, as well as electric motors. Galvanometer movement is the result of interaction of the magnetic field of a permanent magnet which is directed perpendicular to a current carried by a loop or coil of wire.

ferromagnetic materials, such as iron, can be permanently magnetized. Each atom has a net magnetic effect and the atoms tend to align their magnetic fields in arrangements known as **domains**. Each domain contributes to the overall magnetic field of the piece of iron. In an ordinary piece of iron or other ferromagnetic material, the magnetic fields produced by the individual domains cancel out so that the object is not a magnet. In a magnet, the domains are larger in one direction than in any other and a net magnetic effect is produced.

Curie temperature refers to the temperature where it is no longer possible to magnetize an object and a permanent magnet loses its magnetic effect.

SUMMARY OF MATHEMATICAL FORMULAS

magnetic field		
long, straight wire	$B = (\mu_o I)/(2 \pi r)$	The magnitude of the magnetic field strength (B) a perpendicular distance (r) from a long, straight wire carrying a current (I). The SI unit of B is the tesla (T) and $\mu_o = 4\pi \times 10^{-7}$ T/A.
loop of wire	$B = \mu_o I/2r$	the magnitude of the magnetic field strength (B) at the center of a loop of wire of radius (r) which carries a current (I).
force on a current carrying wire in a magnetic field	$F = I \ell B \sin \theta$	The force (F) on a wire carrying a current depends on the magnitude of the current (I), the length (ℓ), magnetic field strength (B), and the angle (θ) between the directions of the current in the wire and the magnetic field.
force on a charged particle traveling through a magnetic field	$F = q v B \sin \theta$	The force (F) on a charged particle traveling through a magnetic field depends on the magnitude of the charge (q), the velocity of the charge (v), the magnetic field strength (B), and the angle (θ) between the direction of motion of the particle and the direction of the magnetic field.
radius of the circular path followed by an ion in a mass spectrograph	$r = (m E)/(q B B')$	The radius of the path depends on the mass of the particle (m), the electric field strength (E) and the magnetic field strength (B) in the velocity selector, and the magnetic field strength (B') which causes the circular motion.

galvanometer movement	$\tau = N I A B$	If the coil of the galvanometer is pivoted in the center, a torque is produced which depends on the number of turns of wire in the coil (N), the current (I), the area of the coil (A), and the magnetic field strength (B). The quantity N I A is the magnetic moment of the coil.
force between two parallel conducting wire.	$F/\ell = (\mu_o I_1 I_2)/(2 \pi L)$	Two parallel, current-carrying conductors produce magnetic fields which result in a force between the conductors. The force per unit length (F/ℓ) depends on the product of the currents (I_1) and (I_2) and is inversely related to the separation between the wires (L).

CONCEPT SUMMARY

Magnets and Magnetic Fields

Two **bar magnets** exert a force on one another. If two **north poles** (or **south poles**) are brought near, a repulsive force is produced. If a north pole and a south pole are brought near, then a force of attraction results. Thus, "Like poles repel, unlike poles attract."

The concept of a field is applied to magnetism as well as gravity and electricity. A **magnetic field** surrounds every magnet and is also produced by a charged particle in motion relative to some reference point. The presence of the magnetic field about a bar magnet can be seen by placing a piece of paper over the bar magnet and sprinkling the paper with iron filings. The magnetic field produced by certain arrangements of bar magnets are represented in the diagrams shown below.

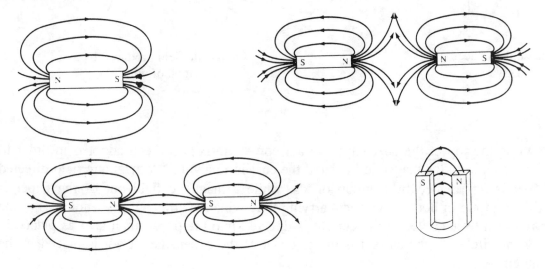

Electric Currents Produce Magnetism

A wire carrying a current (I) produces a magnetic field. The magnitude of the magnetic field strength (B) a perpendicular distance r from a **long, straight wire** is given by

$$B = (\mu_o I)/(2 \pi r)$$ B is measured in teslas (T) and $\mu_o = 4\pi \times 10^{-7}$ T/A, where μ_o is known as the permeability of free space

The direction of the magnetic field produced by a current-carrying wire can be predicted by using the **right-hand rule**. The thumb of the right hand points in the direction of the conventional current in the wire. The fingers encircle the wire in the direction of the magnetic field.

The magnitude of the strength of the magnetic field (B) at the center of a **loop of wire** of radius r which carries a current I is

$$B = \mu_o I/2r$$

The direction of the magnetic field at the center of the loop can again be predicted by using the right-hand rule. The thumb is placed tangent to a point on the loop and is directed in the same direction as the current in the loop at that point. The fingers encircle the wire in the same direction as the magnetic field.

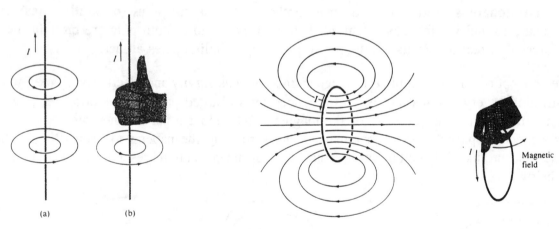

| magnetic field produced by a straight wire | magnetic field produced by a loop of wire |

Conventions

As shown at the top of the next page, certain **conventions** have been adopted in order to represent the direction of the magnetic field and the current in a wire. A magnetic field directed into the paper is represented by a group of x's, while a magnetic field out of the paper is represented by a group of dots. A current-carrying wire which is arranged perpendicular to the page is represented by a circle. If the current is directed into the paper, then an x is placed in the center of the circle. If the current is directed out of the paper, then a dot is placed in the center of the circle.

x x x x x	. . . . .		
x x x x x	. . . . .	⊙	⊗
x x x x x	. . . . .		
B field into paper	B field out of paper	wire with current out of paper	wire with current out of paper

Force on a Current-Carrying Wire in a Magnetic Field

A current (I) in a wire consists of moving electrical charges and a force (F) may be produced when a current-carrying wire of length ℓ is placed in a magnetic field. The magnitude of the force is given by the equation

$$F = I \, \ell \, B \sin \theta$$

where B is the **magnetic field strength** in teslas (T). Other units for magnetic field strength include newtons per ampere meter (N/A m), newtons per coulomb meters per second (N/C m/s), webers per square meter (wb/m^2), and gauss (G).

$$1 \text{ T} = 1 \text{ N/A m} = 1 \text{ N/C m/s} = 1 \text{ wb/m}^2 = 10^4 \text{ G}$$

θ is the angle between the directions of the current in the wire and the magnetic field. The force on the wire is zero if $\theta = 0°$ (sin 0° = 0) and is a maximum if $\theta = 90°$ (sin 90° = 1.0).

As described in the text and shown in the diagram, a second **right-hand rule** is used to predict the direction of the force on the wire: "First you orient your right hand so that the outstretched fingers point in the direction of the (conventional) current; from this position when you bend your fingers they should then point in the direction of the magnetic field lines; if they do not, rotate your hand and arm about the wrist until they do, remembering that your straightened fingers must point in the direction of the current. When your hand is oriented in this way, then the extended thumb points in the direction of the force on the wire."

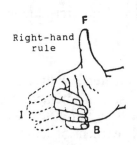

Force on a Charged Particle Moving in a Magnetic Field

An electrically charged particle (q) moving through a magnetic field (B) at speed v may be acted upon by a force (F). The magnitude of the force (F) on the particle is

$$F = q \, v \, B \sin \theta$$

where θ is the angle between the direction of motion of the particle and the direction of the magnetic field. If $\theta = 0°$, then the particle is traveling parallel to the field and no force exists on the particle (sin 0° = 0). If $\theta = 90°$, then sin 90° = 1, the particle is traveling perpendicular to the magnetic field, and the force is a maximum.

The direction of the force can be predicted by again using the right-hand rule: the outstretched fingers of your right hand point along the direction of motion of the positively charged particle (v) and when you bend your fingers they must point along the direction of B; then your thumb points in the direction of the force. If the particle is charged negatively, then the force is directed opposite from the direction of the thumb.

The Mass Spectrograph

The velocity selector (see chapter 27) allows only charged particles which have a particular velocity to pass undeflected regardless of the magnitude of their charge or their mass. Because of this it can be used along with a second magnetic field (B') to determine the mass of an ion. As shown in the diagram, the particle passes through the velocity selector into magnetic field (B') arranged perpendicular to its path. The mass of the particle is given by

$$m = (q\ B'r)/v \quad \text{and since} \quad v = E/B, \quad \text{then} \quad m = (q\ B\ B'\ r)/E$$

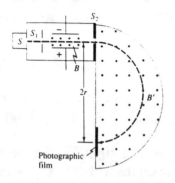

Ions of different elements will not follow the same path because they differ in mass. Even if a pure substance is used, particles having different mass are found. This is because elements contain isotopes, i.e., particles which have the same chemical and physical properties but which have a different number of neutrons in the nucleus. Two isotopes of the same element do not have the same mass; thus their ions will follow paths of different radius.

Galvanometer Movement

Galvanometer movement is the basis of most meters, i.e., ammeters, voltmeters, and ohmeters, as well as electric motors. Galvanometer movement is the result of interaction of the magnetic field of a permanent magnet which is directed perpendicular to a current carried by a loop or coil of wire. Based on the diagram, the force on each vertical segment of length a is

$$F = N\ I\ a\ B \sin 90°$$

where N refers to the number of turns of wire in the coil. If the coil is pivoted in the center, a torque is produced which equals

$$\tau = N\ I\ a\ b\ B$$

where b is the distance between the vertical segments. Since (a)(b) equals the area (A) of the coil, then

$$\tau = N\ I\ A\ B$$

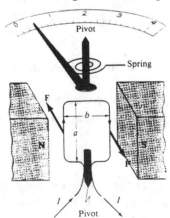

and the quantity N I A is the **magnetic moment** of the coil.

Force between Two Parallel Conductors

Two parallel, current-carrying conductors produce magnetic fields which result in a force between the conductors. If the currents are in the same direction, the force is of one attraction. A repulsive force results if the currents are in opposite directions. The force per unit length (F/ℓ) exerted by conductor 1 on conductor 2 and vice versa is given by

$F/\ell = (\mu_o I_1 I_2)/(2 \pi L)$ where L is the separation between the wires in meters and I_1 and I_2 represent the magnitude of the current in each conductor

Ferromagnetism; Domains

If an object is made from a **ferromagnetic material**, such as iron, each atom has a net magnetic effect. The atoms tend to align their magnetic fields in arrangements known as **domains** and each domain contributes to the overall magnetic field of the piece of iron. In an ordinary piece of iron or other ferromagnetic material, the magnetic field produced by the individual domains cancel out so that the object is not a magnet. In a magnet, the domains are larger in one direction than in any other and a net magnetic effect is produced.

If an unmagnetized ferromagnetic object is placed in a magnetic field, the domains which are aligned with the direction of the external magnetic field tend to grow. The increase in the size of the domains comes at the expense of neighboring domains which are not aligned with the external field. The result is that the unmagnetized object is attracted to the external magnetic field and the object is said to exhibit **ferromagnetism**.

When the object is removed from the external field the domains may remain aligned and the object retains a net magnetic effect. The object may lose this net magnetic effect if it is struck, dropped, or heated. Above a certain temperature, known as the **Curie temperature**, it is not possible to magnetize an object and a permanent magnet loses its magnetic effect.

ANSWERS TO SELECTED TEXT QUESTIONS

QUESTION 5. What kind of field or fields surround a moving electric charge?

ANSWER: A moving electric charge produces a magnetic field. An electric charge, whether moving or not, has an electric field in the region surrounding the charge. However, the electric charge has mass and must be surrounded by a gravitational field. Therefore, an electric field, a magnetic field, and a gravitational field all surround a moving electric charge.

QUESTION 9. Suppose you have three iron rods, two of which are magnets but the third is not. How would you determine which two are the magnets without using any additional objects?

ANSWER: The two north poles of the bar magnets will repel as will the two south poles. The two rods which exhibit repulsion when the ends are oriented one way but attraction when the orientation is reversed must be the bar magnets. Either end of the unmagnetized iron rod is attracted to north pole or a south pole. The unmagnetized rod will never be repelled by either end of a bar magnet.

QUESTION 23. Charged cosmic ray particles from outside the Earth tend to strike the Earth more frequently near the poles than at lower altitudes. Explain.

ANSWER: Figure 20-6 in the text shows that the Earth's magnetic field is similar to the magnetic field produced by a bar magnet (see text, figure 20-3b). Cosmic rays approaching the Earth's equatorial regions tend to enter at a large angle to the magnetic field and be deflected ($F = q v B \sin \theta$). At the polar regions they approach at a small angle to the magnetic field and there is little deflection. This heavy bombardment at the polar regions gives rise to the Aurora Borealis in the northern hemisphere and Australis Borealis in the southern hemisphere.

PROBLEM SOLVING SKILLS

For problems involving a current-carrying wire in a magnetic field:

1. Draw an accurate diagram showing the orientation of the wire in the magnetic field. Use the adopted conventions to indicate the direction of the current and the magnetic field.
2. Use the right-hand rule to determine the direction of the force on the wire.
3. Use $F = I \ell B \sin \theta$ to determine the magnitude of the force.

For problems involving a charged particle traveling through a magnetic field:

1. Draw an accurate diagram showing the motion of the particle through the magnetic field. Use the adopted conventions to indicate the direction of the magnetic field and the direction of motion of the particle.
2. Take note of whether the particle is positively charged or negatively charged and then use the right-hand rule to determine the direction of the force on the particle.
3. If the particle is deflected into circular motion, the magnetic force produces a centripetal acceleration. Use Newton's second law to determine the magnitude of the acceleration.
4. If the speed of the particle as it enters the magnetic field is given or can be calculated, it is possible to determine the radius of circle in which the particle travels.

For problems involving the magnetic field produced by a current-carrying wire:

1. Draw a diagram showing the orientation of the straight wire or the loop of wire and the direction of the current in the wire.
2. Use the second right-hand rule to determine the direction of the magnetic field. Use the adopted conventions to indicate the direction of the magnetic field.
3. Use the appropriate formula to determine the magnitude of the magnetic field.

PROGRAMMED PROBLEMS

PROBLEM 1. A wire carrying 4.0 A of current is placed in a uniform magnetic field of strength 2.0 T as shown in the diagram. a) Use the right-hand rule to determine the direction of the force on the wire. b) Determine the magnitude of the force on a 0.010 meter section of the wire.

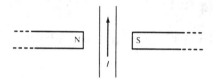

a. 1	Solution: (Section 20-3)
Use the right-hand rule to determine the direction of the force.	Based on the diagram, the current is directed toward the top of the page. Using the right-hand rule, your fingers point in the direction of the current. Your hand is then arranged so that when you bend your fingers, your fingers bend toward the direction of B, which is towards the right side of the page. Your thumb is directed into the paper, which means that the force is directed into the page.
b. 1 Determine the magnitude of the force.	The current is at right angles to the direction of the magnetic field, $\theta = 90°$. Also, 1.0 T = 1.0 N/A m. $$F = I \ell B \sin \theta$$ $$= (4.0 \text{ A})(0.010 \text{ m})(2.0 \text{ T})(\sin 90°)$$ $$F = 0.080 \text{ N}$$

PROBLEM 2. An alpha particle has mass 6.68×10^{-27} kg and charge 3.2×10^{-19} C. The particle is traveling at 3.0×10^6 m/s and enters a magnetic field of magnitude 0.10 T. The magnetic field is directed at right angles to the direction of motion of the particle. a) Determine the radius of the circle in which the particle travels. b) Based on the diagram, determine the direction of motion in which the particle travels, i.e., either clockwise or counterclockwise as viewed from above.

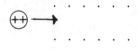

a. 1	Solution: (Section 20-4)
Determine the radius of the circle in which the travels.	The alpha particle is deflected by a magnetic force and travels in a circular path. The magnetic force provides the centripetal particle acceleration (a_c), thus $$\text{net } F = m a_c \quad \text{but} \quad a_c = v^2/r$$

$F_{magnetic} = m\, v^2/r$

$q\, v\, B \sin \theta = m\, v^2/r$

but $\theta = 90°$ and $\sin 90° = 1$ Solving for r gives

$r = (m\, v)/(q\, B)$

$= [(6.68 \times 10^{-27}\ kg)(3.0 \times 10^{6}\ m/s)]/[(3.2 \times 10^{-19}\ C)(0.10\ T)]$

$r = 0.63$ m

---

b. 1

Use the right-hand rule to determine the direction of motion of the particle.

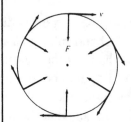

The fingers point toward the right side of the page. Your hand is then arranged so that when you bend your fingers they point out of the paper in the direction of B. Your thumb points toward the bottom of the page. At the point where it enters the magnetic field, the particle is deflected downward. Applying the right-hand rule at several more points indicates that the particle travels in a clockwise circle as viewed from above.

PROBLEM 3. An electron is accelerated through a potential difference of 30.0 volts into a magnetic field of 5.49×10^{-4} T. Determine the a) velocity of the particle as it enters the field, b) radius of the circle in which it travels, and c) period of its motion. d) Based on the diagram shown at the right, determine the direction of motion of the electron in the magnetic field, i.e., either clockwise or counterclockwise as viewed from above. Assume that the initial velocity of the electron is zero. Note: $m_e = 9.1 \times 10^{-31}$ kg and $q_e = 1.6 \times 10^{-19}$ C.

x x x x x

$\Longleftarrow\ominus$

x x x x x

x x x x x

a. 1

Determine the electron's velocity as it enters the magnetic field.

Solution: (Section 20-4)

The electron is accelerated through a potential difference and gains kinetic energy. The work done on the particle is equal to the product of the charge on the particle and the potential difference.

$W = q\, V = \tfrac{1}{2} m\, v^2 - \tfrac{1}{2} m\, v_o^2$

but $v_o = 0$ m/s; thus

$q\, V = \tfrac{1}{2} m\, v^2$ and rearranging gives

$v = (2\, q\, V/m)^{1/2}$

	$v = [2\ (1.6 \times 10^{-19}\ \text{C})(30.0\ \text{V})/(9.1 \times 10^{-31}\ \text{kg})]^{\frac{1}{2}}$
	$v = 3.25 \times 10^{6}\ \text{m/s}$
b. 1 Determine the radius of the circle in which the electron travels.	The magnetic force provides the centripetal acceleration and causes the electron's to follow a circular path. $F_{\text{magnetic}} = m\ a_{\text{centripetal}}$ $q\ v\ B\ \sin\theta = m\ v^2/r$ but $\theta = 90°$ and $\sin 90° = 1$ Solving for r gives $r = (m\ v)/(q\ B)$ $\quad = [(9.1 \times 10^{-31}\ \text{kg})(3.25 \times 10^{6}\ \text{m/s})]/[(1.6 \times 10^{-19}\ \text{C})(5.49 \times 10^{-4}\ \text{T})]$ $r = 3.36 \times 10^{-2}\ \text{m}$
c. 1 Determine the period of the motion. Hint: the period (T) of the motion refers to the time required for the electron to complete one revolution.	The distance traveled in one revolution equals the circumference $(2\ \pi\ r)$ of the circle. The period can be determined by dividing the circumference of the circle by the velocity (v). $T = (2\ \pi\ r)/v$ $\quad = [2\ \pi\ (3.36 \times 10^{-2}\ \text{m})]/(3.25 \times 10^{6}\ \text{m/s})$ $T = 6.50 \times 10^{-8}\ \text{s}$
d. 1 Determine the direction of motion of the particle. Note: remember that an electron carries a negative charge.	Based on the diagram, your fingers point toward the left of the page. Your hand is then arranged so that when you bend your fingers they point into the paper in the direction of B. Your thumb points downward. However, since the particle carries a negative charge, the force is in the opposite direction. Therefore, the force on the particle is upward. Arbitrarily selecting a few more points along the path shows that the electron will travel in a clockwise circle as viewed from above.

PROBLEM 4. Determine the magnetic field strength a) at a point 1.0 meter from a long straight wire which carries a 3.0 A current and b) at the center of a loop of wire 0.10 m in radius that contains 100 turns and carries a 5.0 A current. c) Use the right-hand rule and the conventions discussed in the chapter summary to indicate the direction of the magnetic field produced in each of the situations shown.

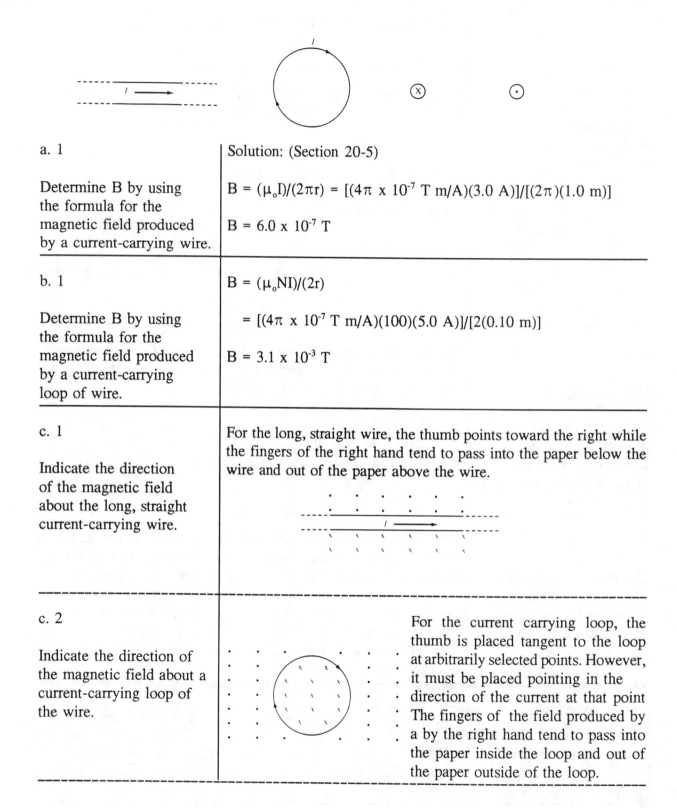

a. 1	Solution: (Section 20-5)
Determine B by using the formula for the magnetic field produced by a current-carrying wire.	$B = (\mu_o I)/(2\pi r) = [(4\pi \times 10^{-7}$ T m/A$)(3.0$ A$)]/[(2\pi)(1.0$ m$)]$ $B = 6.0 \times 10^{-7}$ T
b. 1 Determine B by using the formula for the magnetic field produced by a current-carrying loop of wire.	$B = (\mu_o NI)/(2r)$ $= [(4\pi \times 10^{-7}$ T m/A$)(100)(5.0$ A$)]/[2(0.10$ m$)]$ $B = 3.1 \times 10^{-3}$ T

c. 1

Indicate the direction of the magnetic field about the long, straight current-carrying wire.

For the long, straight wire, the thumb points toward the right while the fingers of the right hand tend to pass into the paper below the wire and out of the paper above the wire.

c. 2

Indicate the direction of the magnetic field about a current-carrying loop of the wire.

For the current carrying loop, the thumb is placed tangent to the loop at arbitrarily selected points. However, it must be placed pointing in the direction of the current at that point The fingers of the field produced by a by the right hand tend to pass into the paper inside the loop and out of the paper outside of the loop.

c. 3 Indicate the direction of the magnetic field produced by a long, straight, current-carrying wire if the current is directed into the paper.	The thumb of the right hand is directed perpendicular to the plane of the paper and into the paper. The fingers encircle the wire in a clockwise direction.
c. 4 Indicate the direction of the magnetic field produced by a long, straight, current-carrying wire if the current is directed out of the paper.	The thumb of the right hand is directed perpendicular to the plane of the paper and out of the paper. The fingers encircle the wire in a counterclockwise direction.

PROBLEM 5. Two long parallel wires carry currents of 2.0 A and 5.0 A, respectively. The wires are 0.20 m apart and the currents flow in the same direction. Determine the a) magnitude and direction of the force on a 0.50 m segment of the wire carrying the 2.0 A current and b) force on a 0.50 m segment of the wire carrying the 5.0 A current.

a. 1 Determine the magnitude of the magnetic field acting on the wire carrying the 2.0 A current.	Solution: (Section 20-6) The magnetic field acting on the wire is due to the magnetic field produced by the wire carrying the 5.0 A current. Therefore, $B = [(4\pi \times 10^{-7} \text{ T m/A})(5.0 \text{ A})]/[(2\pi)(0.20 \text{ m})] = 5.0 \times 10^{-6}$ T
a. 2 Determine the magnitude of the force acting on the wire carrying the 2.0 A current.	The force on a current-carrying wire is given by the equation $F = I \ell B \sin \theta$. $F = (2.0 \text{ A})(0.50 \text{ m})(5.0 \times 10^{-6} \text{ N/A m}) \sin 90°$ $F = 5.0 \times 10^{-6}$ N

a. 3

Determine the direction of the force acting on the 2.0 A current.

Based on the diagram shown below and using the right-hand rule, the magnetic field produced by the 5.0 A current encircles the wire such that it is vertically into the page at the position of the 2.0 A current. Using the right hand rule for the force produced by a magnetic field acting on a current-carrying wire, the fingers point toward the bottom of the page in the direction of the 2.0 A current. The magnetic field is inward, perpendicular to the plane of the paper. The fingers bend until they point inward. The thumb points to the right. The force is therefore toward the right.

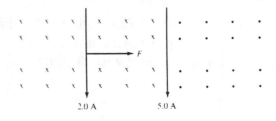

2.0 A   5.0 A

b. 1

Determine the magnitude of the magnetic field produced by the 2.0 A current.

The magnetic field acting on the 5.0 A current is due to the magnetic field produced by the 2.0 A current.

$B = [(4 \pi \times 10^{-7} \text{ T m/A})(2.0 \text{ A})]/[(2\pi)(0.20 \text{ m})]$

$B = 2.0 \times 10^{-6}$ N/A m

b. 2

Determine the force acting on the wire carrying the 5.0 A current.

$F = I \ell B \sin \theta$ but $\theta = 90°$

$F = (5.0 \text{ A})(0.50 \text{ m})(2.0 \times 10^{-6} \text{ N/A m})(\sin 90°)$

$F = 5.0 \times 10^{-6}$ N

b. 3

Determine the direction of the force acting on the wire carrying the 5.0 A current.

The magnetic field produced by the 2.0 A current is directed out of the plane of the paper perpendicular to the position of the 5.0 A current. The fingers point along the direction of the 5.0 A current. The fingers bend to point in the direction of the magnetic field. The thumb points toward the left in the direction of the wire carrying the 2.0 A current. Therefore, the two wires are attracted to one another.

Note: the answer to part b can be predicted by using Newton's third law of motion. The magnetic field produced by one wire interacts with the current in the other wire. Each wire is acted upon by a force due to the other wire. The forces should be equal in magnitude but opposite in direction.

Alternate solution: As described in section 20-6 of the text, the force per unit length between two long, parallel, current-carrying wires is

$$F/\ell = (\mu_o\, I_1\, I_2)/(2\,\pi\,L)$$

where L is the separation between the wires and I_1 and I_2 are the currents.

Substituting the data given in the problem gives

$$F/(0.50\ m) = [(4\pi \times 10^{-7}\ T\ m/A)(2.0\ A)(5.0\ A)]/[2\pi\,(0.20\ m)]$$

$$F = 5.0 \times 10^{-6}\ N$$

PRACTICE PROBLEMS

PROBLEM 1. A 0.10 meter long wire segment oriented in the east-west direction is attached by thin, flexible wires to a battery. The wire segment has a mass of 0.010 kg and the electrical resistance in the circuit is 10.0 Ω. The Earth's magnetic field is directed toward the north and has a magnitude of 5.0 x 10^{-5} T. Determine the magnitude and direction of the current in the wire if the wire remains motionless in the magnetic field.

ANS. 2.0 x 10^4 A toward the east

PROBLEM 2. A proton is accelerated through a potential difference of 75.0 volts into a magnetic field of strength 1.00 x 10^{-3} T. The magnetic field is directed perpendicular to the direction of motion of the proton. a) Determine the radius of the circle in which the proton travels. b) Based on the diagram shown below, use the right-hand rule to determine the direction of the circle in which the proton travels, i.e., either clockwise or counterclockwise as viewed from above the paper. Note: the mass of a proton is 1.67 x 10^{-27} kg and the charge on a proton is 1.6 x 10^{-19} C.

ANS. a) 1.25 m,   x x x x x x

b) counterclockwise   x x x x x x

                        x x x x x x
                                ↑
                                ⊕

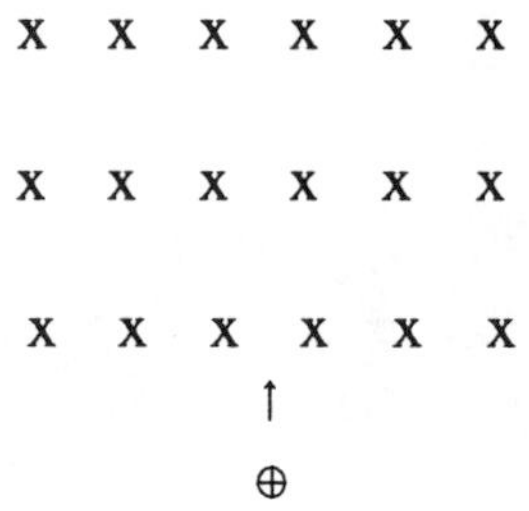

PROBLEM 3. Hydrogen and deuterium nuclei are accelerated from rest through a potential difference of 100 V. A magnetic field of strength 0.10 T is directed perpendicular to their path. Determine the radius of curvature of the orbit of each isotope. The mass of the hydrogen nucleus is 1.67×10^{-27} kg and the mass of the deuterium nucleus is 3.34×10^{-27} kg. The charge on each ion is 1.6×10^{-19} C.

ANS. $r_H = 1.4 \times 10^{-2}$ m, $r_D = 2.0 \times 10^{-2}$ m

PROBLEM 4. A current loop contains 100 turns of wire and has dimensions of 0.020 m by 0.030 m. The current through the loop is 0.10 A and the external magnetic field is directed to produce maximum torque. If the magnitude of the magnetic field is 0.030 T, determine the a) torque on the loop and b) magnetic moment of the coil.

ANS. a) 1.8×10^{-4} m N, b) 6.0×10^{-3} A m

CHAPTER 21

ELECTROMAGNETIC INDUCTION AND FARADAY'S LAW; AC CIRCUITS

KEY TERMS AND PHRASES

magnetic flux is the product of the magnetic field strength (flux density), the area of the plane of the loop through which the magnetic field passes, and the cosine of the angle that the magnetic field makes with a line drawn normal to the plane of the loop.

induced emf is described by Faraday' law which states that voltage is produced in a conductor when the magnetic flux through the conductor changes. The induced current is described by Lenz's law which states that an induced emf produces a current whose magnetic field always opposes the change in magnetic flux which caused it.

electric motor is essentially a galvanometer movement that is arranged so that a coil of wire, referred to as the armature, runs continuously. The electric current through the armature interacts with the external field and the resulting torque causes the armature to rotate. Depending on the design of the motor, the motor can be arranged to run on direct current (dc) or alternating current (ac) electricity.

electric generator uses mechanical work to produce electric energy. The armature of the generator is turned by an external torque and an emf is induced as the coil passes through an external magnetic field.

back or **counter emf** is produced as the armature of a motor rotates in the external magnetic field. This induced emf produces a torque which opposes the motion of the armature.

eddy current is an induced current produced by a changing magnetic flux in a piece of metal. The direction of the induced current is such as to oppose the magnetic field which caused the current.

mutual induction occurs when a changing current in one circuit induces a current in another circuit.

self-induction occurs when a changing current in a circuit induces a back emf in the circuit.

transformer is a device used to increase or decrease an ac voltage.

LRC series circuit consists of an inductor, capacitor, and a resistor connected in series with an alternating current source of voltage.

phasor is a rotating vector. A phasor is used to represent either current or voltage in an ac circuit. Vector algebra is used to analyze an LRC circuit.

inductive reactance is a measure of the effect an inductor has on the current through an ac circuit. It is analogous to the effect a resistor has to the flow of current through a dc circuit.

capacitative reactance is a measure of the effect a capacitor has on the current through an ac circuit. It is analogous to the effect a resistor has to the flow of current through a dc circuit.

impedance in an ac circuit is analogous to resistance in a dc circuit. Impedance results from the combination of inductive reactance, capacitative reactance, and resistance.

resonant frequency is the frequency at which an LRC circuit has minimum impedance to current flow. At this frequency current flow is a maximum and the energy transferred to the system is a maximum.

SUMMARY OF MATHEMATICAL FORMULAS

magnetic flux	$\Phi_B = B \, A \, \cos \theta$	Magnetic flux (Φ_B) is the product of the magnetic field strength (B), the area (A) of the plane of the loop through which the magnetic field passes, and the cosine of the angle ($\cos \theta$) that the magnetic field makes with a line drawn normal to the plane of the loop.
Faraday's law	$\xi = - N \, \Delta \Phi_B / \Delta t$	The magnitude of the induced emf (ξ) in a coil depends on the number of turns (N) in the loop and the rate of change of flux ($\Delta \Phi_B / \Delta t$) through the loop.
EMF induced in a moving conductor	$\xi = v \, \ell \, B \, \sin \theta$	The emf (ξ) induced in a conducting rod or wire which moves through a magnetic field depends on the velocity (v) of the rod relative to the magnetic field, the length (ℓ) of the rod, and the angle (θ) between the direction of motion of the rod and the direction of the magnetic field.

emf produced by an electric generator	$\xi = N A B \omega \sin \omega t$ $\xi_o = N A B \omega$	The magnitude of the emf (ξ) produced by an generator rotating continuously with a constant angular velocity (ω) depends on the number of turns in the armature (N), the area of the loop (A), the magnitude of the magnetic field (B), and the angle (ωt) that the face of the loop makes with the direction of the magnetic field at time t. ξ_o represents the peak voltage of the generator.
"ideal" transformer	$V_s/V_p = N_s/N_p$ $I_p V_p = I_s V_s$ $I_s /I_p = N_s /N_p$	The voltage in the secondary coil (V_s) depends on the voltage in the primary coil (V_p) and the number of turns in each coil, N_s and N_p. In an ideal transformer, the power in the primary ($P_{input} = I_p V_p$) equals the power in the secondary ($P_{output} = I_s V_s$) The ratio of the currents is related to the ratio of the turns.
induced voltage due to mutual inductance	$\xi = M \, \Delta I/\Delta t$	The induced voltage (ξ) in one coil depends on the mutual inductance (M) and is the rate of change of the current in a second coil ($\Delta I/\Delta t$).
induced voltage due to self-inductance	$\xi = - L \, \Delta I/\Delta t$ $L = \mu_o \, N^2 A/\ell$	The induced voltage in a single coil depends on the self-inductance (L) in a single coil and the rate of change of current in the coil ($\Delta I/\Delta t$). the self-inductance (L) of a long coil, called a solenoid, of length ℓ, cross-sectional area A and N turns

energy stored in an inductor	energy = ½ L I²	The energy stored in a coil is related to the self-inductance (L) and the current (I).
	energy = ½ B² A ℓ/μ_o	The energy stored in the inductor's magnetic field is related to the magetic field strength (B) and the volume enclosed by the windings (A ℓ).
energy density	energy density = ½ B²/μ_o	the energy stored per unit volume
current rise in an LR circuit	$I = \xi/R \, (1 - e^{-t/\tau})$	The current (I) in an LR circuit is related to the emf of the source (ξ), the resistance (R), the time constant (τ), and time (t).
	$\tau = L/R$	the inductive time constant of an L circuit
current decay in an LR circuit	$I = \xi/R \; e^{-t/\tau}$	The current (I) in an LR decay circuit depends on the emf of the source (ξ), the resistance (R), the time constant (τ), and time (t).
inductive reactance	$X_L = 2 \pi f L = \omega L$	The inductive reactance (X_L) is related to both the frequency of the source (f) and the self-inductance of the coil (L).
capacitative reactance	$X_C = 1/2\pi fC = 1/\omega C$	The capacitive reactance (X_C) is inversely proportional to the frequency of the source and the capacitance of the capacitor.
LRC series circuit peak voltage impedance	$V_o = [(V_{Ro})^2 + (V_{Lo} - V_{Co})^2]^{½}$	The peak voltage (V_o) across the source is related to the peak voltage across the resistor (V_{Ro}), the inductor (V_{Lo}), and the capacitor (V_{Co}).
	$Z = [(X_L - X_C)^2 + R^2]^{½}$	The impedance (Z) is related to the inductive reactance (X_L), the capacitative reactance (X_C), and the resistance (R).

phase angle	$\tan \phi = (X_L - X_C)/R.$	The phase angle (ϕ) between the voltage and current equals the ratio of the difference between the inductive and capacitative reactance ($X_L - X_C$) and the resistance (R).
power dissipated by the impedance	$\bar{P} = I\,V \cos \phi$ $\bar{P} = I^2\,Z \cos \phi$	Power dissipated in the form of heat is related to the current (I), voltage (V), impedance (Z), and the power factor cos ϕ.
power factor	$\cos \phi = R/Z$	cos ϕ is defined as the power factor of the LRC circuit. For a resistor, $\phi = 0°$ and cos $0° = 1$. Therefore, all of the power is dissipated in the form of joule heating. For a capacitor or an inductor, $\phi = 90°$ and cos $90° = 0$. Therefore, no heat is dissipated by a capacitor (or inductor); all of the energy is stored in the electric (or magnetic) field.
resonant frequency for an LRC circuit	$f_o = 1/[2\pi\,(L\,C)^{1/2}]$	Resonance for an LRC circuit occurs when the impedance is a minimum. At resonance, the energy transferred to the system from the source of emf is a maximum. The resonant frequency is inversely related to the square root of the inductance and the capacitance.

CONCEPT SUMMARY

Faraday's Law of Induction

If a bar magnet is moved toward a coil of wire, an emf will be induced in the wire. The magnitude of the induced emf depends on the magnetic field strength (or **flux density**), the area of the loop, and the time required for the change in **magnetic flux** through the area of the loop to occur. The product of the magnetic flux density (B) and the area of the plane of the loop through which it passes is known as the magnetic flux (Φ_B).

If the magnetic field is given in wb/m^2 and the area of the plane of the loop in m^2, then the flux has the unit of webers (Wb). The formula for the magnetic flux is

$\Phi_B = B\,A \cos \theta$

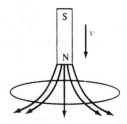
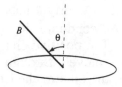

θ is the angle between the direction of B and a line drawn perpendicular to the plane of the loop. If the angle is 0°, the flux passing through the loop is a maximum, cos 0° = 1.0. If the angle is 90°, then no flux passes through the loop, cos 90° = 0.

The magnitude of the induced emf (ξ) in the coil is given by **Faraday's law**:

$$\xi = - N \, \Delta \Phi_B / \Delta t$$

N is the number of turns in the loop and $\Delta \Phi_B / \Delta t$ is the rate of change of flux through the loop.

Lenz's Law

The minus sign in Faraday's law formula indicates that the induced emf opposes the change in flux through the loop. This opposition is described by **Lenz's law,** which states that an induced emf produces a current whose magnetic field always opposes the change in magnetic flux which caused it.

An example of Lenz's law would be that of a north pole of a bar magnet being inserted into a coil of wire as shown in the figure below. The x's represent the magnetic field of the bar magnet. According to Lenz's law, an induced current will be produced in the counterclockwise direction through the coil. Using the right-hand rule, it can be determined that this induced current produces a magnetic field (dots in figure) which opposes the magnetic field of the bar magnet.

If the bar magnet is held motionless, the flux is no longer changing ($\Delta \Phi_B / \Delta t$ = 0) and the induced current disappears. However, if the magnet is withdrawn, an induced current will be produced in the clockwise direction. This current produces a magnetic field which attempts to maintain the magnetic flux through the loop at the same magnitude as when the bar magnet was motionless in the loop. Since the loop contains electrical resistance, this current will quickly be reduced to zero.

It is the relative motion between the magnetic field and the loop which causes the induced emf. Thus, it is possible to induce an emf in a loop by 1) holding the magnetic field constant and changing the area of the loop through which the magnetic flux passes, 2) holding the loop motionless and changing the magnitude of the flux through the loop, or 3) changing the orientation of the loop in the magnetic field by rotating the loop in the field.

EMF Induced in a Moving Conductor

An emf is induced in a conducting rod or wire which moves through a magnetic field. This can be predicted by the right-hand rule. In the diagram shown below, a conducting rod is moved along a U-shaped conductor in a uniform magnetic field that points 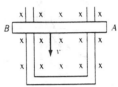 into the paper. The positive charges in the rod are moving with the rod as the rod moves through the magnetic field. Using the right-hand rule, the outstretched fingers point in the direction that the rod is moving. When the fingers bend they must point in the direction of the magnetic field, i.e., into the paper. The thumb points toward point A, indicating that the positive charges tend to move away from B and toward A.

A difference in potential exists between the ends of the rod. This difference in potential is the induced emf. The induced emf in the wire is given by the formula

$$\xi = v \, \ell \, B \sin \theta$$

where v is the velocity of the rod relative to the magnetic field, ℓ is the length in meters of the segment of the rod, and θ is the angle between the direction of motion of the rod and the direction of the magnetic field.

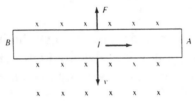

Since the rod is part of a closed circuit, a current will flow through the circuit. As shown in the diagram, the current flows from B toward point A and then clockwise through the circuit. The magnitude of this induced current is determined from Ohm's law, $I = \xi/R$.

The direction of the induced current is such that a force is produced that opposes the external force that causes the rod to move through the magnetic field. This also can be predicted by using the right-hand rule. The outstretched fingers point in the direction of the induced current. When the fingers bend to point in the direction of the magnetic field, the thumb of the right hand points toward the top of the page. The force F, which is produced by the interaction of the induced current and the magnetic field, opposes the motion of the wire. This is another example of Lenz's law.

If the rod travels at constant speed and friction between the rod and the U-shaped conductor is negligible, the induced force is equal in magnitude but opposite in direction to the external force moving the rod through the field.

Electric Motor and Generator

The **electric motor** is essentially a galvanometer movement that is arranged so that a coil of wire, referred to as the **armature**, runs continuously. The electric current through the armature interacts with the external field and the resulting torque causes the armature to rotate. Depending on the design of the motor, the motor can be arranged to run on direct current (dc) or alternating current (ac) electricity.

The **electric generator** uses mechanical work to produce electric energy. The armature is turned by an external torque and an emf is induced as the coil passes through an external magnetic field.

If the loop rotates continuously with a constant angular velocity (ω), then the magnitude of the induced emf is given by

$$\xi = N A B \omega \sin \omega t$$

where ωt is the angle that the face of the loop makes with the direction of the magnetic field at time t and A is the area of the loop.

The output emf of the generator is sinusoidally alternating. However, depending on the design, the generator can be made to produce either dc or ac current.

Counter EMF and Torque

As the armature of a motor rotates in the external magnetic field, an induced emf is produced. This induced emf, which is called a **back** or **counter emf**, produces a torque which opposes the motion of the armature (Lenz's law).

The magnitude of the back emf is proportional to the speed of the armature of the motor. When a motor is turned on, the magnitude of the back emf increases until a balance point is reached and the armature rotates at a constant speed. If the motor speed increases, the back emf increases until a balance point is again achieved.

The above also applies to a generator. The external torque causing the armature to rotate is opposed by a counter torque produced by the counter emf. If this did not occur, it would violate the law of conservation of energy, since it would then be possible to start the armature rotating and produce electrical energy without expending mechanical energy.

Eddy Currents

An induced current, called an **eddy current**, is produced by a changing magnetic flux in a piece of metal. The direction of the induced current is such as to oppose the magnetic field which caused the current (Lenz's law). Eddy currents can be produced by moving the metal through a magnetic field or by allowing a changing magnetic field to pass through a stationary piece of metal.

Eddy currents can be beneficial, e.g., electromagnetic damping in certain analytical balances and the braking system of some electric transit cars. However, eddy currents are often undesirable. For example, the coils of wire which make up the armature of a motor or generator as well as the primary and secondary of a transformer are often wound on an iron core. Eddy currents produced in the iron core dissipate energy in the form of "**joule heating**" ($P = I \xi = I^2 R$). To avoid this problem the core is laminated, which means that it is made of very thin sheets insulated from one another. This insulation causes a large electrical resistance along the path of the eddy current and the magnitude of the eddy current is small. The result is negligible energy losses due to joule heating.

Transformers

A **transformer** is a device used to increase or decrease an ac voltage. It consists of two coils of wire, known as the **primary** and **secondary coils**. The primary is connected to a source of emf and the secondary to a device usually referred to as the load (R_L). A schematic of a simple transformer is shown at right. When the current changes in the primary, the magnetic field it produces changes. As the changing field produced by the primary passes through the secondary, an induced emf is produced. If the secondary is part of a closed circuit, an induced current will

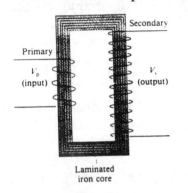

Secondary

Primary

V_p
(input)

V_s
(output)

Laminated
iron core

pass through it. The advantage of a transformer is that it is possible to produce a higher or lower voltage in the secondary as compared to the primary. This is accomplished by having different numbers of turns of wire in the two coils. If the number of turns of wire in the primary is greater in the primary than the secondary, then the voltage in the secondary is lower than in the primary and the transformer is a **step-down** transformer. If the number of turns in the primary is less than the secondary, then the voltage in the secondary is higher than in the primary and the transformer is a **step-up** transformer.

V_s and V_p are the voltages in the secondary and primary coils and N_s and N_p are the number of turns in each coil. The following equation reflects the theoretical limit for an "ideal" transformer:

$$V_s / V_p = N_s / N_p$$

Based on the law of conservation of energy, the power output in the secondary cannot be greater than the power input in the primary. In the "ideal" transformer they would be equal; thus

$$P_{input} = P_{output}$$

$$I_p V_p = I_s V_s \quad \text{and} \quad I_s / I_p = V_p / V_s$$

Power losses in transmission lines are due to joule heating, $P = I^2 R$. Because of joule heating, there is considerable advantage to transmitting power over long distances at high voltage but low current. Thus transformers can step the voltage up at the source and step the voltage down at the output. The ratio of the currents is related to the ratio of the turns as follows:

$$I_s / I_p = N_s / N_p$$

21-9

Mutual Inductance

If two coils are placed near one another, a changing current in coil one creates a changing magnetic field which passes through coil 2. The changing current in coil 1 creates a changing magnetic field in coil 2 and according to Faraday's law, an induced emf is produced in coil 2. The induced emf in coil 2 is given by

$$\xi_2 = M \, \Delta I_1 / \Delta t$$

where M is a proportionality constant called the **mutual inductance** and $\Delta I_1 / \Delta t$ is the rate of change of the current in the first coil. If the current is changing in coil 2 rather than coil 1; then the induced emf occurs in coil 1 and

$$\xi_1 = M \, \Delta I_2 / \Delta t$$

The value of M is the same whether the emf occurs in coil 1 or coil 2. The unit of mutual inductance is the **Henry (H)** where 1 H = 1 ohm second. The value of M depends on whether or not iron is present, the size of the coils, number of turns, and the distance between the coils.

Self-Inductance

Self- inductance appears in a single coil when the current in the coil changes. The changing current produces a changing magnetic field in the coil, and this induces a back emf in the coil. The back emf tends to retard the flow of current if the current in the coil is increasing and induces an emf in the same direction as the current flow if the current is decreasing. The average induced emf is

$$\xi = - L \, \Delta I / \Delta t$$

where $\Delta I / \Delta t$ is the rate of change of current in the coil and L is a proportionality constant measured in henries (H).

The self-inductance (L) of a long coil, called a solenoid, of length ℓ, and cross-sectional area A which contains N turns is $L = \mu_o \, N^2 A / \ell$.

Energy Stored in a Magnetic Field

The energy stored in a coil of inductance L, carrying a current I, is energy $= \frac{1}{2} L \, I^2$. The energy stored in the inductor's magnetic field is energy $= \frac{1}{2} B^2 A \, \ell / \mu_o$. The volume enclosed by the windings of the coil equals the product of A and ℓ. The energy stored per unit volume, or **energy density**, is given by energy density $= \frac{1}{2} B^2 / \mu_o$.

LR Circuit

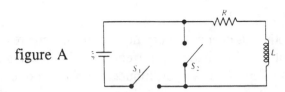

figure A

Figure A is a schematic drawing of an **LR circuit**. If switch 1 (S_1) is closed, the source of emf ξ is connected to the inductor L which has resistance R. The current through the circuit increases according to the formula

$$I = \xi/R \, (1 - e^{-t/\tau})$$

t is the time in seconds since the switch was closed, and τ is the inductive time constant of the circuit, $\tau = L/R$. It can be shown that when $t = \tau$, the current in the circuit has reached 63% of its maximum value, i.e., $I = 0.63 \, I_{max}$. Figure B, shown below, shows the growth of current in the circuit as a function of time.

figure B   figure C

If switch 1 is disconnected after a steady current I is reached, the battery is then removed from the circuit and no current flows. If switch 2 is now connected, the energy stored in the inductor L causes a current to flow as follows:

$$I = \xi/R \, e^{-t/\tau}$$

This decay current is represented graphically in figure C. It can be shown that for a decay current, the current reaches 37% of its initial value ($0.37 \, I_{max}$) after 1 time constant ($t = \tau$).

AC Circuits and Impedance

Resistor

An ac source connected to a resistor produces a current (I) and a voltage drop (V) across the resistor given by the following equations:

$$I = I_0 \cos 2 \pi f t \quad \text{and} \quad V = V_0 \cos 2 \pi f t$$

where I_0 and V_0 are the peak values, f is frequency of the source in hertz, and t is the time in seconds.

The instantaneous voltage and current are said to be in phase since both are zero at the same moment of time and both reach their maximum values in either direction at the same time. The graphs shown below represent the voltage and current through the resistor as a function of time.

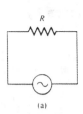

(a)

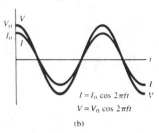

$I = I_0 \cos 2\pi ft$

$V = V_0 \cos 2\pi ft$

(b)

The relationship between V and I follows Ohm's law I = V/R and the average power dissipated is given by

$$P = \overline{I\,V} = (I_{rms})^2\,R = (V_{rms})^2/R \quad \text{where rms refers to root-mean-square.}$$

and $I_{rms} = 0.707\,I_0$, $V_{rms} = 0.707\,V_0$

Inductor

The current and voltage in an inductor connected to an ac source are given by

$$I = I_0 \cos 2\,\pi\,f\,t \quad \text{and} \quad V = -V_0 \sin 2\,\pi\,f\,t$$

The current "lags" behind the voltage and is out of phase by 90°, i.e., the current reaches its maximum and minimum values ¼ cycle after the voltage.

An inductor produces a back emf $V = L\,\Delta I/\Delta t$ and therefore resists the flow of current through it. Energy from the source is momentarily stored in the magnetic field and as the field decreases, the energy is transferred back to the source. Thus no power is dissipated in the inductor.

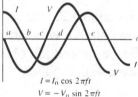

$I = I_0 \cos 2\pi ft$

$V = -V_0 \sin 2\pi ft$

The magnitude of the current is related to the applied voltage at a given frequency by the equation

$$V_L = I\,X_L \quad \text{where} \quad X_L = 2\,\pi\,f\,L$$

X_L is the **inductive reactance** of the inductor and has units of ohms. The inductive reactance is related to both the frequency of the source and the magnitude of the self-inductance of the coil.

The current and the voltage are not in phase, i.e., the peak current and peak voltage in the inductor are not reached at the same time. The equation $V = I\,X_L$ is valid on the average but not at a particular instant of time.

Capacitor

The current and voltage in a capacitor connected to an ac source are given by

$$I = I_o \cos 2\pi ft \quad \text{and} \quad V = V_o \sin 2\pi ft$$

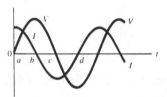

The current leads the voltage and is out of phase by 90°, i.e., the current reaches maximum and minimum values ¼ cycle before the voltage.

The average power dissipated in the capacitor is zero. Energy from the source is stored in the electric field between the plates of the capacitor. As the electric field decreases the energy is transferred back to the source.

The magnitude of the current is related to the applied voltage at a given frequency by the equation

$$V = I X_C \quad \text{where} \quad X_C = 1/2\pi fC \quad \text{where } X_C \text{ is the } \textbf{capacitive reactance} \text{ and has units of ohms}$$

Capacitive reactance is inversely proportional to the frequency of the source and the capacitance of the capacitor. As in the case of the inductor, the current and voltage are not in phase, and while the equation $V = I X_C$ is valid on the average, it is not valid at a particular instant of time.

LRC Series Circuit

An LRC circuit (see diagram A) can be analyzed by using a phasor diagram. In a phasor diagram, vector-like arrows are drawn in an xy coordinate system to represent V_{Ro}, V_{Lo}, and V_{Co}. As shown in diagram B, the magnitude of V_{Ro} is represented by an arrow drawn along the +x axis, V_{Lo} along the +y axis, and V_{Co} along the -y axis.

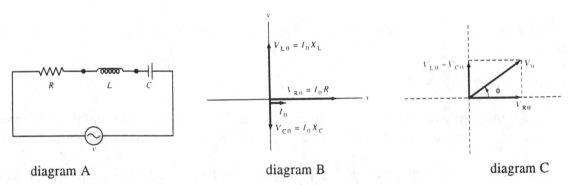

diagram A   diagram B   diagram C

Assuming that V_{Lo} is greater than V_{Co}, then diagram B reduces to diagram C. The resultant of ($V_{Lo} - V_{Co}$) and V_{Ro} is V_o. The peak voltage from the source (V_o) is out of phase with V_{Ro} and the current by the angle ϕ, where ϕ is the phase angle. Using the Pythagorean theorem, it can be shown that

$$V = [(V_{Ro})^2 + (V_{Lo} - V_{Co})^2]^{1/2}$$

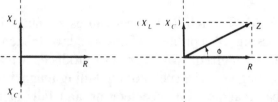

The **impedance** Z of an LRC circuit can be determined in a like manner.

$$Z = [(X_L - X_C)^2 + R^2]^{1/2} \text{ where Z is measured in ohms}$$

The phase angle between the voltage and current can be determined from $\cos \phi = R/Z$ or $\tan \phi = (X_L - X_C)/R$. The power dissipated by the impedance is given by

$$P = I_{rms} V_{rms} \cos \phi$$

where $\cos \phi$ is the power factor. If the circuit contains only resistance, $\phi = 0°$, $\cos 0° = 1$, and $P = I_{rms} V_{rms}$.

If the circuit contains a pure capacitor and/or inductor, $\phi = 90°$ and $\cos 90° = 0$, $P = 0$, and no power is dissipated.

Resonance in AC Circuits

Since $Z = [(X_L - X_C)^2 + R^2]^{1/2}$, the impedance in an ac circuit is a minimum when $X_L = X_C$. At this point $I_{rms} = V_{rms}/R$ and I_{rms} is a maximum. Also, the energy transferred to the system is a maximum and a condition known as **resonance** occurs.

For example, a resonant circuit in a radio is used to tune in a particular station. A range of frequencies reach the circuit; however, a significant current flows only at or near the resonant frequency. At resonance $X_L = X_C$; therefore,

$$2\pi f_o L = 1/(2\pi f_o C) \quad \text{and rearranging gives} \quad f_o = 1/[2\pi (L C)^{1/2}]$$

where f_o is the **resonant frequency** of the circuit in hertz.

SELECTED TEXT QUESTIONS WITH ANSWERS

QUESTION 9. Explain why, exactly, the lights may dim briefly when a refrigerator motor starts up. (Hint: consider "terminal voltage.")

ANSWER: When a motor starts up, the current through the armature is large because the induced counter emf is small and the electrical resistance to current flow is small. Electrical outlets on the same circuit as the motor will have a voltage drop and any house lights on the same circuit with the motor will dim. As the armature speeds up, it produces an increasingly large counter emf which reduces the flow of current through the motor. The voltage at the electrical outlets rises and the lights return to their former brightness.

QUESTION 16. A bar magnet falling inside a vertical metal tube reaches a terminal velocity even if the tube is evacuated so that there is no air resistance. Explain.

ANSWER: The falling magnet causes a change in flux in the region of the tube through which it is passing. According to Faraday's law, this change of flux induces an emf. The induced emf produces a current, which in turn produces a magnetic field which opposes the motion of the falling magnet. The speed of the falling magnet increases until the fields are equal and opposite. The bar magnet stops accelerating and falls at a constant speed which is called the terminal velocity.

QUESTION 27. Under what circumstances is the impedance in an LRC circuit a minimum?

ANSWER: For an LRC circuit $Z = [R^2 + (X_L - X_C)^2]^{\frac{1}{2}}$. The impedance is a minimum when $X_L = X_C$. At this point $Z = R$, the impedance is a minimum, and a condition known as resonance occurs. The resonant frequency (f_o) is inversely related to both L and C, i.e.,

$$f_o = 1/[2\pi (L\ C)^{\frac{1}{2}}].$$

PROBLEM SOLVING SKILLS

For a straight wire moving at speed v through a uniform magnetic field:

1. Use $\xi = v\ \ell\ B \sin \theta$ to determine the magnitude of the induced emf.
2. If the wire is part of a closed circuit, use Ohm's law to determine the current through the wire and the equation $F = I\ \ell\ B \sin \theta$ to determine the magnitude of the force which opposes the mechanical force acting on the wire.
3. The power dissipated in the form of Joule heating is given by $P = I^2 R$.

For problems involving a changing magnetic flux through a loop of wire:

1. Determine the rate of change of flux ($\Delta \Phi / \Delta t$).
2. Apply Faraday's law to determine the magnitude of the induced emf.
3. Apply Ohm's law to determine the magnitude of the induced current and Lenz's law to determine its direction.

For "ideal" transformer problems:

1. Complete a data table listing the voltage, current, and number of turns of wire in both the primary and secondary coils.
2. Use the mathematical equation which relates the voltage and number of turns in the primary to the voltage and number of turns in the secondary.
3. An ideal transformer is 100% efficient. The power in the primary equals the power in the secondary. Use this concept to solve for the current in the secondary.

For problems involving ac circuits and impedance:

1. Use the formulas which relate the frequency to reactance to determine the inductive reactance and capacitive reactance.
2. Draw phasor diagrams and determine the voltage drop across each circuit element as well as the impedance in the circuit.
3. Use a phasor diagram to determine the phase angle. Determine the power factor.
4. If requested, determine the rms current and the average power dissipated.
5. If requested, determine the resonant frequency.

PROGRAMMED PROBLEMS

> **PROBLEM 1.** The magnitude of the magnetic field through a 100 turn loop of wire changes from 0.00 wb/m² to 3.00 wb/m² in 0.100 s. The radius of the loop is 0.0500 m and its electrical resistance is 10.0 Ω. a) Determine the magnitude of the average induced emf and the current in the loop. b) Determine the direction of the induced current in the loop if the loop is in the plane of the paper and the magnetic field is directed out of the paper.

a. 1 Determine the area of of the loop of wire.	Solution: (Section 21-2) $A = \pi \, r^2 = \pi (0.0500 \text{ m})^2$ $A = 7.85 \times 10^{-3} \text{ m}^2$
a. 2 Determine the change in magnetic flux through the loop.	The magnetic field is directed perpendicular to the coil. Using the convention discussed in the text, the angle θ is 0°. $\Delta \Phi = \Phi_f - \Phi_i$ $\quad = B_f \, A \cos \theta - B_i \, A \cos \theta$ $\quad = (3.00 \text{ wb/m}^2)(7.85 \times 10^{-3} \text{ m}^2)(\cos 0°)$ $\qquad\qquad - (0.00 \text{ wb/m}^2)(7.85 \times 10^{-3} \text{ m}^2)(\cos 0°)$ $\Delta \Phi = 2.36 \times 10^{-2} \text{ wb}$
a. 3 Use Faraday's law to determine the magnitude of the induced emf and Ohm's law to determine the induced current.	$\xi = - \text{N} \, \Delta \Phi / \Delta t = - (100)(2.36 \times 10^{-2} \text{ wb})/0.10 \text{ s})$ $\xi = - 23.6 \text{ V}$ $I = \xi/R = - 23.6 \text{ V}/10.0 \ \Omega = - 2.36 \text{ A}$
b. 1 Draw a diagram showing the loop and the external magnetic field.	. external magnetic field ← loop
b. 2 Use Lenz's law to predict the direction of the induced current.	Based on Lenz's law, the direction of the induced current is such that it must oppose the external magnetic field that caused it. Since the external magnetic field is directed out of the paper, the current in the loop must produce a magnetic field which is directed into the paper (see diagram at top of next page). The right-hand rule indicates that a clockwise current in the loop will produce a magnetic field that is directed into the paper.

. external magnetic field

x magnetic field produced by
the induced current

PROBLEM 2. A 4.00 wb/m² magnetic field is directed perpendicular to the plane of a 100 turn loop which has a radius of 0.100 m. The loop is rotated through 180° in 0.200 s. Determine the magnitude of the induced emf in the loop.

a. 1	Solution: (Section 21-2)
Determine the change in flux through the loop.	The area of the loop and the strength of the external magnetic field are not changing. However, the angle between the plane of the loop and the magnetic field changes from 0° to 180° as the coil is rotated. An emf will be induced in the loop because as the loop rotates it cuts through lines of magnetic flux. The change in flux can be determined as follows:

$$\Delta \Phi = \Phi_f - \Phi_i$$

$$= B_f \, A_f \cos 180° - B_i \, A_i \cos 0°$$

$$= B \, A \, (\cos 180° - \cos 0°) \text{ but } \cos 180° = -1 \text{ and } \cos 0° = +1$$

$$= (4.00 \text{ wb/m}^2)[\pi (0.100 \text{ m}^2)](-1.0 - +1.0)$$

$$\Delta \Phi = -0.251 \text{ wb}$$

a. 2	$\xi = -N \, \Delta\Phi/\Delta t$
Use Faraday's law to determine the magnitude of the induced emf.	$= -(100)(-0.251 \text{ wb}/0.20 \text{ s})$
	$\xi = +126 \text{ V}$

PROBLEM 3. The metal rails of the U-shaped conductor shown in the diagram are 1.00 m apart and the magnetic field is 0.200 wb/m². The magnetic field is directed perpendicular to the paper. The electrical resistance of the closed circuit is 1.00 Ω and is assumed to be constant. Determine the a) magnitude and direction of the induced current in the circuit when the rod is moving at 3.00 m/s, b) magnitude and direction of the external force required to keep the rod moving at a constant speed, and c) power required to keep the rod moving at constant speed as compared to the rate of Joule heating in the circuit.

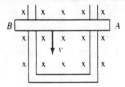

a. 1 Determine the magnitude of the induced emf.	Solution: (Section 21-3) Since every point in the wire is moving at the same speed, use $\xi = v \ell B \sin \theta$ $\xi = (3.00 \text{ m/s})(1.00 \text{ m})(0.200 \text{ wb/m}^2) \sin 90° = 0.600 \text{ volts}$
a. 2 Determine the magnitude of the induced current.	$I = \xi/R$ $I = 0.600 \text{ V}/1.00 \ \Omega = 0.600 \text{ A}$
a. 3 Use Lenz's law to determine the direction of the induced current.	Lenz's law predicts that the induced current must oppose the change that causes it. The current opposes the change by interacting with the external magnetic field to produce a force which opposes the rod's motion. Using the right-hand rule, the current can be shown to be flowing from B to A and therefore in a clockwise direction through the U-shaped conductor.
b. 1 Determine the magnitude of the external force.	Note: 1 wb/m^2 = 1 T = 1 N/A m $F = I \ell B \sin \theta = (0.600 \text{ A})(1.00 \text{ m})(0.200 \text{ N/A m}) \sin 90°$ $F = 0.120 \text{ N}$
c. 1 Determine the mechanical power required to keep the rod moving at constant speed.	The external force is moving the wire through the magnetic field, the angle between the force and the direction of motion is 0°. The wire travels at constant speed; therefore, $P = F v \cos 0°$ $P = (0.120 \text{ N})(3.00 \text{ m/s})(\cos 0°) = 0.36 \text{ watt}$
c. 2 Determine the rate of joule heating.	$P = I^2 R = (0.600 \text{ A})^2(1.00 \ \Omega) = 0.36 \text{ watt}$ The power supplied by the external force is dissipated in the form of heat in the circuit.

PROBLEM 4. A wire segment 0.0500 meters long is oriented in the east-west direction. The wire is attached by a thin, flexible wire to a galvanometer. The wire segment has a mass of 0.010 kg and the electrical resistance in the circuit is 1.00 Ω. The wire is dropped from rest into a magnetic field of strength 2.00 T which is directed toward the north. Determine the a) maximum velocity achieved by the wire and b) direction of the induced current in the circuit. Assume that air resistance is negligible.

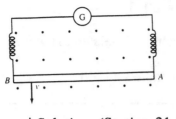

a. 1	Solution: (Section 21-3)
Locate the forces acting on the wire. What is the net force on the wire when it reaches its maximum velocity?	Gravity causes the wire to accelerate downward. As it accelerates, the induced current gradually increases and the force produced by the interaction of the induced current and the external magnetic field increases until it is equal but opposite to the wire's weight. The net force on the wire when it reaches maximum speed is zero.
a. 2 Determine the magnitude of the current flowing through the wire.	$F_{grav} = F_{magnetic}$ $mg = I \ell B \sin \theta$ but $\theta = 90°$ and $\sin 90° = 1.0$ $I = (mg)/(\ell B) = [(0.010 \text{ kg})(9.8 \text{ m/s}^2)]/[(0.050 \text{ m})(2.00 \text{ N/A m})]$ $I = 0.980$ A
a. 3 Use Ohm's law to determine the magnitude of the induced emf.	$\xi = I R$ $= (0.980 \text{ A})(1.00 \text{ } \Omega)$ $\xi = 0.980$ volt
a. 4 Determine the maximum velocity of the wire.	$\xi = v \ell B \sin \theta$ $0.980 \text{ volt} = v (0.050 \text{ m})((2.00 \text{ N/A m})(\sin 90°)$ $v = 9.80$ m/s
b. 1 Use the right-hand rule to determine the direction of the induced current.	Since the force produced by the induced current is directed upward, the thumb points upward. The magnetic field is directed outward; therefore, when you bend your fingers they must point outward. When your hand is properly oriented you will note that the straightened fingers point toward B. Therefore, the current flows from A toward B and clockwise around the closed path.

PROBLEM 5. An ac generator consists of an 800 turn flat rectangular coil 0.0600 m long and 0.0400 m wide. The coil rotates at 1200 rpm in a magnetic field of 0.400 T. Determine the a) angular velocity of the coil in radians per second and b) peak output voltage of the generator.

a. 1	Solution: (Section 21-5)
Determine the angular angular velocity of the coil.	The angular velocity ω is related to the frequency (f) by the formula $\omega = 2\pi f = (2\pi \text{ rad/rev})(1200 \text{ rev/min})(1 \text{ min/60 sec}) = 126$ rad/s
b. 1	The peak voltage is given by $\xi_o = N B A \omega$
Determine the peak output voltage.	where $A = (0.0600 \text{ m})(0.0400 \text{ m}) = 2.40 \times 10^{-3} \text{ m}^2$ $\xi_o = (800)(0.400 \text{ T})(2.4 \times 10^{-3} \text{ m}^2)(126 \text{ rad/s})$ $\xi_o = 96.5$ V

PROBLEM 6. The primary winding of a transformer contains 300 turns and draws 4.0 A from a 120 V source of emf. If the secondary contains 600 turns, determine the emf and current in the secondary. Assume that the transformer is 100% efficient.

a. 1	Solution: (Section 21-7)
Complete a data table.	V_p = 120 volts $\qquad V_s$ = ? N_p = 300 turns $\qquad N_s$ = 600 turns I_p = 4.0 A $\qquad I_s$ = ?
a. 2	$V_s / V_p = N_s / N_p$
Determine the emf in the secondary.	$V_s / 120 \text{ V} = 600/300$ $V_s = 240$ V
a. 3	The transformer is 100% efficient; therefore, the power output in the secondary equals the power input in the primary.
Determine the current in the secondary.	$P_{input} = P_{output}$ $I_p V_p = I_s V_s$ $(4.0 \text{ A})(120 \text{ V}) = I_s (240 \text{ V})$ $I_s = 2.0$ A

PROBLEM 7. The current in a 100 mH coil changes from -40.0 mA to +40.0 mA in 5.00 ms. Determine the magnitude of the induced emf.

a. 1	Solution: (Section 21-9)
Use the formula for the induced emf in a coil of self-inductance L.	The induced emf in a coil in which the current is changing is given by $$\xi = -L\, \Delta I/\Delta t$$ but $\Delta I = I_f - I_i = 40.0\text{ mA} - -40.0\text{ mA} = +80.0\text{ mA}$ $$\xi = -[(100 \times 10^{-3}\text{ H})(+80.0 \times 10^{-3}\text{ A})]/(5.00 \times 10^{-3}\text{ s})]$$ $$\xi = 1.60\text{ V}$$

PROBLEM 8. A 2000 turn, air-filled coil is 2.00 m long, 0.100 m in diameter and carries a current of 1.00 A. Determine the a) self-inductance of the coil and b) energy stored in the coil's magnetic field.

a. 1	Solution: (Sections 21-9 and 21-10)
Determine the cross-sectional area of the coil.	$A = \pi\, d^2/4 = \pi\, (0.100\text{ m})^2/4 = 7.85 \times 10^{-3}\text{ m}^2$
a. 2 Determine the self-inductance of the coil.	$L = \mu_0\, N^2\, A/\ell$ $= (4\pi \times 10^{-7}\text{ T m/A})(2000)^2(7.85 \times 10^{-3}\text{ m}^2)/(2.00\text{ m})$ $L = 0.0197\text{ H} = 19.7\text{ mH}$
b. 1 Determine B if the magnitude of the field produced by a long coil is given by $B = \mu_0\, N\, I/\ell$.	$B = \mu_0\, N\, I/\ell$ $= (4\pi \times 10^{-7}\text{ T m/A})(2000)(1.00\text{ A})/(2.00\text{ m})$ $B = 1.26 \times 10^{-3}\text{ T}$
b. 2 Determine the energy stored in the magnetic field.	energy $= \frac{1}{2}(B^2/\mu_0)\, A\, \ell$ $= \frac{1}{2}[(1.26 \times 10^{-3}\text{ T})^2/(4\pi \times 10^{-7}\text{ T m/A})](7.85 \times 10^{-3}\text{ m}^2)(2.00\text{ m})$ energy $= 9.86 \times 10^{-3}\text{ J}$

PROBLEM 9. a) Determine the reactance of a 100.0 mH inductor connected to a 120 Hz ac line. b) The inductor is replaced with a 12.5 μF capacitor. Determine the reactance.

a. 1	Solution: (Section 21-12)
Determine the inductive reactance.	$X_L = 2 \pi f L$
	$= 2 \pi (120 \text{ Hz})(100.0 \times 10^{-3} \text{ H})$
	$X_L = 75.4 \ \Omega$
b. 1	$X_C = 1/(2 \pi f C)$
Determine the reactance of the capacitor.	$X_C = 1/[2\pi (120 \text{ Hz})(12.5 \times 10^{-6} \text{ F})]$
	$X_C = 106 \ \Omega$

PROBLEM 10. An LRC circuit has a 200.0 mH inductor with 40.0 Ω of resistance connected in series with a 100 μF capacitor and a 60 Hz, 120 volt ac source. Determine the a) total impedance, b) phase angle, c) rms current, and d) power dissipated in the circuit.

a. 1	Solution: (Section 21-13)
Determine the inductive reactance, capacitive reactance, and total impedance.	$X_L = 2 \pi f L$
	$X_L = 2 \pi (60 \text{ Hz})(200 \times 10^{-3} \text{ H}) = 75.4 \ \Omega$
	$X_C = 1/(2\pi f C) = 1/[2\pi (60 \text{ Hz})(100 \times 10^{-6} \text{ F})]$
	$X_C = 26.5 \ \Omega$
	$Z = [(X_L - X_C)^2 + R^2]^{1/2}$
	$= [(75.4 \ \Omega - 26.5 \ \Omega)^2 + (40.0 \ \Omega)^2]^{1/2}$
	$Z = 63.2 \ \Omega$
b. 1	$\tan \phi = (X_L - X_C)/R$
Determine the phase angle.	$\tan \phi = (75.4 \ \Omega - 26.5 \ \Omega)/(40.0 \ \Omega) = 1.22$
	$\phi = 50.7°$
c. 1	$I_{rms} = V_{rms}/Z = (120 \text{ V})/(63.2 \ \Omega)$
Determine the rms current.	$I_{rms} = 1.90 \text{ A}$
d. 1	$P = I V \cos \phi$
Determine the power dissipated in the circuit.	$P = (1.90 \text{ A})(120 \text{ V}) \cos 50.7° = 144 \text{ watts}$ or

21-23

$$P = (I_{rms})^2 R = (1.90 \text{ A})^2 (40.0 \ \Omega)$$

$$P = 144 \text{ watts}$$

PROBLEM 11. An LRC circuit has a 1.50 mH inductor with 100 Ω of resistance connected in series with a 100 μF capacitor and a 120 volt ac source. Determine the a) resonant frequency and b) peak current for the circuit.

a. 1	Solution: (Section 21-14)
Derive a formula for the resonant frequency.	The impedance reaches its minimum value at resonance. $X_L = X_C$, and $Z = R$.
	$X_L = X_C$
	$2\pi \ f_o \ L = 1/(2\pi \ f_o \ C)$ Rearranging gives
	$f_o^2 = 1/(4\pi^2 \ L \ C)$
	$\quad = 1/[4 \ \pi^2 \ (1.50 \times 10^{-3} \text{ H})(100 \times 10^{-6} \text{ F})]$
	$f_o^2 = 1.69 \times 10^5 \text{ Hz}^2$
	$f_o = 411 \text{ Hz}$
b. 1	$V_{rms} = 120 \text{ volts}$
Determine the peak voltage.	The peak voltage $V_o = (2)^{1/2} \ V_{rms}$.
	Therefore, $V_o = (2)^{1/2} (120 \text{ V}) = 170 \text{ V}$
b. 2	The peak current occurs at the resonant frequency when the impedance to current flow is at a minimum. At this frequency, $Z = R$, thus
Determine the peak current.	
	$I = V_o/R = 170 \text{ V}/100 \ \Omega = 1.70 \text{ A}$

PRACTICE PROBLEMS

PROBLEM 1. A uniform magnetic field of strength 0.50 T is directed at an angle of 90° to the plane of a 100 turn, rectangular copper coil of length 0.040 m and width 0.050 m. The resistivity of copper is 1.7×10^{-8} Ω m and the diameter of the copper wire which makes up the coil is 1.0 mm. The magnetic field through the loop decreases to zero in 0.30 s. Determine the a) initial flux through the coil, b) average induced emf in the coil as the field decreases to zero, c) amount of charge that passes through the coil as the field decreases.

ANS. a) 1.0×10^{-3} wb, b) 0.33 V, c) 0.25 coulomb

PROBLEM 2. The magnitude of the magnetic field through a 100 turn loop of wire changes from 3.00 wb/m^2 out of the plane of the loop to 2.00 wb/m^2 into the plane of the loop in 0.100 s. The radius of the loop is 0.0500 m and its electrical resistance is 10.0 ohms. a) Determine the magnitude of the average induced emf and the current in the loop. b) Determine the direction of the induced current in the loop if the loop is in the plane of the paper and the magnetic field is initially directed out of the paper.

ANS. a) 39.3 V,   b) counterclockwise as viewed from above the paper

PROBLEM 3. An ac generator consists of a 400 turn flat rectangular coil 0.0800 m long and 0.0250 m wide. The coil rotates at a constant angular velocity in a magnetic field of 0.400 T. Determine the angular velocity required in order to produce a peak output voltage of 120 volts.

ANS. 375 rad/s

PROBLEM 4. A transformer has 1000 turns in the primary which draws 3.0 A from a 120 volt source. The current in the secondary is 9.0 A. Determine the a) potential difference across the secondary and b) number of turns in the secondary.

ANS. a) 40 V, b) 330 turns

CHAPTER 22

ELECTROMAGNETIC WAVES

KEY TERMS AND PHRASES

Maxwell's equations consist of four basic equations that describe all electric and magnetic phenomena.

electromagnetic waves are transverse waves which have both an electric and a magnetic component. Maxwell, based on his equations, concluded that a changing magnetic field (B) will produce a changing electric field (E) which is at right angles to the magnetic field and that changing electric field will produce a changing magnetic field. The net result of the interaction of the changing E and B fields is the production of an electromagnetic wave which propagates away from the wave source at the speed of light.

electromagnetic spectrum includes radio waves, microwaves, infrared radiation, visible light, x rays, and gamma rays. They differ in frequency (f) and wavelength (λ) but all travel at the speed of light.

Poynting vector describes the instantaneous energy transported per unit time per unit area perpendicular to the wave direction. The vector is in the direction in which the wave is moving.

SUMMARY OF MATHEMATICAL FORMULAS

speed of electromagnetic waves in a vacuum	$v = (\epsilon_o \mu_o)^{-\frac{1}{2}}$	The speed of EM waves in a vacuum is inversely related to the square root of the permittivity of free space (ϵ_o), where $\epsilon_o = 8.85 \times 10^{-12}$ C^2/N m^2 and the permeability of free space (μ_o), where $\mu_o = 4 \pi \times 10^{-7}$ T m/A.
speed of electro-magnetic waves	$v = f \lambda$	The speed (v) of electromagnetic waves equals the product of the frequency (f) and the wavelength (λ). The speed of light (v) in a vacuum equals 3.00×10^8 m/s.

energy density	$u = \frac{1}{2}\,\epsilon_o E^2 + \frac{1}{2}\,B^2/\mu_o$ $u = (\epsilon_o/\mu_o)^{\frac{1}{2}}\,E\,B$	The total energy stored per unit volume (u) at any particular instant. E and B represent the instantaneous magnitudes of the electric and magnetic fields at a particular point.
Poynting vector	$\bar{S} = E_o\,B_o/2\mu_o$	The average energy transported per unit time per unit area perpendicular to the wave direction is represented by the Poynting vector (S). The vector is in the direction in which the wave is moving. The units of S are joules per second per square meter ($J/s\ m^2$). E_o represents the maximum values of the electric field vector while B_o represents the maximum value of the magnetic field vector.

CONCEPT SUMMARY

Maxwell's Equations

All electric and magnetic phenomena can be described in four basic equations known as **Maxwell's equations**. The following is a summary of each equation:

1) Gauss' law for electricity is a generalized form of Coulomb's law and relates electric charge to the electric field that the charge produces.
2) Gauss' law of magnetism describes magnetic fields and predicts that magnetic fields are continuous, i. e. they have no beginning or ending point. The equation predicts that there are no magnetic monopoles.
3) Faraday's law of induction describes the production of an electric field by a changing magnetic field.
4) Maxwell's extension of Ampere's law describes the magnetic field produced by a changing electric field or by an electric current.

Production of Electromagnetic Waves

Maxwell concluded that a changing magnetic field (B) will produce a changing electric field (E) and the changing electric field will produce a changing magnetic field. The net result of the interaction of the changing E and B fields is the production of a wave which has both an electric and a magnetic component and travels through empty space. This wave is referred to as an electromagnetic wave (EM).

Electromagnetic waves are produced by accelerated electric charges. For example, accelerated electrons in atoms give off visible light while oscillating electric charges in an antenna are undergoing acceleration and produce radio waves. The following diagram represents the field strengths of an electromagnetic wave produced by a sinusoidally varying source of emf.

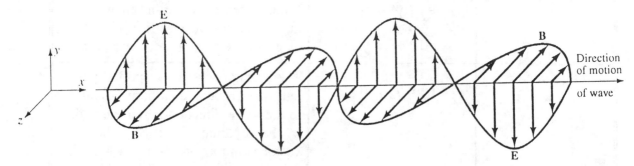

From the figure it should be noted that electromagnetic waves are transverse waves with E and B at right angles to one another and both are perpendicular to the direction of travel.

The speed of EM waves in a vacuum is given by $v = (\epsilon_o \mu_o)^{-\frac{1}{2}}$ where

ϵ_o is the permittivity of free space, $\epsilon_o = 8.85 \times 10^{-12}$ C^2/N m^2 and

μ_o is the permeability of free space, $\mu_o = 4\pi \times 10^{-7}$ T m/A

Electromagnetic Spectrum

Radio waves and visible light are only a small part of what is known as the electromagnetic spectrum (EM). The spectrum includes radio waves, microwaves, infrared radiation, visible light, x rays, and gamma rays. They differ in frequency (f) and wavelength (λ). The velocity, frequency, and wavelength are related by the following equation which represents the speed of all EM waves in free space.

$v = f\lambda$ where the speed (v) of electromagnetic waves in a vacuum = 3.00×10^8 m/s

The following table gives the approximate range of wavelengths for each portion of the EM spectrum:

radio waves 10^4 m to 1 m   ultraviolet 10^{-8} m to 10^{-9} m

microwaves 1 m to 10^{-4} m   x rays 10^{-9} m to 10^{-11} m

infrared 10^{-4} m to 10^{-7} m   gamma rays less than 10^{-12} m

visible light 4×10^{-7} m to 7×10^{-7} m

Energy Density of an EM Wave

Electromagnetic waves transport energy from one region of space to another. This energy is associated with the electric and magnetic fields of the wave. The energy per unit volume (**energy density**) stored in the electric field is given by

$u = \frac{1}{2} \epsilon_o E^2$ where u is the energy density and has units of joules per cubic meter (J/m^3)

The energy stored per unit volume in the magnetic field is given by $u = \frac{1}{2} B^2/\mu_o$.

The total energy stored per unit volume at any particular instant is given by

$$u = \frac{1}{2} \epsilon_o E^2 + \frac{1}{2} B^2/\mu_o$$

E and B represent the instantaneous magnitudes of the electric and magnetic fields at a particular point.   The total energy density can be stated in terms of the electric field only as $u = \epsilon_o E^2$ or the magnetic field only as $u = B^2/\mu_o$ or in terms of both E and B as

$$u = (\epsilon_o/\mu_o)^{\frac{1}{2}} E B$$

Poynting Vector

The instantaneous energy transported per unit time per unit area perpendicular to the wave direction is represented by the **Poynting vector (S)**. The vector is in the direction in which the wave is moving and the magnitude of the vector is given by

$$S = \epsilon_o c E^2 \quad \text{or} \quad S = E B/\mu_o$$

The units of S are joules per second per square meter (J/s m^2).   The average energy transported per unit area per unit time is given by

$$\bar{S} = \frac{1}{2} \epsilon_o c E^2 = \frac{1}{2} c/\mu_o B_o^2 = E_o B_o/2\mu_o$$

where E_o and B_o represent the maximum values of E and B.

SELECTED TEXT QUESTIONS WITH ANSWERS

QUESTION 1. The electric field in an EM wave traveling north oscillates in an east-west plane. Describe the direction of the magnetic field vector in this wave.

ANSWER: The plane of magnetic field vector is perpendicular to both the direction of travel of the wave and the plane of the electric field vector. Therefore, the magnetic field vector is in a vertical plane which is perpendicular to the surface of the Earth.

QUESTION 5. Are the wavelengths of radio and television signals longer or shorter than those detectable by the human eye?

ANSWER: Based on the table given in figure 22-10, radio and television signals have a longer wavelength than visible light.

QUESTION 8. In the electromagnetic spectrum, what type of EM wave would have a wavelength of 10^3 km? 1 km? 1 m? 1 cm? 1 mm? 1 μm?

ANSWER: Based on figure 22-10, 10^3 km, 1 km, and 1 m are in the radio wave region of the EM spectrum; 1 cm and 1 mm are in the microwave region; 1 μm is in the microwave region.

PROBLEM SOLVING SKILLS

For problems involving the velocity, frequency, or wavelength of a periodic electromagnetic wave in a vacuum:

1. Express the wavelength of the wave in meters and the frequency of the wave in hertz.
2. Use the equation $c = f \lambda$ to solve the problem where $c = 3.0 \times 10^8$ m/s.

For problems related to the energy in electromagnetic waves:

1. Note the instantaneous magnitude of the electric and/or magnetic field vector.
2. Solve for the energy density stored in the electric field using the equation $u = \frac{1}{2} \epsilon_o E^2$. Solve for the energy density stored in the magnetic field using the equation $u = \frac{1}{2} B^2/\mu_o$.

For problems related to the time average value of the Poynting vector:

1. Determine the time average value of the Poynting vector by using the equation

$$\overline{S} = E_o B_o/2\mu_o$$

For problems related to the time average power a specified distance from a source of electromagnetic waves which radiates uniformly in all directions:

1. Determine the surface area of an imaginary sphere through which the energy radiates.

2. Determine the time average power using the equation $\overline{P} = \overline{S} A$.

PROGRAMMED PROBLEMS

PROBLEM 1. A radar pulse travels from the Earth to the Moon and back in 2.60 seconds. Calculate the distance from the Earth to the Moon in a) meters and b) miles.

a. 1	Solution: (Section 22-4)
Calculate the distance distance from the Earth to the Moon in meters.	A radar pulse is an electromagnetic wave which travels at 3.00×10^8 m/s. The time required for the pulse to travel from the Earth to the moon is equal to $(\frac{1}{2})(2.60 \text{ s}) = 1.30$ s. Since it travels in a straight line at a constant speed, the distance traveled can be determined as follows: $$d = c\, t = (3.00 \times 10^8 \text{ m/s})(1.30 \text{ s})$$ $$d = 3.90 \times 10^8 \text{ m}$$
b. 1	1.000 miles = 1609 m Therefore,
Calculate the distance from the Earth to the Moon in miles.	$$d = 3.90 \times 10^8 \text{ m} \times (1.000 \text{ mile}/1609 \text{ m})$$ $$d = 2.42 \times 10^5 \text{ miles}$$ Note: the Moon follows an elliptical path in its orbit about the Earth; therefore, the value obtained in this problem would be the distance at a certain point in the Moon's orbit.

PROBLEM 2. Determine the frequency of yellow light of wavelength 6.00×10^{-7} meters.

a. 1	Solution: (Section 22-4)
Use $c = f\, \lambda$ to calculate the frequency of yellow light.	The frequency, wavelength, and velocity of light are related by the equation $c = f\, \lambda$, where c refers to the speed of light in a vacuum. $$f = c/\lambda = (3.00 \times 10^8 \text{ m/s})/(6.00 \times 10^{-7} \text{ m})$$ $$f = 5.00 \times 10^{14} \text{ Hz}$$

PROBLEM 3. The energy density of an EM wave is 5.00×10^5 J/m³. Determine the peak magnitude of the a) electric field strength and b) magnetic field strength.

a. 1	Solution: (Section 21-7)
Determine the peak magnitude of the electric field strength.	The energy density is related to the electric field strength by the equation

$u = \epsilon_o E^2$

but $\bar{E}^2 = E_o^2/2$ Therefore,

$u = \frac{1}{2} \epsilon_o E_o^2$ Rearranging gives

$E_o = (2u/\epsilon_o)^{\frac{1}{2}}$

$E_o = [(2)(5.00 \times 10^{-5} \text{ J/m}^3)/(8.85 \times 10^{-12} \text{ C}^2/\text{N m}^2)]^{\frac{1}{2}}$

$E_o = 3.36 \times 10^3 \text{ N/C}$

---

b. 1

Determine the peak magnitude of the magnetic field strength.

The energy density can be expressed in terms of the magnetic field strength as follows:

$u = B^2/\mu_o$ but $\bar{B}^2 = B_o^2/2$ Therefore,

$u = \frac{1}{2} B_o^2/\mu_o$

$5.00 \times 10^{-5} \text{ J/m}^3 = \frac{1}{2}B_o^2/(4\pi \times 10^{-7} \text{ N s}^2/\text{C}^2)$

and solving for B_o:

$B_o = (3.36 \times 10^3 \text{ N/C})/(3.0 \times 10^8 \text{ m/s}) = 1.12 \times 10^{-5} \text{ T}$

Alternate solution: It can be shown that

$c = E_o/B_o$ Therefore,

$B_o = E_o/c = (3.36 \times 10^3 \text{ N/c})/(3.00 \times 10^8 \text{ m/s})$

$B_o = 1.12 \times 10^{-5} \text{ T}$

PROBLEM 4. A source of EM waves radiates uniformly in all directions. The amplitude of the electric field 1.00×10^4 m from the source is 200 V/m. Determine the time average value of a) the Poynting vector and b) time average power radiated by the source.

a. 1

Determine the time average value of the Poynting vector.

Solution: (Section 22-7)

The time average value of the Poynting vector is given by

$\bar{S} = \frac{1}{2} \epsilon_o c E_o^2$

E_o is the peak value (amplitude) of the electric field; thus

$\bar{S} = \frac{1}{2}(8.85 \times 10^{-12} \text{ C}^2/\text{Nm}^2)(3.00 \times 10^8 \text{m/s})(200 \text{ V/m})^2$

$\bar{S} = 53.1 \text{ J/s m}^2$

b. 1

Determine the time
average power radiated
by the source.

Because the EM wave radiates uniformly in all directions, it is possible to assume that all of the energy passes through an imaginary sphere of surface area $4\pi r^2$. The radius of this sphere is taken to be 1.00×10^4 m with the source at the center.

The time average power ($\bar{P}$) radiated by the source is related to the Poynting vector as follows:

$$\bar{P} = \bar{S} \ A = \bar{S} \ (4\pi \ r^2)$$

$$= (53.1 \text{ J/s m}^2)(4\pi)(1.00 \times 10^4 \text{ m})^2$$

$$\bar{P} = 6.67 \times 10^{10} \text{ J/s} = 6.67 \times 10^{10} \text{ watt}$$

PRACTICE PROBLEMS

PROBLEM 1. Suppose the Sun exploded without warning. How long would it take light from the event to reach the Earth?   Hint: the average distance between the Earth and the Sun is 93,000,000 miles or 1.5×10^{11} m.

ANS. 500 seconds or 8.3 min

PROBLEM 2. A radio station is assigned a frequency of 60 kHz. The amplitude of the electric field at a point 1.0 km from the transmitter is 10.0 V/m. Determine the a) wavelength of the EM wave and b) peak value of the magnetic field at this point.

ANS. a) 5.0×10^3 m, b) 3.33×10^{-8} T

PROBLEM 3. Calculate the instantaneous energy density in a region of space where the magnetic field produced by an EM wave has an instantaneous value of 1.0×10^{-5} T.

ANS. 8.0×10^{-5} J/m^3

PROBLEM 4. A 25 watt light bulb radiates uniformly in all directions. Assume that the bulb acts as a point source of energy and determine the a) time average value of the Poynting vector at a point 1.0 meter from the bulb and b) peak values of both the electric field and magnetic field at this same point.

ANS. a) 2.0 J/m^2 s,   b) 39 V/m, 1.3×10^{-7} T

CHAPTER 23

LIGHT: GEOMETRICAL OPTICS

KEY TERMS AND PHRASES

ray of light is a single beam of light which travels in a straight line until it strikes an obstacle. If the ray strikes a highly reflective surface, then **specular** (or mirror) reflection occurs. If the ray strikes a rough surface, then **diffuse** reflection results.

laws of specular reflection 1) The incident ray, reflected ray, and the normal to the surface are all in the same plane and 2) the angle of incidence equals the angle of reflection.

virtual image refers to an image that only "appears" to be behind the surface of a mirror (or lens). The image is not actually located at its apparent position and cannot be formed on a screen.

real image of an object refers to an image that can be formed on a screen placed at the position where rays from a point on the object pass through a common point.

spherical mirrors are curved mirrors which are sections of a sphere. The surface of a convex mirror curves toward the observer while the surface of a concave mirror curves away from the observer.

focal point of a spherical mirror is the point where rays parallel and very close to the principal axis all pass through (or appear to come from) after reflecting from the mirror.

focal point of a lens is the point where rays parallel and very close to the principal axis all pass through (or appear to come from) after refraction by the lens.

focal length of a mirror (or lens) is the distance from the center of the mirror (or lens) to the focal point.

linear magnification refers to the ratio of the height of the image to the height of the object.

refraction occurs when light changes direction as it passes from one transparent substance into another. The light changes direction at the interface between the two substances.

Snell's law describes the relationship between the incident ray and the refracted ray during refraction.

index of refraction equals the ratio of the speed of light in a vacuum to the speed of light in a transparent substance. The indices of refraction, along with the angle of incidence, determine the angle of bending of the light at the interface between two transparent substances.

total internal reflection occurs when light traveling a medium of higher index of refraction is completely reflected at the interface of a medium of lower index of refraction.

critical angle is the minimum angle of incidence at which total internal reflection occurs.

convex lens causes light to converge as it passes through the lens. The surfaces of a convex lens curve outward toward the observer.

concave lens causes light to diverge as it passes through the lens. The surfaces of a concave lens curve away from the observer.

SUMMARY OF MATHEMATICAL FORMULAS

second law of specular reflection	$\angle i = \angle r$	The angle of incidence ($\angle i$) equals the angle of reflection ($\angle r$).
mirror and lens equations	$1/d_o + 1/d_i = 1/f$	Equation relates the distance from the center of the mirror (lens) to the object (d_o) and the image (d_i) to the focal length (f) of the mirror (lens).
	$f = r/2$	The focal length (f) equals one-half the radius of curvature (r) of the mirror. The focal point is real for a concave mirror but virtual for a convex mirror.
	$m = h_i/h_o = -d_i/d_o$	The linear magnification (m) refers to the ratio of the size of the image (h_i) to the size of the object (h_o). The magnification equals the ratio of the image distance (d_i) to the object distance(d_o).
Snell's law	$n_1 \sin \theta_1 = n_2 \sin \theta_2$	The law of refraction states the relationship between the angle of incidence (θ_1) and the angle of refraction (θ_2). n_1 is the index of refraction of the incident medium while n_2 is the index of refraction of the medium into which the light passes.

critical angle for total internal reflection	$\sin \theta_c = n_2/n_1$	At the critical angle (θ_c) and at all angles greater than this angle, light is totally reflected back into the medium of the incident ray.

CONCEPT SUMMARY

Laws of Reflection

A **ray** of light is a single beam of light which travels in a straight line until it strikes an obstacle. If the ray strikes a highly reflective surface, then **specular** (or mirror) reflection occurs. If the ray strikes a rough surface, then **diffuse** reflection results. In specular reflection, it is found that 1) the **incident ray (i), reflected ray (r)**, and the **normal (N)** to the surface are all in the same plane and 2) the **angle of incidence ($\angle i$)** equals the **angle of reflection ($\angle r$)**.

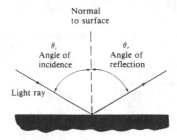

Plane Mirrors

Using the laws of specular reflection it is possible to show that the image formed by an object placed in front of a plane (flat) mirror has the following characteristics: it is erect or upright, **virtual**, and the same size as the object. Also, the apparent distance from the mirror to the image is equal to the actual distance from the mirror to the object.

The word **virtual** refers to the fact that the image of the object only "appears" to be behind the surface of the mirror but the light does not pass through the mirror and the image is not actually located at its apparent position. It is not possible to form a virtual image on a piece of paper or photographic film placed at the image position.

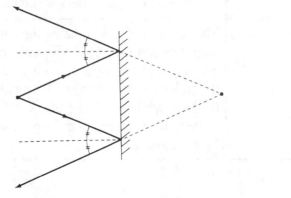 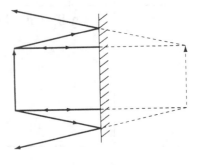

In order to locate the position of the image of any point on an object placed in front of a mirror, it is necessary to draw at least two rays of light which leave that point and reflect from the mirror. The figures at the bottom of the previous page show the use of ray diagrams in locating the image produced by 1) a point and 2) an arrow placed in front of a plane mirror.

Spherical Mirrors

Spherical mirrors are curved mirrors which are sections of a sphere. Convex and concave mirrors are two types of spherical mirrors.

The **principal focus** or **focal point** of a spherical mirror is the point where rays parallel and very close to the **principal axis** all pass through (or appear to come from) after reflecting from the mirror. As shown in the following diagrams, the focal point (F) of the mirror is located halfway between the center of curvature and the center of the mirror. The focal length (f) equals one-half the radius of curvature of the mirror, i.e., $f = r/2$. The focal point is real for a concave mirror but virtual for a convex mirror.

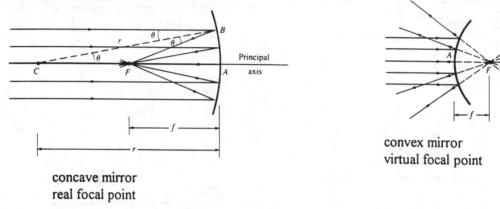

concave mirror
real focal point

convex mirror
virtual focal point

Images Formed by Spherical Mirrors

The characteristics of the images formed by spherical mirrors depend on the distance from the mirror to the object (d_o), the focal length of the mirror (f), and whether the mirror is concave or convex. The following summarizes the possibilities:

type of mirror	object distance as compared to the focal length	characteristics of the image
concave	$d_o > 2f$	real, inverted, diminished
	$d_o = 2f$	real, inverted, same size
	$f < d_o < 2f$	real, inverted, magnified
	$d_o = f$	no image formed
	$d_o < f$	virtual, erect, magnified

Convex mirrors produce only virtual, erect, and diminished images.

The images formed by spherical mirrors may be located by using two rays. 1) A ray from the top of the object reflects from the center of the mirror. The principal axis is the normal to the mirror at this point. The angles of incidence and reflection are easily measured so that the reflected ray may be drawn. 2) A ray from the top of the object, which is parallel to the principal axis, will reflect through the focal point of the concave mirror and appear to be coming from the focal point of a convex mirror.

The image of the top of the object is located at the point where the rays cross in the case of a real image and appear to cross after being traced behind the mirror in the case of a virtual image. The following diagrams represent each of the possibilities discussed in the table on the previous page. Note: d_i is the distance from the image to the mirror.

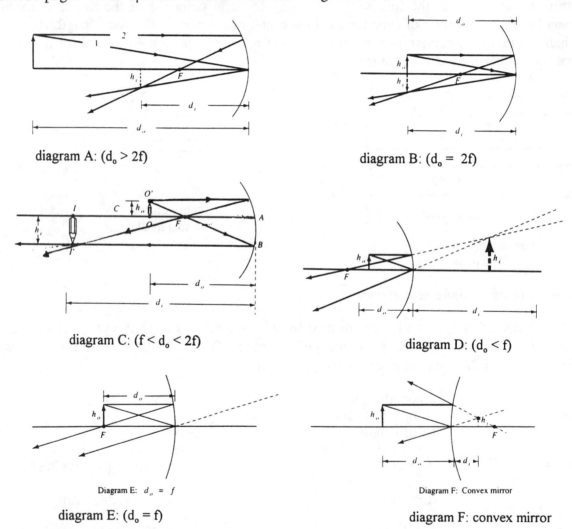

diagram A: ($d_o > 2f$)

diagram B: ($d_o = 2f$)

diagram C: ($f < d_o < 2f$)

diagram D: ($d_o < f$)

diagram E: ($d_o = f$)

diagram F: convex mirror

Mirror Equation

The mirror equation for both concave and convex spherical mirrors is as follows:

$1/d_o + 1/d_i = 1/f$ where $f = r/2$ and r is the radius of curvature of the mirror.

23-5

Magnification and Size of Image

The linear magnification (m) refers to the ratio of the size of the image (h_i) to the size of the object (h_o). The magnification produced by a curved mirror is given by

$$m = h_i/h_o = - d_i/d_o \quad \text{and} \quad h_i = m \, h_o$$

Sign Conventions

The following sign conventions are given in the text to be used with the mirror equations: When the object, image, or focal point is on the reflecting side of the mirror (on the left side in all of the drawings), the corresponding distance is considered positive. If any of these points are behind the mirror (on the right), the corresponding distance is considered to be negative. The object height and the image height are considered positive if they lie above the principal axis and negative if they lie below the principal axis. A negative sign is inserted in the magnification equation so that for an upright image the magnification is positive and for an inverted image it is negative.

Refraction

As light passes from one medium into another it changes direction at the interface between the two media. This change of direction is known as **refraction**. The relationship between the **angle of incidence** and the **angle of refraction** is given by **Snell's law**:

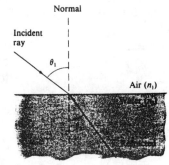

$$n_1 \sin \theta_1 = n_2 \sin \theta_2$$

θ_1 is the angle of incidence and is measured from the normal to the surface to the line of the incident ray.

θ_2 is the angle of refraction and is measured from the normal to the surface to the line of the refracted ray.

n_1 is the index of refraction of the medium in which the light is initially traveling. The index of refraction is a dimensionless number which varies from a low of 1.00 for a vacuum to 2.42 for diamond.

n_2 is the index of refraction of the medium into which the light passes.

Total Internal Reflection

As light passes from a medium of higher index of refraction into a medium of lower index of refraction an angle of incidence is reached at which the angle of refraction is 90°. This angle is known as the **critical angle** (θ_c) and at all angles greater than this angle, the light is totally reflected back into the medium of the incident ray. The equation for the critical angle can be determined by using Snell's law.

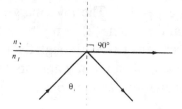

$$n_1 \sin \theta_c = n_2 \sin 90° \quad \text{where} \quad \sin \theta_c = n_2/n_1$$

Thin Lenses

Rays of light parallel to the principal axis of a **convex lens** converge at the focal point (F) of the lens after refraction. A convex lens, which is also known as a **converging lens**, has a real focal point.

Rays of light parallel to the principal axis of a **concave lens** diverge after passing through the lens. If the refracted rays are traced on a straight line back through the lens, they "appear" to converge at a focal point. A concave lens, which is also known as a **diverging lens**, has a virtual focal point.

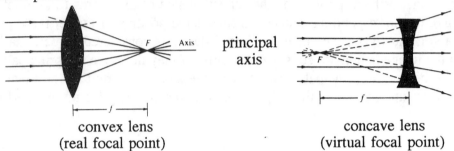

convex lens
(real focal point)

concave lens
(virtual focal point)

Images Formed by a Thin Lens

As in the case of images formed by spherical mirrors, the characteristics of images formed by thin lenses depend on the object distance (d_o), focal length of the lens (f), and whether or not the lens is convex or concave. The following table summarizes the possibilities.

type of lens	object distance as compared to the focal length	characteristic of image
convex	$d_o > 2f$	real, inverted, diminished
	$d_o = 2f$	real, inverted, same size
	$f < d_o < 2f$	real, inverted, magnified
	$d_o = f$	no image formed
	$d_o < f$	virtual, erect, magnified

Concave lenses produce only virtual, erect, and diminished images.

The images formed by thin lenses may be located by using the following two rays. 1) A ray from the top of the object which strikes the center of the lens. At this point the two sides of the lens are parallel. The ray emerges from the lens slightly displaced but traveling parallel to its original direction. 2) A ray from the top of the object which is parallel to the principal axis of a convex lens passes through the focal point of the side of the lens opposite the object. If the lens is concave, then the ray appears to have come from the virtual focal point located on the same side of the lens as the object.

For a real image, the image of any point on the object is located where the two rays from the point intersect after refraction. For a virtual image it is necessary to trace the path of the refracted rays back through the lens. The image is located at the point where the rays "appear" to cross.

The following diagrams represent each of the situations described in the table.

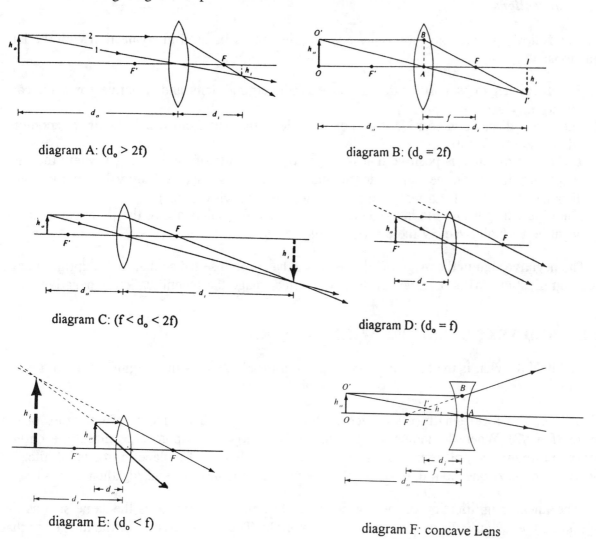

diagram A: ($d_o > 2f$)

diagram B: ($d_o = 2f$)

diagram C: ($f < d_o < 2f$)

diagram D: ($d_o = f$)

diagram E: ($d_o < f$)

diagram F: concave Lens

Lens Equation

The lens equation for both convex and concave lenses is

$$1/d_o + 1/d_i = 1/f$$

where d_o is the object distance from the lens, d_i is the image distance from the lens, and f is the focal length of the lens.

Magnification and Size

The linear magnification (m) of a lens is given by

$$m = h_i/h_o = -d_i/d_o \quad \text{and} \quad h_i = m\, h_o$$ where h_i is the image height and h_o is the object height.

Sign Conventions

The following sign conventions are given in the text to be used in connection with the lens equations:

1. The focal length is positive for a convex (converging) lens and negative for a concave (diverging) lens.
2. The object distance is positive if it is on the side of the lens from which the light is coming; otherwise it is negative.
3. The image distance is positive if it is on the opposite side of the lens from where the light is coming; if it is on the same side, the image distance is negative. Equivalently, the image distance is positive for a real image and negative for a virtual image.
4. The object height and the image height are positive for points above the principal axis and negative for points below the principal axis.

The negative sign in the magnification equation has been inserted so that for an upright image the magnification (m) is positive and for an inverted image the magnification is negative.

SELECTED TEXT QUESTIONS WITH ANSWERS

QUESTION 8. What is the focal length of a plane mirror? What is the magnification of a plane mirror?

ANSWER: The focal length of a curved mirror equals one-half the radius of curvature of the mirror ($f = r/2$). When the radius of curvature is very large, i.e., approaches infinity, a curved mirror approximates a plane mirror. Therefore, the focal length of a plane mirror is infinite. As shown in the next question, $d_i = -d_o$ for a plane mirror and since $m = -d_i/d_o$, then $m = +1$.

The linear magnification of the mirror is +1. Therefore, the image is the same size as the object and the + sign indicates that the image is upright. The characteristics of the image are that it is virtual, upright, and the same size as the object.

QUESTION 9. Does the mirror equation, eq. 23-3, hold for a plane mirror? Explain.

ANSWER: The mirror equation does hold for a flat mirror. The focal length of a flat mirror equals infinity. Substituting $f = \infty$ into the mirror equation gives

$$1/d_o + 1/d_i = 1/\infty \quad \text{but} \quad 1/\infty = 0 \quad \text{then} \quad 1/d_o + 1/d_i = 0$$

$$1/d_o = -1/d_i \quad \text{and} \quad d_i = -d_o$$

The image distance equals the object distance from the mirror and the negative sign indicates that the image is virtual.

QUESTION 12. What is the angle of refraction when a light ray meets the boundary between two materials perpendicularly?

ANSWER: When a light ray is perpendicular to the boundary between two materials the angle of incidence is $0°$. According to Snell's law :

$n_1 \sin \theta_1 = n_2 \sin \theta_2 \quad$ but $\quad \theta_1 = 0° \quad$ and $\quad \sin \theta_1 = 0$

$n_1 (0) = n_2 \sin \theta_2 \quad$ and $\quad \sin \theta_2 = 0 \quad$ and $\quad \theta_2 = 0°$

In this special case, the angle of refraction is zero and the light ray is not bent as it passes from one medium to the other.

QUESTION 19. Where must the film be placed if a camera lens is to make a sharp image of an object very far away?

ANSWER: For an ordinary fixed focal length camera, any distance beyond approximately 6 feet is very far away. Thus 6 feet or more approximates infinity in equation 23-3.

$1/d_o + 1/d_i = 1/f \quad$ but $\quad d_o = \infty \quad$ and $\quad 1/d_o = 1/\infty = 0$

$0 + 1/d_i = 1/f, \quad 1/d_i = 1/f, \quad$ and $\quad d_i = f$

The film must be placed at the focal point of the lens in order for a clear image to be formed.

PROBLEM SOLVING SKILLS

For problems involving image formation by a convex or a concave mirror:

1. Identify whether the mirror is concave or convex. The focal length is positive for a concave mirror and negative for a convex mirror.
2. Choose an appropriate scale factor to represent the focal length, object distance and height of the object.
3. Use the two suggested rays to draw an accurate ray diagram. Draw in the image at the point where the two rays cross after reflection from the mirror.
4. Use the mirror equations and sign conventions to mathematically determine the image distance, image height, and magnification.
5. State the characteristics of the image: real or virtual; erect or inverted; magnified, diminished, or same size as the object.

For problems involving refraction of light as it passes from one medium to another:

1. Draw an accurate diagram locating the incident ray and normal to the surface. Determine the angle of incidence.
2. Complete a data table using the information given in the problem.
3. Use Snell's law to solve the problem.
4. If the problem involves total internal reflection, then at the critical angle, the angle of refraction is 90°. Use Snell's law to determine the magnitude of the critical angle.

For problems involving image formation by a concave or a convex lens:

1. Identify whether the lens is concave or convex. The focal length is positive for a convex lens and negative for a concave lens.
2. Lens problems involve refraction of light while mirror problems involve reflection. Apply the steps listed above for spherical mirror problems to solve problems involving lenses.

PROGRAMMED PROBLEMS

a. 1	Solution: (Section 23-3)
Choose an appropriate scale factor and draw an accurate ray diagram.	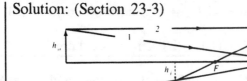

b. 1	$1/d_o + 1/d_i = 1/f$
Mathematically determine the image distance from the mirror, magnification, and height of the image.	$1/30 \text{ cm} + 1/d_i = 1/10 \text{ cm}$
	$1/d_i = 1/10 \text{ cm} - 1/30 = 2/30 \text{ cm}$
	$d_i = 15 \text{ cm}$
	$m = -d_i/d_o = -15 \text{ cm}/30 \text{ cm} = -0.50$
	$h_i = m\, h_o = (-0.50)(3.0 \text{ cm}) = -1.5 \text{ cm}$
	Note: recall that the negative sign for both the magnification and image height indicates that the object is real and inverted.

c. 1	The image is real, inverted, and diminished.
State the characteristics of the image.	

a. 1	Solution: (Section 23-3)
Choose an appropriate scale factor and draw an accurate ray diagram.	

b. 1	$1/d_o + 1/d_i = 1/f$
Mathematically determine the image distance from the mirror, magnification, and height of the image.	$1/60$ cm $+ 1/d_i = 1/30$ cm
	$1/d_i = 1/30$ cm $- 1/60$ cm $= (2 - 1)/60$ cm
	$1/d_i = 1/60$ cm and $d_i = 60$ cm
	$m = -d_i/d_o = -(60$ cm$)/(60$ cm$) = -1.0$
	$h_i = (-1.0)(2.0$ cm$) = -2.0$ cm

c. 1	The image is real, inverted, and the same size as the object.
State the characteristics of the image.	

PROBLEM 3. An object 1.0 cm high is placed 20 cm from a concave mirror of focal length 15 cm. a) Draw a ray diagram and locate the position of the image formed. Draw in the image. b) Mathematically determine the image distance from the mirror, magnification, and height of the image. c) State the characteristics of the image.

a. 1	Solution: (Section 23-3)
Choose an appropriate scale factor and draw an accurate ray diagram.	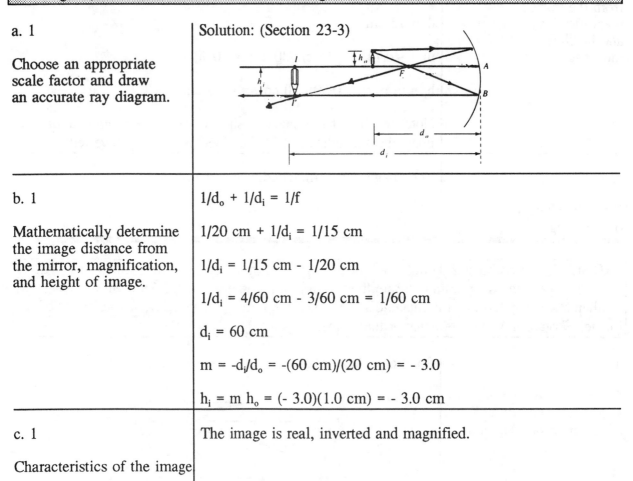

b. 1	$1/d_o + 1/d_i = 1/f$
Mathematically determine the image distance from the mirror, magnification, and height of image.	$1/20$ cm $+ 1/d_i = 1/15$ cm
	$1/d_i = 1/15$ cm $- 1/20$ cm
	$1/d_i = 4/60$ cm $- 3/60$ cm $= 1/60$ cm
	$d_i = 60$ cm
	$m = -d_i/d_o = -(60$ cm$)/(20$ cm$) = -3.0$
	$h_i = m\, h_o = (-3.0)(1.0$ cm$) = -3.0$ cm

c. 1	The image is real, inverted and magnified.
Characteristics of the image	

> **PROBLEM 4.** An object 3.0 cm high is placed 5.0 cm from a concave mirror of focal length 10 cm. a) Draw a ray diagram and locate the position of the image formed. Draw in the image. b) Mathematically determine the image distance from the mirror, magnification, and height of the image. c) State the characteristics of the image.

a. 1 Choose an appropriate scale factor and draw an accurate ray diagram.	Solution: (Section 23-3) 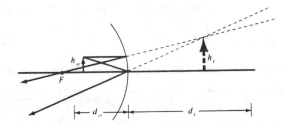

b. 1 Mathematically determine the image distance from the mirror, magnification, and height of image.	$1/d_o + 1/d_i = 1/f$ $1/5.0 \text{ cm} + 1/d_i = 1/10 \text{ cm}$ $1/d_i = 1/10 \text{ cm} - 1/5.0 \text{ cm} = 1/10 \text{ cm} - 2/10 \text{ cm}$ $d_i = -10 \text{ cm}$ $m = -d_i/d_o = -(-10 \text{ cm})/(5.0 \text{ cm}) = +2.0$ $h_i = m\, h_o = (+2.0)(3.0 \text{ cm}) = +6.0 \text{ cm}$
c. 1 State the characteristics of the image.	The positive sign for both the magnification and the image height indicates that the image is virtual and erect. The image is erect and magnified.

> **PROBLEM 5.** An object 3.0 cm high is placed 15 cm from a concave mirror of focal length 15 cm. a) Draw a ray diagram and locate the position of the image formed. Draw in the image. b) Mathematically determine the image distance from the mirror, magnification, and height of the image. c) State the characteristics of the image.

a. 1 Choose an appropriate scale factor and draw an accurate ray diagram.	Solution: (Section 23-3) 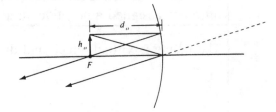

b. 1	$1/d_o + 1/d_i = 1/f$
Mathematically determine the image distance from the mirror, magnification, and height of image.	$1/15 \text{ cm} + 1/d_i = 1/15 \text{ cm}$
	$1/d_i = 1/15 \text{ cm} - 1/15 \text{ cm}$
	$1/d_i = 0$; therefore, d_i is undefined

c. 1	No image, either real or virtual, is formed.
State the characteristics of the image.	

PROBLEM 6. An object 3.0 cm high is placed 30 cm from a convex mirror of focal length 20 cm. a) Draw a ray diagram and locate the position of the image formed. Draw in the image. b) Mathematically determine the image distance from the mirror, magnification, and height of the image. c) State the characteristics of the image.

a. 1	Solution: (Section 23-3)
Choose an appropriate scale factor and draw an accurate ray diagram.	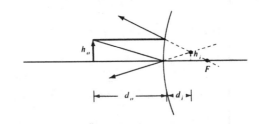

b. 1	$1/d_o + 1/d_i = 1/f$
Mathematically determine the image distance from the mirror, magnification, and height of the image.	$1/30 \text{ cm} + 1/d_i = 1/(-20 \text{ cm})$
	$1/d_i = -1/20 \text{ cm} - 1/30 \text{ cm} = -3/60 \text{ cm} - 2/60 \text{ cm} = -5/60 \text{ cm}$
	$d_i = -12 \text{ cm}$
	$m = -d_i/d_o = -(-12 \text{ cm})/(30 \text{ cm}) = +0.40$
	$h_i = (+0.40)(3.0 \text{ cm}) = +1.2 \text{ cm}$
	Note: the sign convention states that the focal length of a convex mirror is assigned a negative value.

c. 1	The image is virtual, erect, and diminished.
State the characteristics of the image.	

PROBLEM 7. A ray of light strikes the surface of a flat glass plate at an incident angle of 30°. The index of refraction of air and glass are 1.00 and 1.50, respectively. Determine the a) angle of reflection and b) angle of refraction.

a. 1	Solution: (Section 23-5)
Draw an accurate diagram showing the incident ray, the reflected ray, and the approximate path of the refracted ray.	The reflected ray follows the laws of specular reflection; thus' the angle of reflection is 30°.

b. 1

Use Snell's law to determine the angle of refraction.

$n_1 \sin \theta_1 = n_2 \sin \theta_2$

$(1.00) \sin 30° = (1.50) \sin \theta_2$

$\sin \theta_2 = (1.00)(0.50)/(1.50) = 0.33$

$\theta_2 = 19°$

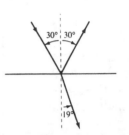

PROBLEM 8. Water fills the space between two parallel glass plates as shown in the figure below. If the angle of incidence of the ray of light entering the top glass plate is 60°, determine the angle of refraction of the light in the a) top glass plate and   b) water. c) Draw the approximate path taken by the ray of light as it passes through each medium.

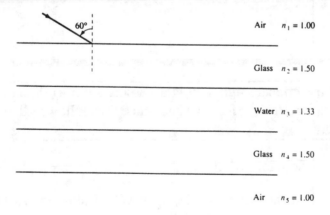

Air $n_1 = 1.00$

Glass $n_2 = 1.50$

Water $n_3 = 1.33$

Glass $n_4 = 1.50$

Air $n_5 = 1.00$

a. 1

Determine the angle of refraction in the top glass plate.

Solution: (Section 23-5)

$n_1 \sin \theta_1 = n_2 \sin \theta_2$

$(1.00) \sin 60° = (1.50) \sin \theta_2$

$\sin \theta_2 = (1.00)(0.87)/(1.50) = 0.58$

$\theta_2 = 36°$

b. 1

Determine the angle of incidence as the light strikes the glass-water interface. Next, determine the angle of refraction in the water.

Because the interface between the media is parallel, the 36° angle of refraction of the light as it passes through the top glass plate is equal to the angle of incidence as the light strikes the interface between the glass and the water.

$$n_2 \sin \theta_2 = n_3 \sin \theta_3$$

$$(1.50) \sin 36° = (1.33) \sin \theta_3$$

$$\sin \theta_3 = (1.50)(0.58)/(1.33) = 0.654$$

$$\theta_3 = 41°$$

---

c. 1

Draw the approximate path taken by the ray of light as it passes through each medium.

Applying geometry and Snell's law, it is possible to show that the angle of refraction in the lower glass plate is 36° and the angle in the air is 60°.

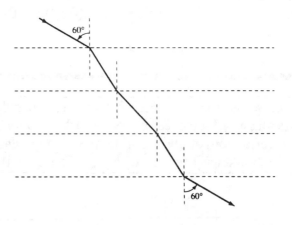

PROBLEM 9. Determine the critical angle for light passing from a) diamond into air and b) glass into water. The index of refraction for each substance is as follows: diamond 2.42, air 1.00, glass 1.50, and water 1.33.

---

a. 1

Use Snell's law to determine the critical angle for an air-diamond interface.

Solution: (Section 23-6)

At the critical angle, $\theta_{air} = 90°$. Using Snell's law,

$$n_{diamond} \sin \theta_{diamond} = n_{air} \sin \theta_{air}$$

$$(2.42) \sin \theta_c = (1.00) \sin 90°$$

$$\sin \theta_c = (1.00)(1.00)/(2.42) = 0.41$$

$$\theta_c = 24°$$

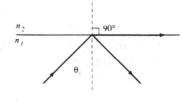

b. 1	$n_{glass} \sin \theta_{glass} = n_{water} \sin \theta_{water}$
Use Snell's law to determine the critical angle angle for a water-glass interface.	$(1.50) \sin \theta_c = (1.33) \sin 90°$
	$\sin \theta_c = (1.33)(1.00)/(1.50) = 0.89$
	$\theta_c = 63°$
	Note: total internal reflection occurs only when light is traveling from a medium of higher index of refraction into a medium of lower index of refraction.

PROBLEM 10. An object 3.0 cm high is placed 30 cm from a convex lens of focal length 10 cm. a) Draw a ray diagram and locate the position of the image formed. Draw in the image. b) Mathematically determine the image distance from the lens, magnification, and height of the image. c) State the characteristics of the image.

a. 1	Solution: (Sections 23-7 and 23-8)
Choose an appropriate scale factor and draw an accurate ray diagram.	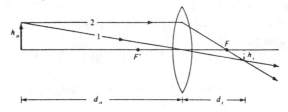

b. 1	$1/d_o + 1/d_i = 1/f$
Mathematically determine the image distance from the lens, magnification, and height of the image.	$1/30$ cm $+ 1/d_i = 1/10$ cm
	$1/d_i = 1/10$ cm $- 1/30 = 2/30$ cm
	$d_i = 15$ cm
	$m = - d_i /d_o = - 15$ cm$/30$ cm $= - 0.50$
	$h_i = m \, h_o = (- 0.50)(3.0$ cm$) = - 1.5$ cm
	Note: Recall that the negative sign for both the magnification and image height indicates that the object is real and inverted.

c. 1	The image is real, inverted, and diminished.
State the characteristics of the image.	

PROBLEM 11. An object 2.0 cm high is placed 60 cm from a convex lens of focal length 30 cm. a) Draw a ray diagram and locate the position of the image formed. Draw in the image. b) Mathematically determine the image distance from the lens, magnification, and height of the image. c) State the characteristics of the image.

a. 1	Solution: (Sections 23-7 and 23-8)
Choose an appropriate scale factor and draw an accurate ray diagram.	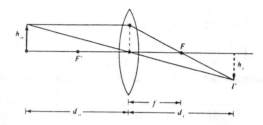

b. 1	$1/d_o + 1/d_i = 1/f$
Mathematically determine the image distance from the lens, magnification, and height of the image.	$1/60 \text{ cm} + 1/d_i = 1/30 \text{ cm}$
	$1/d_i = 1/30 \text{ cm} - 1/60 \text{ cm} = (2 - 1)/60 \text{ cm}$
	$1/d_i = 1/60 \text{ cm}$ and $d_i = 60 \text{ cm}$
	$m = -d_i/d_o = -(60 \text{ cm})/(60 \text{ cm}) = -1.0$
	$h_i = (-1.0)(2.0 \text{ cm}) = -2.0 \text{ cm}$

c. 1	The image is real, inverted, and the same size as the object.
State the characteristics of the image.	

PROBLEM 12. An object 5.0 cm high is placed 30 cm from a convex lens of focal length 20 cm. a) Draw a ray diagram and locate the position of the image formed. Draw in the image. b) Mathematically determine the image distance from the lens, magnification, and height of the image. c) State the characteristics of the image.

a. 1	Solution: (Sections 23-7 and 23-8)
Choose an appropriate scale factor and draw an accurate diagram.	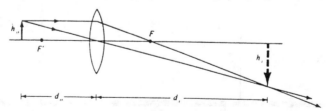

b. 1 Mathematically determine the image distance from the lens, magnification, and height of the image.	$1/d_o + 1/d_i = 1/f$ $1/30$ cm $+ 1/d_i = 1/20$ cm $1/d_i = 1/20$ cm $- 1/30$ cm $1/d_i = (3 - 2)/60$ cm $= 1/60$ cm $d_i = 60$ cm $m = -d_i/d_o = -(60 \text{ cm})/(30 \text{ cm}) = -2.0$ $h_i = m\ h_o = (-2.0)(5.0 \text{ cm}) = -10.0$ cm
c. 1 State the characteristics of the image.	The image is real, inverted, and magnified.

PROBLEM 13. An object 1.0 cm high is placed 10 cm from a convex lens of focal length 10 cm. a) Draw a ray diagram and locate the position of the image formed. Draw in the image. b) Mathematically determine the image distance from the mirror, magnification, and height of the image. c) State the characteristics of the image.

a. 1 Choose an appropriate scale factor and draw an accurate ray diagram.	Solution: (Sections 23-7 and 23-8) 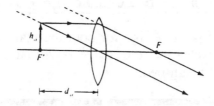
b. 1 Mathematically determine the image distance from the mirror, magnification, and height of the image.	$1/d_o + 1/d_i = 1/f$ $1/10$ cm $+ 1/d_i = 1/10$ cm $1/d_i = 1/10$ cm $- 1/10$ cm $1/d_i = 0, \quad d_i$ is undefined
c. 1 State the characteristics of the image.	No image, either real or virtual, is formed.

PROBLEM 14. An object 3.0 cm high is placed 5.0 cm from a convex lens of focal length 10 cm. a) Draw a ray diagram and locate the position of the image formed. Draw in the image. b) Mathematically determine the image distance from the lens, magnification, and height of the image. c) State the characteristics of the image.

a. 1	Solution: (Sections 23-7 and 23-8)
Choose an appropriate scale factor and draw an accurate ray diagram.	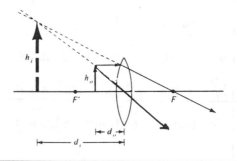

b. 1	$1/d_o + 1/d_i = 1/f$
Mathematically determine the image distance from the lens, magnification, and height of image.	$1/5.0$ cm $+ 1/d_i = 1/10$ cm
	$1/d_i = 1/10$ cm $- 1/5.0$ cm $= 1/10$ cm $- 2/10$ cm
	$d_i = -10$ cm
	$m = -d_i/d_o = -(-10$ cm$)/(5.0$ cm$) = +2.0$
	$h_i = m\, h_o = (+2.0)(3.0$ cm$) = +6.0$ cm

c. 1	The positive sign for both the magnification and the image height indicates that the image is virtual and erect. The image is erect and magnified.
State the characteristics of the image.	

PROBLEM 15. An object 3.0 cm high is placed 30 cm from a concave lens of focal length 20 cm. a) Draw a ray diagram and locate the position of the image formed. Draw in the image. b) Mathematically determine the image distance from the lens, magnification, and height of the image. c) State the characteristics of the image.

a. 1	Solution: (Section 23-4)
Choose an appropriate scale factor and draw an accurate ray diagram.	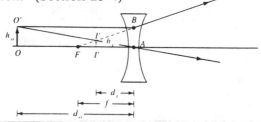

b. 1	$1/d_o + 1/d_i = 1/f$
Mathematically determine the image distance from the lens, magnification, and height of the image.	$1/30$ cm $+ 1/d_i = 1/(-20$ cm$)$
	$1/d_i = -1/20$ cm $- 1/30$ cm $= -3/60$ cm $- 2/60$ cm $= -5/60$ cm
	$d_i = -12$ cm
	$m = -d_i/d_o = -(-12$ cm$)/(30$ cm$) = +0.40$
	$h_i = (+0.40)(3.0$ cm$) = +1.2$ cm
	Note: the sign convention states that the focal length of a concave lens is assigned a negative value.
c. 1	The image is virtual, erect, and diminished.
State the characteristics of the image.	

PRACTICE PROBLEMS

PROBLEM 1. An object 5.0 cm tall is placed 25 cm from a spherical mirror and a real image 10.0 cm high is observed. a) Is the mirror concave or convex? b) Determine the distance from the mirror to the image. c) Determine the focal length of the mirror. d) Draw a ray diagram and locate the position of the image. Draw in the image. e) State the characteristics of the image.

ANS. a) concave, b) 50 cm, c) 17 cm, e) real, inverted, magnified

PROBLEM 2. The figure below represents the path of a ray of light as it passes from air into glass. Determine the index of refraction of the glass.

ANS. 1.46

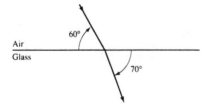

PROBLEM 3. The critical angle for light as it passes from glass into oil is 60.0°. If the index of refraction of the glass is 1.50, determine the index of refraction of the oil.

ANS. 1.30

PROBLEM 4. An object is placed 10.0 cm from a lens and a real image is formed on a screen placed 50.0 cm from the lens. Determine the a) focal length of the lens and b) magnification of the image.

ANS. a) 8.33 cm, b) - 5.0

CHAPTER 24

THE WAVE NATURE OF LIGHT

KEY TERMS AND PHRASES

interference patterns are produced when two (or more) coherent sources producing waves of the same frequency and amplitude superimpose.

sources are coherent when there is a fixed phase between the waves emitted by the sources.

crest is the highest point of that portion of a transverse wave above the equilibrium position.

trough is the lowest point of that portion of a transverse wave below the equilibrium position.

destructive interference occurs if the amplitude of the resultant of two interfering waves is smaller than the displacement of either wave.

constructive interference occurs if the amplitude of the resultant of two interfering waves is larger than the displacement of either wave.

diffraction occurs when waves bend and spread out as they pass an obstacle or narrow opening. The amount of diffraction depends on the wavelength of the waves and the size of the obstacle. In the case of a narrow opening, the amount of bending increases as the size of the opening decreases.

diffraction grating consists of a large number of closely spaced parallel slits which diffract light incident on the grating. The diffracted light will exhibit a pattern of constructive interference and destructive interference.

thin film interference patterns occur when light waves are reflected at both the top and bottom surfaces of the film. The pattern of dark and bright lines observed results from light reflected from the top surface interfering with the light reflected from the bottom surface.

polarization is a property of light that indicates light is a transverse wave phenomenon. A transverse wave is a wave in which the particles of the medium move at right angles to the direction of motion of the wave. An electromagnetic wave in which the electric vector is vibrating in only one plane is said to be plane polarized.

Brewster's angle is the angle of incidence at which maximum polarization of the reflected light occurs.

double refraction occurs when light passes through certain transparent crystals and the incident ray is separated into two refracted rays. One ray, known as the ordinary ray (or o ray), follows Snell's law. The second ray, known as the extraordinary ray (or e ray), does not follow Snell's law.

SUMMARY OF MATHEMATICAL FORMULAS

speed of light and the wavelength in a medium other than a vacuum.	$v = c/n$	The speed of light (v) in a medium is the ratio of the speed of light in a vacuum (c) to the medium's index of refraction (n).
	$\lambda_1 = \lambda_{vacuum}/n_1$	The wavelength of the light in a medium (λ_1) is the ratio of the wavelength in a vacuum to the index of refraction of the medium (n_1).
double slit interference	$m\lambda = d \sin \theta$	The relationship between the wavelength (λ), the distance between the slits (d), and θ is the grating angle for constructive interference. m is the order of the interference fringe, m = 0, 1, 2, 3, etc.
	$(m + \frac{1}{2})\lambda = d \sin \theta$	The relationship between the wavelength (λ), the distance between the slits (d), and θ is the grating angle for destructive interference.
single slit diffraction bright fringes dark fringes	$(m + \frac{1}{2})\lambda = D \sin \theta$ $m\lambda = D \sin \theta$	The pattern for bright or dark fringes in a single slit diffraction pattern depends on the wavelength (λ), the slit width (D), and the diffraction angle (θ). m = 1, 2, 3, etc.

thin film interference	$t = [(2m + 1)/4] \lambda_{film}$ or $4t = (2m + 1) \lambda_{film}$ where $\lambda_{film} = \lambda_{air}/n_{film}$	If the indices of refraction of the media above and below the thin film are lower (or higher) than the index of refraction of the medium of the thin film, then maximum reflection of light of a particular wavelength occurs if the thickness (t) of the film is an odd-number multiple of quarter wavelengths. Note: m = 0, 1, 2, etc.
	$t = m \lambda_{film}/2$ or $2t = m \lambda_{film}$	Minimum reflection of the light occurs if the thickness of the film is a whole-number multiple of half wavelengths. Note: m = 1, 2, etc.
		Note: if the media above and below the thin film are such that a phase change occurs at both the top and bottom surfaces, then the equation for maximum reflection is given by $2t = m \lambda_{film}$ and for minimum reflection by $4t = (2m + 1) \lambda_{film}$.
Michelson interferometer	$t = m \lambda/2$	The Michelson interferometer is a device which uses wave interference to determine the wavelength of light. By measuring the distance that one of the mirrors moves and the number of interference fringes (m) that pass the observer's field of view, the wavelength (λ) of the light can be determined.
Brewster's angle	$\tan \theta_p = n_2/n_1$	The angle of incidence at which maximum polarization (θ_p) of the reflected light occurs is known as Brewster's angle. The tangent of Brewster's angle (θ_p) is related to the ratio of the index of refraction (n_2) of the substance from which the light is reflected to the index of refraction (n_1) of the substance in which the light is initially traveling.

CONCEPT SUMMARY

Reflection

In chapter 11 it was observed that a **wave front** striking a straight barrier follows the law of reflection (see page 11-9). This law was found to apply to light in chapter 23.

Refraction

Water waves undergo **refraction** as they travel from deep to shallow water because the speed of the wave changes (see page 11-10). Refraction of light occurs as light travels from one medium to another and this was discussed in terms of Snell's law in chapter 23. The wave theory proposed by Christian Huygens (1629-1695) predicts that the speed of light is less in water or glass than in air. Measurements of the speed of light in various materials agree with the wave theory. The speed of light in a medium (v) is inversely related to the medium's index of refraction (n), i.e., $v = c/n$ where c is the speed of light in a vacuum and $c = 3.0 \times 10^8$ m/s.

For light traveling from medium 1 into medium 2

$n_2/n_1 = v_1/v_2 = \lambda_1/\lambda_2$ where λ is the wavelength of the light in the medium.

Interference

Two **coherent** sources producing waves of the same frequency and amplitude produce an interference pattern. The following diagram was used in chapter 12 (page 12-5) to demonstrate the pattern produced by sound waves. s_1 and s_2 represent the sources of the waves while point A is a point of **constructive interference** and point B is a point of **destructive interference**.

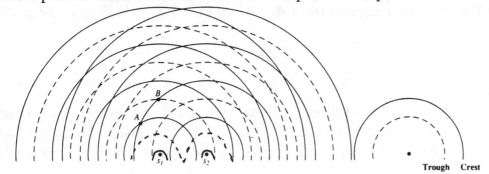

Trough   Crest

Young's Double Slit Experiment

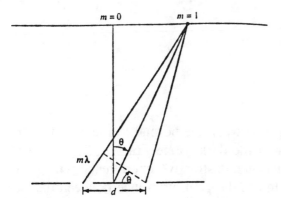

Young's double slit experiment shows that mono-chromatic light passing through two openings also produces an interference pattern. The position of lines of constructive interference can be determined from the following equation:

$$m \lambda = d \sin \theta$$

where d is the distance between the slits, and θ is the grating angle. This angle is equal to the angle formed by a line drawn from the center point between the slits to the region of constructive interference and the normal drawn to the center of the line which connects the slits. m is the order of the interference fringe. m is a dimensionless number which takes on integer values starting with zero, i.e., m = 0, 1, 2, 3, etc. λ is the wavelength of the monochromatic light incident on the double slit. $m\lambda$ is the path difference between the sources and the region of constructive interference. The path difference is a whole number of wavelengths for constructive interference.

Regions of destructive interference alternate with regions of constructive interference in an interference pattern. The position of lines of destructive interference can be found from

$(m + \frac{1}{2})\lambda = d \sin \theta$

where m, λ, d, and θ are the same as described previously and $(m + \frac{1}{2})\lambda$ is the path difference between the sources to the points of destructive interference.

Diffraction

The wave theory predicts and experiments confirm that waves will bend and spread out as they pass an obstacle or narrow opening. In the case of a narrow opening, the amount of bending increases as the size of the opening decreases.

Because the wavelength of light is much shorter than that of water waves or sound waves, the **diffraction** of light is only noticeable when the size of the opening is comparable to the wavelength of the light.

Single Slit Diffraction

Due to the combined effects of diffraction and interference, monochromatic light passing through a single slit of width D produces an interference pattern of alternating bright and dark lines. The positions of lines of destructive interference can be determined by the following equation:

$m\lambda = D \sin \theta$

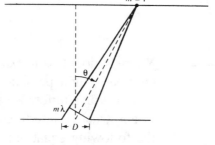

where D is the width of the opening and θ is the angle between the normal to the center of the slit and the center of the dark fringe. m is the order of the dark fringe, m = 1, 2, etc. For m = 0, the wave theory predicts that constructive rather than destructive interference will occur. λ is the wavelength of the light incident on the slit. $m\lambda$ is the path difference between the two edges of the slit and the position of the dark fringe.

Diffraction Grating

A **diffraction grating** consists of a large number of closely spaced parallel slits which diffract light incident on the grating. The diffracted light will exhibit constructive interference at points given by the equation

$$m\lambda = d \sin \theta$$

where m is the order of the bright fringe, m = 0, 1, 2, etc. λ is the wavelength of the incident light. d is the distance between the centers of the adjacent slits. The grating angle (θ) is measured from a line drawn normal to the center of the grating and a line drawn from the center of the grating to the position of the bright fringe.

The grating equation has the same form as the equation found in the Young double slit experiment. In the double slit experiment the distance d was measured between the centers of the two slits. In the diffraction grating equation d refers to the distance between the centers of adjacent slits. The amount of light passing through the diffraction grating is much greater than in the case of the double slit. As a result, the intensity of the bright lines is much greater.

A diffraction grating is particularly useful in separating the component wavelengths of the light incident on the grating. Because of this, it is frequently used in the analysis of spectrum produced by various gases, e.g., mercury, hydrogen, and helium.

Interference by Thin Films

On page 11-9, it was noted that a transverse wave pulse traveling through a light section of rope will undergo partial transmission and partial reflection at a point where the pulse enters the heavier section. If the wave pulse is a crest, then the transmission will be a crest but the reflection will be a trough (see diagram). A 180° phase change occurs for the reflected part of the wave. If the pulse is traveling in a heavy section and enters a light section, no phase change occurs for either the transmitted or reflected wave pulse.

Light waves traveling from one medium to another undergo partial reflection and partial transmission at the interface of the two mediums. By analogy with the rope of chapter 11, the medium with the lower index of refraction is analogous to the light section of the rope. The medium with the higher index of refraction is analogous to the heavy section of rope.

light medium   heavy medium   before   after   heavy medium   light medium

As shown in the diagram, reflection and transmission of light waves in **thin films** occurs at both the top and bottom surfaces of the film. The light reflected back to the observer is the result of light reflected from the top surface interfering with the reflection from the bottom surface.

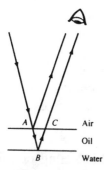

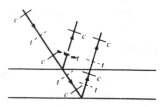

Light waves reflecting from both the top and bottom surfaces of a thin film.

Crests and troughs are represented by a solid line for a crest and a dotted line for a trough. Pulses are shown reflecting from both top and and bottom surfaces of the thin film.

If the indices of refraction of the media above and below the thin film are lower (or higher) than the index of refraction of the medium of the thin film, then **maximum** reflection of light of a particular wavelength occurs if the thickness (t) of the film is an odd-number multiple of quarter wavelengths. The equation for thickness of the film in terms of wavelength is

$$t = [(2m + 1)/4]/\lambda_{film} \quad \text{or} \quad 4t = (2m + 1) \lambda_{film} \quad \text{where} \quad \lambda_{film} = \lambda_{air}/n_{film} \quad \text{and} \quad m = 0, 1, 2, \text{etc.}$$

Minimum reflection of the light occurs if the thickness of the film is a whole number multiple of half wavelengths. The equation for minimum reflection is given by

$$t = m \lambda_{film}/2 \quad \text{or} \quad 2t = m \lambda_{film} \quad \text{where} \quad m = 0, 1, 2, \text{etc.}$$

If the media above and below the thin film are such that a phase change occurs at both the top and bottom surfaces, then the equation for **maximum reflection** is given by $2t = m \lambda_{film}$ and for **minimum reflection** by $4t = (2m + 1) \lambda_{film}$.

Since ordinary white light is made up of colors of varying wavelengths (400 nm - 700 nm), then maximum reflection of certain colors will occur while there will be no reflection of other colors. This and the fact that the thickness of the thin film usually varies across the surface of the film results in the spectrum of colors seen reflected from soap bubbles and oil films.

Michelson Interferometer

The **Michelson interferometer** is a device which uses wave interference to determine the wavelength of light. As shown in the diagram located at the top of the next page, the light from source S strikes a half-silvered mirror (M_s). Part of the light is reflected to a movable mirror (M_1) while part of the light is transmitted and reflects from a fixed mirror (M_2). After reflection, the beams arrive at the observer's eye.

Because the speed of light is less in glass than in air, a glass plate called the compensator (C) is placed in the path of beam 2. Thus, both beams travel through the same thickness of glass before arriving at the observer's eye.

Assume that light reaching the observer's eye from mirrors M_1 and M_2 are in phase. Constructive interference occurs and the observer sees light. If mirror M_1 is moved $\frac{1}{4}\lambda$ toward

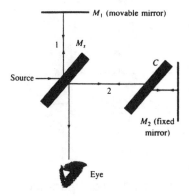

the observer, the path difference changes by $\frac{1}{4}\lambda + \frac{1}{4}\lambda = \frac{1}{2}\lambda$ and destructive interference occurs. If the mirror is moved another $\frac{1}{4}\lambda$, then the total path difference has changed by $\frac{1}{2}\lambda + \frac{1}{2}\lambda = 1\lambda$ from its original setting and constructive interference again occurs.

The distance (t) that the mirror M_1 moves can be determined by a device known as a micrometer. The wavelength of the monochromatic light from source S can be determined by counting the number of bright (or dark) fringes (m) which pass the field of view as M_1 is moved. The equation used is $t = m\lambda/2$.

Polarization of Light

The interference phenomena previously studied in this chapter can be produced by longitudinal as well as transverse waves. **Polarization** is a property of light that indicates light is a **transverse wave** phenomena.

An electromagnetic wave in which the electric vector is vibrating in only one plane is said to be **plane polarized**. Ordinary light is not polarized, which means that its electric vector is vibrating in many planes at the same time.

There are several ways to polarize light. One method is to use a material such as polaroid that removes all of the electric vectors except those in a particular plane. The polaroid filter that causes the electric vector to vibrate in only one plane is known as the polarizer. The axis of a second polaroid filter, known as the analyzer, can be crossed with the axis of the polarizer and block all light to the observer.

A second method to polarize unpolarized light is to reflect ordinary light from materials such as glass or water. The angle of incidence at which maximum polarization of the reflected light occurs is known as **Brewster's angle**. Brewster's angle θ_p can be found by using the equation

$$\tan \theta_p = n_2/n_1$$

where n_1 is the index of refraction of the medium in which the light is initially traveling and n_2 is the index of refraction of the medium from which the light is reflected.

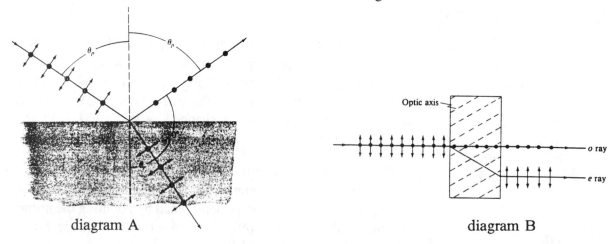

diagram A   diagram B

As shown in diagram A, at Brewster's angle the reflected light is plane polarized parallel to the surface. The angle between the refracted ray and the reflected ray is 90°.

A third method to polarize unpolarized light is by **double refraction**. When light passes through a crystal such as calcite, the incident ray is separated into two refracted rays (see diagram B). One ray, known as the ordinary ray (or o ray), follows Snell's law. The second ray, known as the extraordinary ray (or e ray), does not follow Snell's law. The two emerging beams are both plane polarized. The plane of polarization of the o ray is perpendicular to the plane of polarization of the e ray.

SELECTED TEXT QUESTIONS WITH ANSWERS

QUESTION 4. Two rays of light from the same source destructively interfere if their path lengths differ by how much?

ANSWER: In order for destructive interference to occur, the path length must differ by an odd-numbered multiple of one-half wavelength, i.e., path difference = $(m + \frac{1}{2})\lambda$ where m = 0, 1, 2, etc.

QUESTION 9. Why doesn't the light from two headlights of a distant car produce an interference pattern?

ANSWER: The light from the two headlamps is not coherent. The light produced by one light is produced at random with respect to the second light. In order to produce an interference pattern, the light waves must be coherent, i.e., the emitted waves maintain the same phase relationship to one another at all times.

QUESTION 22. What does polarization tells us about the nature of light?

ANSWER: Polarization indicates that light must be a transverse wave rather than a longitudinal wave. As discussed in chapter 12, sound waves are longitudinal waves which exhibit interference when emitted by two sources which are in phase. Sound waves are longitudinal waves; the molecules of the medium vibrate only along the direction of motion of the wave. There is no way that a longitudinal wave can be plane polarized.

QUESTION 24. How can you tell if a pair of sunglasses is polarizing or not?

ANSWER: Look for glare reflecting from the floor, a lake, or even the top of a car. Make sure that the angle between the normal to the surface and your eye is 45° or more. Hold one lens in front of your eye and rotate the lens. If the lens is made of polarizing material, the intensity of the light reaching your eye will change as the lens rotates.

PROBLEM SOLVING SKILLS

For problems involving double slit interference, single slit diffraction, or a diffraction grating:

1. Determine whether the light is passing through a double slit, a single slit or a diffraction grating.
2. Draw an accurate diagram labeling d or D, ΔX, L, the order of the image (m) and the path difference $m\lambda$.
3. Complete a data table based on the information given in the problem.
4. Note whether the problem involves constructive interference (bright fringes) or destructive interference (dark fringes).
5. Choose the appropriate formula and solve the problem.

For problems involving thin film interference:

1. Assume a crest is incident on the top surface of the film.
2. Determine whether the subsequent reflections and transmissions from each interface will be a crest or trough.
3. Draw a diagram placing a t for trough and c for crest at each position where the light is reflected and transmitted.
4. Determine the minimum thickness of the film in terms of a fraction of a wavelength of the incident light required to produce a) maximum reflection of the light and b) minimum reflection of the light.
5. Based on step 4, write a formula for film thicknesses that will produce a) maximum reflection of the light and b) minimum reflection of the light.
6. If the wavelength of the light in air is given, it is necessary to determine the wavelength of the light in the film.
7. Complete a data table, choose the appropriate formula, and solve the problem.

For problems involving the angle of maximum plane polarization of reflected light:

1. Determine the relative index of refraction of the two mediums, i.e., the ratio of the index of refraction of the medium from which the light is reflected to the index of refraction of the medium in which the light is incident.
2. Use $\tan \theta_p = n_2/n_1$ to solve for Brewster's angle (θ_p).

PROGRAMMED PROBLEMS

a. 1	Solution: (Section 24-2)
Determine the speed of light in water.	The speed of light in a vacuum is 3.00×10^8 meters per second and the index of refraction of a vacuum is 1.00. Assume that medium 1 is the vacuum and medium 2 is the water.
	$n_2/n_1 = v_1/v_2$
	$1.33/1.00 = (3.00 \times 10^8 \text{ m/s})/v_2$
	$v_2 = 2.26 \times 10^8$ m/s
b. 1	1 nm = 1×10^{-9} m; however, it is appropriate to express the wavelength of light in nanometers (nm). There is no need to convert the wavelength to meters.
Determine the wavelength of the light in water.	$n_2/n_1 = \lambda_1/\lambda_2$
	$1.33/1.00 = 450 \text{ nm}/\lambda_2$
	$\lambda_2 = 338$ nm

a. 1	Solution: (Section 24-3)
Convert all of the data to SI units.	0.500 mm = 5.00×10^{-4} m
	0.120 cm = 1.20×10^{-3} m.
a. 2	$\tan \theta = \Delta X/L = (1.20 \times 10^{-3} \text{ m})/(1.0 \text{ m})$
Refer to the diagram at the bottom of page 24-4 and determine the angular deflection for m = 1.	$\tan \theta = 1.20 \times 10^{-3}$
	The angle is small, and if the angle is expressed in radians, then $\sin \theta \approx \tan \theta \approx \theta$
	and $\theta = 1.20 \times 10^{-3}$ rad

a. 3	$m \lambda = d \sin \theta_1$
Determine the wavelength of the light.	$(1)(\lambda) = (5.00 \times 10^{-4} \text{ m})(1.20 \times 10^{-3})$
	$\lambda = 6.00 \times 10^{-7}$ m or 600 nm
b. 1	The maximum value that θ can have is 90°.
Determine the maximum number of bright fringes that can be observed.	$m \lambda = d \sin \theta$
	$m (6.00 \times 10^{-7} \text{ m}) = (5.00 \times 10^{-4} \text{ m})(\sin 90°)$
	$m = (5.00 \times 10^{-4} \text{ m})(1.00)/(6.00 \times 10^{-7} \text{ m})$
	$m = 833$
	833 fringes are formed on each side of the central maximum. Thus, counting the central fringe, the theoretical maximum number of lines that could be observed would be 833 + 833 + 1 = 1667. However, it is necessary to round the number to three significant figures; thus the answer would be 1670.

PROBLEM 3. A laser beam is incident on two slits separated by 0.500 mm. The wavelength of the light produced by the laser is 600 nm. The interference pattern is formed on a screen 1.00 m from the slits. Determine the a) angular deflection of the second-order dark fringe (dark line), and b) distance from the central maximum to this fringe.

a. 1	Solution: (Section 24-3)
Determine the angular deflection of the second order dark fringe.	$(m + \frac{1}{2})\lambda = d \sin \theta_2$
	$(2 + \frac{1}{2})(6.00 \times 10^{-7} \text{ m}) = (5.00 \times 10^{-4} \text{ m}) \sin \theta_2$
	$\sin \theta_2 = 3.00 \times 10^{-3}$
	The angle is small and expressed in radians,
	then $\sin \theta_2 \approx \tan \theta_2 \approx \theta_2$ and
	$\theta_2 = 3.00 \times 10^{-3}$ radians
b. 1	$\tan \theta_2 = \Delta X/L$
Determine the distance from the central maximum to the second order dark fringe.	$\Delta X = L \tan \theta_2 = (1.00 \text{ m})(3.00 \times 10^{-3} \text{ rad})$
	$\Delta X = 3.00 \times 10^{-3}$ m = 3.00 mm

a. 1	Solution: (Section 24-5)
Refer to the diagram on page 24-5 and determine the angular deflection (θ_2) of the second-order dark fringe.	$m\lambda = D \sin \theta_2$ $$(2)(6.30 \times 10^{-7} \text{ m}) = (0.300 \times 10^{-3} \text{ m})(\sin \theta_2)$$ $$\sin \theta_2 = [(2)(6.30 \times 10^{-7} \text{ m})]/(0.300 \times 10^{-3} \text{ m})$$ $$\sin \theta_2 = 4.20 \times 10^{-3}$$ θ is small and expressed in rad; then $\sin \theta \approx \tan \theta \approx \theta$, and $$\theta_2 = 4.20 \times 10^{-3} \text{ radians}$$
a. 2 Determine the distance (ΔX) from the center of the central maximum to the second-order dark fringe.	$\tan \theta_2 = \Delta X/L$ $$\Delta X = L \tan \theta_2 = (1.00 \text{ m})(4.20 \times 10^{-3} \text{ rad})$$ $$\Delta X = 4.20 \times 10^{-3} \text{ m or } 4.20 \text{ mm}$$

a. 1	Solution: (Section 24-6)
Determine the distance between adjacent grooves in the diffraction grating in meters.	There are 15000 grooves per inch; then the distance between adjacent grooves is 1/15000 inch. $$d = (1 \text{ inch}/15000 \text{ grooves})(2.54 \text{ cm}/1 \text{ inch})(1 \text{ m}/100 \text{ cm})$$ $$d = 1.69 \times 10^{-6} \text{ m}$$
a. 2 Determine the first-order angular deflection.	$m\lambda = d \sin \theta_1$ $$(1)(6.30 \times 10^{-7} \text{ m}) = (1.69 \times 10^{-6} \text{ m}) \sin \theta_1$$ $$\sin \theta_1 = 0.372 \quad \text{and} \quad \theta_1 = 21.9°$$

b. 1 Determine the second-order angular deflection.	$m\lambda = d \sin \theta_2$ $(2)(6.30 \times 10^{-7} \text{ m}) = (1.69 \times 10^{-6} \text{ m}) \sin \theta_2$ $\sin \theta_2 = 0.746$ and $\theta_2 = 48.2°$
c. 1 Determine the distance from the central maximum to the second-order bright fringe.	$\tan \theta_2 = \Delta X/L$ $\Delta X = L \tan \theta_2 = (2.00 \text{ m})(\tan 48.2°)$ $\Delta X = 2.24 \text{ m}$
d. 1 Determine the maximum number of bright fringes that can be observed.	$m\lambda = d \sin \theta$ The maximum possible value of θ is 90°; thus $m(6.30 \times 10^{-7} \text{ m}) = (1.69 \times 10^{-6} \text{ m})(\sin 90°)$ $m = 2.7$ Since m must be an integer, only two bright fringes can be observed on either side of the central bright fringe. The maximum number of fringes that can be observed is 2 + 2 + 1 = 5.

PROBLEM 6. A ring of wire is dipped into a soap solution and then held so that the soap film on the ring is vertical. As the soap film gradually drains toward the bottom, a dark band appears at the top with alternating bright and dark bands of light appearing along the length of the film. Determine the thickness of the film at the first three bright bands, as counted from the top, if the incident light has a wavelength of 600 nm and is directed perpendicular to the surface of the film. The index of refraction of the soap solution is 1.33.

a. 1 Assume that a wave crest (c_1) strikes the top interface. Determine whether each subsequent transmission and reflection will be a crest or a trough.	Solution: (Section 24-8) Since $n_{film} > n_{air}$, then a phase change occurs and the reflection of c_1 at the top interface will be a trough (t_1). No phase change occurs for transmitted light so the crest (c_1) will strike the lower interface as a crest. At the lower interface the crest (c_1) will reflect as a crest (c_1'). This is because $n_{film} > n_{air}$. When c_1' strikes the top interface, it will be transmitted as a crest.
a. 2 Draw a diagram using c for crest and t for trough for the situation described in part a. 1.	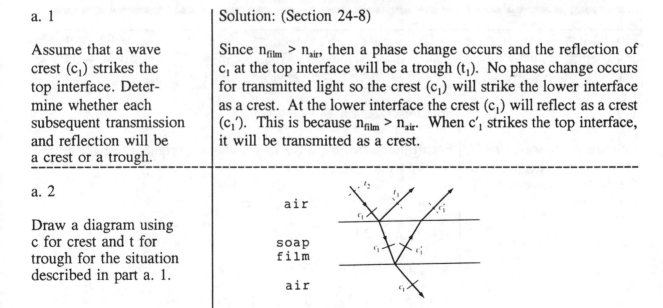

a. 3 Derive a formula for the film thickness that results in maximum reflection reflection of the incident light.	If the thickness of the film is $\frac{1}{4}\lambda$, then c_1' will have traveled $\frac{1}{4}\lambda_{film} + \frac{1}{4}\lambda_{film} = \frac{1}{2}\lambda_{film}$ and c_1' will meet trough t_2 which is reflecting at the top interface as a crest c_2. The superposition of c_1' and c_2 will result in constructive interference and maximum reflection. Thus, maximum reflection occurs if the thickness of the film follows the equation $t = (2m + 1)\lambda_{film}/4$ or $4t = (2m + 1)\lambda_{film}$ where $\lambda_{film} = \lambda_{air}/n_{film} = 600$ nm$/1.33 = 450$ nm
a. 4 Determine the thickness of the film at the position of the first three bright bands as counted from the top of the film.	The minimum thickness for maximum reflection occurs if $m = 0$. Since the film drains toward the bottom, the minimum thickness is located at the position of the first bright band observed near the top of the film. $4t = (2m + 1)\,\lambda_{film}$ If $m = 0$, then $4t = [2(0) + 1](450$ nm$)$ and $t = 113$ nm $= 1.13 \times 10^{-7}$ m Successive bright bands occur for $m = 1$, $m = 2$. If $m = 1$, then $4t = [2(1) + 1](450$ nm$)$ and $t = 338$ nm $= 3.38 \times 10^{-7}$ m If $m = 2$, then $4t = [2(2) + 1](450$ nm$)$ and $t = 563$ nm $= 5.63 \times 10^{-7}$ m

PROBLEM 7. The bright green line produced by the mercury spectrum is used as the source in an experiment involving the Michelson interferometer. When the movable mirror is moved 0.273 mm, 1000 bright fringes are counted. Determine the wavelength of the light.

a. 1 Determine the wavelength of the light. Note: refer to the diagram at the top of page 24-8.	Solution: (Section 24-9) Each time mirror M_1 travels $\frac{1}{2}\lambda$, another bright fringe passes the field of view. Therefore $t = m\,\lambda/2$ and rearranging gives $\lambda = 2t/m = 2(0.273$ mm$)/1000$ $\lambda = 2(0.273 \times 10^{-3}$ m$)/1000$ $\lambda = 0.546 \times 10^{-6}$ m $= 5.46 \times 10^{-7}$ m $= 546$ nm

a. 1	Solution: (Section 24-10)
Determine Brewster's angle for an air-glass interface.	$\tan \theta_p = n_2/n_1$
	$\tan \theta_p = 1.65/1.00$
	$\theta_p = 58.8°$
b. 1	$\tan \theta_p = 1.65/1.33$
Determine Brewster's angle for a water-glass interface.	$\tan \theta_p = 1.24$
	$\theta_p = 51.1°$

PRACTICE PROBLEMS

PROBLEM 1. In a laboratory experiment involving the mercury spectrum, a 600 groove per mm diffraction grating is used to observe a mercury light source. The first-order green fringe is observed to be 34.7 cm to the right of a 1.00 meter long center line which connects the mercury light source and the grating. Determine the wavelength of the green fringe.

ANS. 546 nm

PROBLEM 2. A laser beam of wavelength 630 nm is incident on a single slit of unknown width. The diffraction pattern is formed on a screen 2.0 m from the slit. The distance from the center of the central maximum to the second-order dark fringe is 2.0 mm. Determine the slit width.

ANS. 1.26 mm

PROBLEM 3. Two optically flat glass plates are separated at one end by a wire 0.0273 mm in diameter. Monochromatic light is incident on the plates and 100 dark fringes are observed. Determine the wavelength of the light.

ANS. 546 nm

PROBLEM 4. Light strikes a glass plate of index of refraction 1.50 at Brewster's angle. Determine a) Brewster's angle for the glass, b) the angle of refraction of light in the glass, and c) the angle between the reflected ray and the refracted ray.

ANS. a) 56.3°,   b) 33.7°,   c) 90°

CHAPTER 25

OPTICAL INSTRUMENTS

KEY TERMS AND PHRASES

simple camera consists of a convex lens, light-tight box, photographic plate or film, shutter, and an iris diaphragm or stop. The image of the object is formed on the film. The image is real, inverted, and diminished in size.

human eye functions in a manner similar to a simple camera. Light passes through a transparent outer membrane called the cornea, a clear liquid called the aqueous humor, and the lens into a cavity which is filled with a second clear liquid called the vitreous humor. The image is formed on the back of the eye on the retina and transmitted to the brain via the optic nerve.

rods on the retina are sensitive to the intensity level of light while the **cones** distinguish color.

fovea is the region of the retina where the cones are most densely packed. Vision is most acute in the fovea.

accommodation refers to the ability of the lens of the eye to change shape in order to bring objects into focus.

myopia or nearsightedness is a condition where a person can see nearby objects clearly but objects at a distance are blurred.

hyperopia or farsightedness is a condition where a person can see distant objects clearly but nearby objects are blurred.

diopter is a measure of the refractive power of a lens. The power in diopters is inversely related to the focal length expressed in meters.

magnifying glass is a converging lens which is used to produce a virtual, upright, and enlarged image of an object.

near point is the closest distance from the eye that an object can be placed and still be focused on the retina.

astronomical telescope, a refraction type, consists of a large-diameter convex lens with a long focal length (f_o) called the objective and a small-diameter convex lens with a short focal length (f_e) as the eyepiece. A magnified, inverted, virtual image of an object, e.g., a planet, is observed.

compound microscope consists of two convex lenses, an objective, and an eyepiece separated by a distance. The object is placed just beyond the focal point of the objective. The image observed is virtual, inverted, and greatly magnified.

spherical aberration occurs when rays of light from a point on the axis of the lens do not produce a point image after refraction. Instead, they produce a small circular patch of light.

chromatic aberration occurs because the index of refraction of transparent materials varies with wavelength. As a result, white light from a point source produces an image containing colors spread out over a small region.

achromatic doublet is a lens combination of two lenses used to correct chromatic aberration. One lens is a convex lens and the other a concave lens. The lenses have different indices of refraction and radii of curvature.

Rayleigh criterion states that two images are just resolvable when the center of the diffraction disk of one image is directly over the first minimum of the diffraction pattern of the other.

X-rays are electromagnetic waves of very short wavelength emitted when high energy electrons strike a metal target.

X-ray diffraction patterns are produced as a result of reflection of an X-ray beam from the surface of a crystal. The subsequent reflection of the X-rays from the planes of atoms results in a pattern of constructive and destructive interference.

SUMMARY OF MATHEMATICAL CONCEPTS

f-stop	f-stop = f/D	The f-stop refers to the adjustment to the size of the opening necessary to compensate for the outside brightness and also the shutter speed. f is the focal length of the lens and D is the diameter of the opening.
angular magnification	$M = \theta'/\theta$	The angular magnification or magnifying power (M) is the ratio of the angle the object subtends with the eye (θ') when the magnifier is used to the angle the object subtends with the eye (θ) without the magnifier at a distance of 25 cm.
magnification of a magnifying glass with the image at the near point	$M = 1 + N/f$	The magnification depends on the near point (N) and the focal length of the lens. N is usually taken to be 25 cm.

magnification of a magnifying glass with the image at infinity	$M = N/f$	If the eye is relaxed, the image is at infinity and the object is then precisely at the focal point. The magnification (M) equals the ratio of the near point to the focal length.
magnification of a refracting telescope	$M = -f_o/f_e$	The magnification (M) equals the ratio of the focal length of the objective lens (f_o) to the focal length of the eyepiece (f_e). The negative sign indicates an inverted image.
magnification of a compound microscope	$M = M_e/M_o$ or $M \approx N/(f_e\,f_o)$	The total magnification (M) of a compound microscope equals the product of the magnification of the eyepiece (M_e) and the magnification of the objective (M_o). The approximation is accurate when f_e and f_o are small compared to the distance between the lenses.
Rayleigh criterion	$\theta_{min} = 1.22\ \lambda/D$	The Rayleigh criterion is used to determine when two images can be resolved. The minimum angle (θ_{min}) for resolution depends on the wavelength (λ) of the light and the diameter of the objective lens (D).
resolving power of a microscope	$RP = S = f\,\theta = 1.22\lambda f/D$	For a microscope, the resolution is defined in terms of the resolving power (RP) where S is the minimum separation of two objects that can just be resolved and f is the focal length of the objective lens.
X-ray diffraction Bragg equation	$m\lambda = 2\,d\,\sin\theta$	If an X-ray beam is incident on the crystal, the subsequent reflection of the X-rays from the planes of atoms results in a diffraction pattern. Constructive interference occurs when $m\lambda = 2d\sin\theta$ where $m\lambda$ is a whole number of wavelengths and $m = 1, 2, 3$, etc. d is the distance between the layers of atoms and θ is the angle of incidence of the x-ray beam with the surface of the crystal.

CONCEPT SUMMARY

The Camera

A simple **camera** consists of a convex lens, light-tight box, photographic plate or film, shutter, and an iris diaphragm or stop.

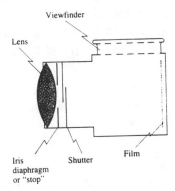

Three main adjustments must be made in order to produce a clear image of the object on the film: shutter speed, f-stop, and distance from lens to film. The shutter speed determines the length of time that the shutter is open to allow light to pass through to the film. Shutter speed varies from a second or more for a long exposure to 1/1000 second or less to capture the image of a moving object and avoid a blurred picture.

The amount of light reaching the film is controlled by the iris diaphragm or stop. The stop blocks light which has passed through the lens from reaching the film. The size of the opening is adjusted to compensate for the outside brightness and also the shutter speed and is specified in terms of the f-stop. The f-stop is given by the equation

f-stop = f/D   where f is the focal length of the lens and D is the diameter of the opening

Depending on the distance from the object to the lens, it may be necessary to adjust the distance between the lens and the film in order to produce a sharp image on the film. This is accomplished by turning a ring on the lens which moves the lens toward or away from the film. On inexpensive cameras, the optics are arranged so that all objects beyond a certain distance, usually 6 feet, are in focus and no adjustment is necessary.

The Human Eye

The basic structure of the **human eye** is shown in the diagram at the top of the next page. Light passes through a transparent outer membrane called the **cornea**, a clear liquid called the **aqueous humor**, and the lens into a cavity which is filled with a second clear liquid called the **vitreous humor**. The image is formed on the back of the eye on the **retina** and transmitted to the brain via the **optic nerve**. The image formed on the retina is received by millions of light-sensitive receptors known as **rods and cones**. The rods are sensitive to the intensity level of light while the cones distinguish color. Vision is most acute at the **fovea**, a region where the cones are most densely packed.

The amount of light passing through the eye is controlled by a diaphragm called the **iris** which opens and closes automatically as the eye adjusts to light intensity. The opening in the center of the iris is called the pupil of the eye.

The lens of the eye changes shape and therefore its focal length in order to focus the image of an object on the retina. For distant objects, the **ciliary muscles** relax, the lens becomes thinner and the focal length greater, and the image is focused on the retina. For nearby objects, the muscles contract causing the curvature of the lens to increase and decreasing the focal length. Again the image comes to a focus on the retina. The ability of the eye to adjust in this manner is called **accommodation**.

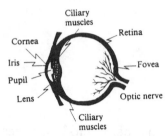

Common defects of the eye include nearsightedness or **myopia** and farsightedness or **hyperopia**. In nearsightedness, a person can see nearby objects clearly but objects at a distance are blurred. This is usually because the eyeball is too long and the image comes to a focus in front of the retina. This condition can be corrected by a concave or diverging lens which causes the rays to come to focus at the retina.

A farsighted person can see distant objects clearly but a nearby object, e.g., the print on this page, is blurred. This is usually caused by an eyeball that is too short. The rays from the object have not yet come to a focus when they strike the retina. A convex or converging lens is used to converge the rays so that the image comes to a focus at the retina. When prescribing eyeglasses, the power (P) of a lens, expressed in **diopters**, is used in place of the focal length. $P = 1/f$ where the focal length (f) is expressed in meters. P is positive for a converging lens and negative for a diverging lens.

The Magnifying Glass

A simple **magnifier** or magnifying glass is a converging lens which is used to produce an enlarged image of an object on the retina of the eye. The diagram shown below shows the virtual image of an object produced when the lens is used as a magnifier compared to the object viewed by the unaided eye focused at its near point.

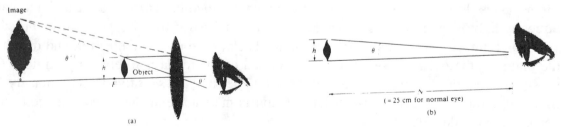

The **angular magnification** or **magnifying power M** is given by $M = \theta'/\theta$. θ' is the angle the object subtends with the eye when the magnifier is used and θ is the angle the object subtends with the eye without the magnifier at a distance of 25 cm. The angular magnification can be written in terms of the focal length of the lens as follows:

$$M = 1 + N/f = 1 + 25 \text{ cm}/f$$

N is the near point, which is the closest distance from the eye that an object can be placed and still be focused on the retina. N is usually taken to be 25 cm. If the eye is relaxed when using the magnifying glass, the image is then at infinity and the object is then precisely at the focal point. In this instance $M = N/f$.

Astronomical Telescope

A refraction-type astronomical **telescope** consists of a large-diameter convex lens with a long focal length (f_o) called the **objective** and a small-diameter convex lens with a short focal length (f_e) as the **eyepiece**. As shown in the diagram, distant objects such as stars and planets can be considered to be an infinite distance from the telescope. As a result, the image (I_1) is located at the focal point of the objective. The eyepiece is positioned so that the image produced by the objective lens is at or just inside the focal point of the eyepiece. A magnified, inverted virtual image (I_2) of the object is observed.

For an object at infinity, the distance between the lenses is $f_o + f_e$ and the magnifying power of the telescope is $M = \theta'/\theta = -f_o/f_e$ where the negative sign indicates an inverted image.

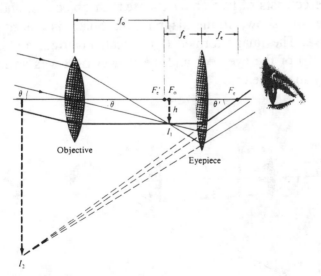

The largest astronomical telescopes employ a concave mirror as the objective. In a **Newtonian focus telescope,** the light from the objective is reflected to a plane mirror. The plane mirror reflects the light to a convex lens which acts as the eyepiece.

In a **Cassegrainian focus**, the light is reflected from the objective to a small convex mirror which then reflects the light through a hole in the objective.

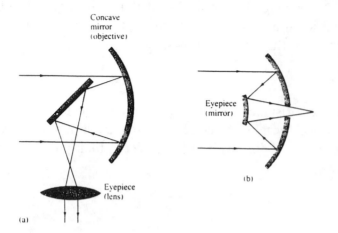

A concave mirror can be used as the telescope of an astronomical telescope. Either a lens or a mirror can be used as the eyepiece. Arrangement (a) is called the Newtonian focus and (b) the Cassegrainian focus. Other arrangements are also possible.

Terrestrial Telescope

A **terrestrial telescope** is designed so that an erect image is observed. The **Galilean telescope** uses a convex lens of long focal length as the objective and a concave lens of short focal length as the eyepiece. The eyepiece is a concave lens, its focal length is negative, and the angular magnification is positive. The angular magnification is given by $M = - f_o / f_e$.

Compound Microscope

A **compound microscope** consists of two convex lenses, an objective, and an eyepiece or ocular separated by a distance. As shown in the diagram, the object is placed just beyond the focal point (f_o) of the objective. The image formed (I_1) is real, inverted, and enlarged. I_1 is formed inside the focal point (f_e) of the eyepiece and the viewer observes an image (I_2) which is virtual, inverted, and greatly magnified.

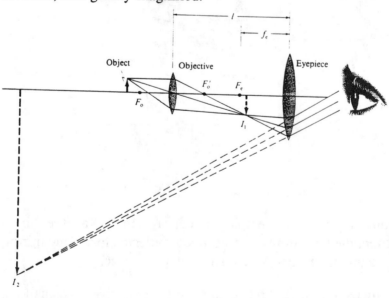

The magnification produced by the objective is given by $M_o = d_i / d_o = (\ell - f_e)/d_o$ where d_i is the distance from the lens to the image, d_o is the distance from the lens to the object, and ℓ is the distance between the two lenses.

The magnification produced by the eyepiece is $M_e = N/f_e$ where N is the near point, which is usually taken to be 25 cm. The total magnification is given by

$$M = M_e / M_o = N/f_e \ \times \ (\ell - f_e)/d_o \approx N/(f_e \ f_o)$$

The approximation is accurate when f_e and f_o are small compared to ℓ.

Lens Aberrations

The ray diagrams drawn to locate images produced by thin lenses are only approximately correct. Only an "ideal" lens gives an undistorted image, the formation of images by real lenses are limited by what are referred to as **lens aberrations**.

In spherical aberrations (diagram a) rays of light from a point on the axis of the lens, which pass through different sections of the lens, do not produce a point image after refraction. Instead, they produce a small circular patch of light. The point at which the circle has its smallest diameter is referred to as the circle of least confusion. Spherical aberrations can be approximately corrected by the expensive method of grinding a nonspherical lens or, more frequently, by using a combination of two or more lenses.

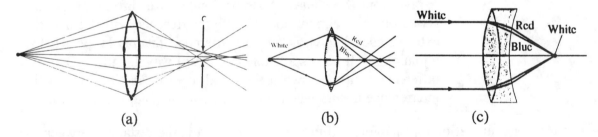

(a)   (b)   (c)

Chromatic aberration (diagram b) occurs because the index of refraction of transparent materials varies with wavelength. For example, blue light is bent more by glass than red light. As a result, white light from a point source produces an image containing colors spread out over a small region.

Chromatic aberration (diagram c) is approximately corrected through use of a combination of two lenses called an **achromatic doublet**. One lens is a convex lens and the other a concave lens. The lenses have different indices of refraction and radii of curvature.

Limits of Resolution

The ability of a lens to produce distinct images of two point objects which are very close together is called the **resolution** of the lens. Two principal factors limit the resolution of a lens. One factor is lens aberration and the second is the wave nature of light.

The magnification produced by a microscope or telescope is limited by diffraction. Magnification beyond a certain point does not lead to an increase in sharpness or resolution of images. The **Rayleigh criterion** states that two images are just resolvable when the center of the diffraction disk of one image is directly over the first minimum of the diffraction pattern of the other. The first minimum is at an angle θ = 1.22 λ/D from the central maximum and therefore the objects are considered to be "just" resolvable at this angle. λ is the wavelength of the incident light and D is the diameter of the objective lens.

For a microscope, the resolution is defined in terms of the **resolving power (RP)** where

$$RP = S = f\,\theta = 1.22\ \lambda f/D$$

where S is the minimum separation of two objects that can just be resolved and f is the focal length of the objective lens.

X-ray Diffraction

X-rays were discovered in 1895 by Wilhelm Roentgen. The nature of x-rays was not determined until 1913 when it was shown by Max Von Laue that X-rays exhibit properties of electromagnetic waves of very short wavelength.

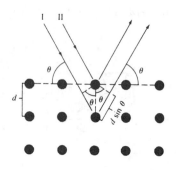

The wavelength of x-rays is comparable to the spacing of atoms in crystals such as sodium chloride (NaCl). If an X-ray beam is incident on the crystal, the subsequent reflection of the X-rays from the planes of atoms results in a diffraction pattern. Based on the diagram, constructive interference will occur if the extra distance, known as the path difference, that ray I travels before rejoining ray II equals a whole number of wavelengths. The path difference can be shown to be equal to 2 d sin θ. Constructive interference occurs when mλ = 2 d sin θ (Bragg equation).

mλ is a whole number of wavelengths and m = 1, 2, 3, etc. d is the distance between the layers of atoms and θ is the angle of incidence of the X-ray beam with the surface of the crystal.

ANSWERS TO SELECTED TEXT QUESTIONS

QUESTION 3. Why must a camera lens be moved farther from the film to focus on a closer object?

ANSWER: The relationship between the object distance, image distance, and focal length of the lens is given by the Gaussian form of the lens equation: $1/d_o + 1/d_i = 1/f$. The focal length of the lens is fixed and the image distance is measured from the lens to the film where the image is formed. If the object is close to the lens, the image distance must be increased so that a clear image will be formed on the film (see page 23-8, diagram C). If the object is far away from the lens, the image distance between the lens and film must be decreased in order to form a clear image (see page 23-8, diagram A). Inexpensive cameras often are made so that the distance from the lens to the film is fixed. In this type of camera an object distance of 6 feet to infinity ensures that a clear image will be formed on the film.

QUESTION 14. Which aberrations present in a simple lens are not present (or are greatly reduced) in the human eye?

ANSWER: The cornea of the human eye is less curved at the edges than at the center and the lens is less dense at the edges than at the center. Both effects cause rays at the outer edges to be bent less strongly and thus help reduce spherical aberration.

There is significant absorption of the shorter wavelengths by the lens of the eye and the retina is less sensitive to the blue and violet wavelengths. This is the region of the spectrum where chromatic aberration is the greatest and as a result chromatic aberration is greatly reduced in the eye.

QUESTION 16. What are the advantages (give at least two) for the use of large reflecting mirrors in astronomical telescopes?

ANSWER: A large objective allows in more light and this is necessary to produce bright images of distant objects. However, the construction and grinding of large lenses is difficult and they tend to sag under their own weight. A mirror has only one surface to be ground and can be supported along the entire area of the nonreflecting side. Also, since mirrors reflect light they do not exhibit chromatic aberration.

PROBLEM SOLVING SKILLS

For problems related to power of a lens:

1. Use the equation $P = 1/f$ to determine either the power or the focal length of the lens.
2. Use the Gaussian form of the lens equation to determine the position of the image.
3. Use the equation $M = 1 + N/f$ to determine the magnification of a converging lens.

For problems related to the final image produced by two converging lenses:

1. Use the Gaussian form of the lens equations to determine the position and size of the image produced by the lens closest to the object.
2. Determine the distance from the second lens to the image produced by the first lens. This distance represents the object distance from the second lens.
3. Use the Gaussian form of the lens equation to determine the position and size of the image produced by the second lens.

To determine the limit of resolution of a lens for two point objects which are close together:

1. Apply the Rayleigh criterion to determine the minimum angle at which the objects are resolvable.
2. If the objects are being viewed through a microscope, use the formula for resolving power to determine the minimum separation of the objects which allows resolution.

For problems related to X-ray diffraction:

1. Apply the Bragg equation to determine either the distance between atoms in the crystal or the wavelength of the incident x-rays.

PROGRAMMED PROBLEMS

PROBLEM 1. A student requires reading glasses of +2.5 diopters to read the print on this page when the page is held 25 cm from her eyes. Determine the minimum distance that she would have to hold the newspaper in order to read the print without her eyeglasses. Assume that the distance from the lens to the eye is negligible.

a. 1 Determine the focal length of the lens.	Solution: (Section 25-2) The power of the lens is related to the focal length by the equation $P = 1/f$. The power of the lens is +2.5 diopters, and the focal length of the lens is $f = 1/+2.5$ diopters $= + 0.40$ meters $= 40$ cm
a. 2. Use the lens equation determine the image distance.	With her glasses on, the print is held 25 cm from her eyes and the image is formed at the near point. Thus $d_o = 25$ cm while d_i is the distance from the lens to the near point of her vision. $1/d_o + 1/d_i = 1/f$ $1/25$ cm $+ 1/d_i = 1/40$ cm $d_i = - 67$ cm Without her glasses, the student would have to hold the page at least 67 cm from her eyes in order to read the print.

PROBLEM 2. The maximum magnification produced by a particular converging lens is 3 times when used by a person whose near point is 20 cm. Determine the a) focal length of the lens and b) magnification when the person's eye is relaxed.

a. 1 Determine the focal of the lens.	Solution: (Section 25-3) The maximum magnification is related to the focal length as length follows: $M = 1 + N/f$, $3 = 1 + 20/f$ and rearranging $20/f = 3 - 1$, $20/f = 2$, $f = 20/2$ $f = 10$ cm
b. 1 Determine the magnification when the eye is relaxed.	When the eye is relaxed, the image is seen at infinity and the object is at the focal point; thus $M = N/f = 20$ cm$/10$ cm $= 2$

PROBLEM 3. An astronomical telescope consists of two convex lenses, an eyepiece of focal length 90 cm and an objective of focal length 3.0 cm. Determine the magnification of the telescope if used to view a distant object.

a. 1	Solution: (Section 25-4)
Determine the magnification of the telescope. Hint: assume that the object is at infinity.	The magnifying power can be found as follows: $M = - f_o/f_e = - 90$ cm/3.0 cm $M = -30$

PROBLEM 4. Two converging lenses are 20 cm apart. The focal lengths of the lenses are 5.0 cm and 10 cm, respectively. An object 3.0 cm high is placed 15 cm in front of the 5.0 cm lens. a) Draw an accurate ray diagram and locate the final image formed by the combination. b) Mathematically determine the position and size of the final image.

a. 1	Solution: (Section 25-4)
Draw an accurate diagram and locate the final image formed by the combination.	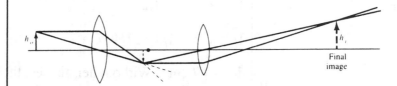

b. 1	
Determine the position and size of the image produced by the lens which is closest to the object.	$1/d_o + 1/d_i = 1/f_1$ $1/15$ cm $+ 1/d_i = 1/5.0$ cm $d_i = 7.5$ cm $h_i /h_o = - d_i/d_o$ $h_i /3.0$ cm $= - 7.5$ cm/15 cm $h_i = - 1.5$ cm　　　The image is real, inverted, and diminished.

b. 2	
Determine the distance from the second lens to the image and also the height of the image. State the characteristics of the final image.	The image produced by the first lens now becomes the object for the second lens. The distance from the second lens to the image produced by the first lens is equal to the distance between the lenses minus the distance from the first lens to the image distance, i.e., 20 cm - 7.5 cm = 12.5 cm. $1/d_o + 1/d_i = 1/f_2$ $1/12.5$ cm $+ 1/d_i = 1/10$ cm

$d_i = 50$ cm

The height of the final image is given by

$h_i / h_o = - d_i / d_o$

$h_i / (-1.5 \text{ cm}) = - (50 \text{ cm})/(12.5 \text{ cm})$

$h_i = + 6.0$ cm

The final image is real, upright, and magnified.

PROBLEM 5. During a laboratory experiment, a student positions two lenses to form a compound microscope. The objective and eyepiece have focal lengths of 0.40 cm and 3.0 cm, respectively. The lenses are 4.9 cm apart and an object is placed 0.50 cm in front of the objective lens. a) Draw a ray diagram and locate the position of the final image. b) Mathematically determine the position of the final image.

a. 1	Solution: (Section 25-5)
Draw an accurate diagram and locate the position of the final image.	The following diagram is not drawn to scale. This is necessary in order to show the final image on the diagram. 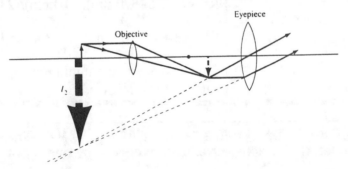

b. 1	Let f_o = focal length of the objective lens.
Mathematically determine the position of the image produced by the eyepiece.	$1/d_o + 1/d_i = 1/f_o$
	$1/0.50 \text{ cm} + 1/d_i = 1/0.40$ cm
	$d_i = 2.0$ cm

b. 2	The image produced by the objective lens becomes the object for the eyepiece. The two lenses are 4.9 cm apart; thus the image produced is 2.9 cm from the eyepiece. Therefore,
Mathematically determine the position of the final image.	$1/d_o + 1/d_i = 1/f_e$ where f_e = focal length of the eyepiece
	$1/2.9 \text{ cm} + 1/d_i = 1/3.0$ cm

$d_i = -87$ cm   The final image is a virtual image formed
87 cm in front of the eyepiece.

PROBLEM 6. The objective lens of a compound microscope is 2.00 cm in diameter and has a 10.0 cm focal length. Light of wavelength 570 nm is used to illuminate objects to be viewed. a) What is the angular separation of nearby objects when they are "just" resolvable? b) What is the resolving power of this lens?

a. 1	Solution: (Section 25-7)
What is the angular separation of nearby objects when they are "just" resolvable?	The minimum angular separation is given by
	$\theta = 1.22\ \lambda/D = (1.22)(570$ nm$)/(2.0$ cm$)$
	$= (1.22)(570 \times 10^{-9}$ m$)/(2.0 \times 10^{-2}$ m$)$
	$\theta = 3.48 \times 10^{-5}$ radians or 1.99×10^{-3} degrees
	This is the limit on resolution for this lens due to diffraction.
b. 1	resolving power = RP = s = f θ = 1.22 λf/D
What is the resolving power of this lens?	RP = $(1.22)(570$ nm$)(10.0$ cm$)/(2.0$ cm$)$
	RP = 3.48×10^{-6} m or 3.48×10^{-3} mm
	This distance is the minimum separation of two point objects that can "just" be resolved.

PROBLEM 7. X-rays having a wavelength of 0.200 nm are scattered by a crystal. Determine the angles for all diffraction maximum produced by one set of layers of ions spaced 0.412 nm apart.

a. 1	Solution: (Section 25-11)
Use the Bragg equation to determine the possible angles.	θ cannot exceed 90°; therefore, the possible angles are 90° or less.
	$m\lambda = 2d \sin \theta_1$ If m = 1, then
	$(1)(0.200$ nm$) = 2(0.412$ nm$) \sin \theta_1$
	$\sin \theta_1 = 0.243$
	$\theta_1 = 14.0°$

If m = 2, then

$(2)(0.200 \text{ nm}) = 2(0.412 \text{ nm}) \sin \theta_2$

$\sin \theta_2 = 0.485$

$\theta_2 = 29.0°$

If m = 3, then

$(3)(0.200 \text{ nm}) = 2(0.412 \text{ nm}) \sin \theta_3$

$\sin \theta_3 = 0.728$

$\theta_3 = 46.7°$

If m = 4, then

$(4)(0.200 \text{ nm}) = 2(0.412 \text{ nm}) \sin \theta_4$

$\sin \theta_4 = 0.971$

$\theta_4 = 76.1°$

If m = 5, then $\sin \theta_5 > 1$, therefore, only four diffraction maxima are produced.

PRACTICE PROBLEMS

PROBLEM 1. A student finds that she is unable to see objects distinctly which are more than 25 cm from her eye. Determine the type and power of the lens required for her to see a distant object clearly. Hint: assume that the distant object is an infinite distance away.

ANS. diverging or concave lens, - 4.0 diopters

PROBLEM 2. A telescope has an objective lens of focal length 90.0 cm and an eyepiece of focal length 5.00 cm. The telescope is used to view an object 50.0 meters from the objective lens. The observer notes a virtual image of the object formed 25.0 cm in front of the eyepiece. Determine the distance between the two lenses.

ANS. 95.8 cm

PROBLEM 3. A microscope has an objective lens of focal length 0.30 cm and an eyepiece of focal length 5.0 cm. The two lenses are 30.0 cm apart and the final image is a virtual image formed 25.0 cm from the eyepiece. Determine the magnification of the instrument.

ANS. -510

PROBLEM 4. A student notes that she is "just" able to distinguish the two headlights of a distant car. Using her knowledge of physics, she makes the following estimates: for night lighting conditions the diameter of one pupil is 0.60 mm, average wavelength of light coming from the headlights is 600 nm, and the distance between the headlights for most cars is about 2.0 m. Assuming that her eyes can attain their theoretical resolving power, what is the distance between the student and the car?

ANS. 1600 m or about 1 mile

CHAPTER 26

SPECIAL THEORY OF RELATIVITY

KEY TERMS AND PHRASES

postulates of the special theory are 1) all inertial frames of reference are equivalent, and 2) observers, regardless of their relative velocity or the velocity of the source, must measure the same value for the speed of light in a vacuum.

frame of reference is a point in space with respect to which the motion of objects can be measured.

inertial frame of reference is one which is either at rest or is moving in a straight line at a constant speed. In all inertial frames of reference, the laws of physics are the same and hold in the same way.

noninertial frame of reference is one which is undergoing accelerated motion. This means a frame in which the speed is changing or the direction of motion is changing or both the speed and the direction of motion are changing.

principle of simultaneity Two events which are observed to occur simultaneously at different points in one frame of reference will not occur simultaneously according to an observer in a second frame of reference which is in motion relative to the first.

proper time is measured by an observer at rest with respect to the timing device.

relativistic time dilation refers to the difference in the measurement of time by an observer in motion relative to the timing device.

proper length is measured by an observer at rest with respect to the object measured.

relativistic length contraction refers to the decrease in measured length of an object when the object is in motion relative to an observer.

proper mass is measured by an observer at rest with respect to the object measured.

relativistic mass refers to the mass of an object, as measured by an observer, when the object is in motion relative to an observer.

rest energy is the energy an object has due to its mass alone.

total energy of an object equals the sum of the object's mechanical energy and rest energy. The total energy is expressed by the formula $E = m c^2$ where m is the relativistic mass and c is the speed of light. According to the equation, it is possible to convert mass into energy and vice versa.

SUMMARY OF MATHEMATICAL FORMULAS

time dilation equation	$\Delta t = \Delta t_o / [1 - (v/c)^2]^{\frac{1}{2}}$	Δt_o is the proper time measured by the observer at rest with the timing device. Δt is the relativistic time measured by the observer in motion relative to the timing device. v is the speed of the timing device relative to observer measuring Δt while c is the speed of light in a vacuum.
length contraction equation	$L = L_o [1 - (v/c)^2]^{\frac{1}{2}}$	L is the relativistic length measured by the observer in motion relative to the object. L_o is the proper length measured by the observer at rest with respect to the object.
relativistic mass	$m = m_o / [1 - (v/c)^2]^{\frac{1}{2}}$	m is the relativistic mass and is measured by an observer in motion relative to the object. The proper mass (m_o) is measured by an observer at rest with respect to the object.
relativistic addition of velocities	$u = (v + u')/(1 + vu'/c^2)$	u is the speed of the object as measured by an observer in motion relative to the frame of reference of the moving object. v is the speed of the moving frame of reference. u' is the speed of the object relative to an observer at rest in the moving frame of reference. If u' is in the same direction as v, then u' is positive. If u' is opposite from v, then u' is negative.
rest energy	$E_o = m_o c^2$	An object's rest energy (E_o) equals the product of the rest mass (m_o) and the speed of light (c) squared.

relativistic total energy	$E = mc^2$	The total energy (E) equals the product of the relativistic mass (m) and the speed of light (c) squared. The total energy of an object equals the sum of the object's mechanical energy and rest energy.
relativistic kinetic energy	$KE = E - E_o$ $KE = (m - m_o) c^2$	For a moving object, the kinetic energy equals the difference between an object's total energy (E) and its rest energy (E_o).

CONCEPT SUMMARY

Postulates of the Special Theory

The **special theory of relativity** is based on two postulates formulated by Albert Einstein:

1) All inertial frames of reference are equivalent.

2) Observers, regardless of their relative velocity or the velocity of the source, must measure the same value for the speed of light in a vacuum.

Inertial Frames of Reference

A **frame of reference** is a point in space with respect to which the motion of objects can be measured. An **inertial frame of reference** is one which is either at rest or is moving in a straight line at a constant speed. In all inertial frames of reference, the laws of physics are the same and hold in the same way.

A **noninertial frame of reference** is one which is undergoing accelerated motion. This means a frame in which the speed is changing or the direction of motion is changing or both the speed and the direction of motion are changing.

As a consequence of the first postulate, it is possible to determine relative motion between objects traveling at constant velocity with respect to one another, but it is not possible to determine absolute motion.

Principle of Simultaneity

Two events which are observed to occur simultaneously at different points in one frame of reference will not occur simultaneously according to an observer in a second frame of reference which is in motion relative to the first.

The following "thought" experiment will be used to clarify this principle. Two railroad cars are directly opposite one another when lightning bolts strike points shown in the diagram located below. The cars are moving relative to one another with observer O_1 in car 1 assuming that he is at rest while car 2 is moving to the left at speed v. Likewise, observer O_2 in car 2 assumes that he is at rest and sees car 1 moving to the right at speed v.

diagram A   diagram B

As shown in diagram A, the lightning bolts strike the front and rear ends of each car simultaneously. As shown in diagram B, a moment later the light from the events simultaneously reach O_2's position. However, in O_1's reference frame, the light from the front end has already reached O_1, while the light from the rear has not yet reached O_1. So from O_1's reference frame the lightning strike at the front of the car must have occurred before the strike at the rear of the car.

Predictions from the Special Theory

Time Dilation

Einstein's two postulates plus the principle of simultaneity lead to the prediction that time is measured differently in frames of reference moving relative to one another.

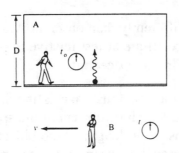

Observer A on a spaceship is moving toward the right at a speed v relative to an Earthbound observer B. A pulse of laser light is fired vertically upward from the floor of the spacecraft just as the spacecraft is passing B's position.

A is moving at a constant speed in a straight line, he is an inertial frame of reference. From A's point of view he is **not** moving but instead B is moving toward the left at speed v. A sees the light travel vertically upward toward the ceiling, a distance D, at the speed of light (c). The time required for the light to travel from the floor to the ceiling, from A's viewpoint, can be determined from the equation

time = distance/velocity or $\Delta t_0 = D/c$ and $D = c \, \Delta t_0$

The time measured by observer A is the proper time Δt_0. D is the distance from the floor to the ceiling. c is the speed of light in a vacuum.

As shown in the diagram 1 at the top of the next page, from B's frame of reference he is at rest and it is A who is moving toward the right at speed v. B sees the beam of light travel from the floor to the ceiling, but A is traveling toward the right, and B observes that the beam does not travel vertically upward. B sees the beam travel at an angle to the vertical. According to the second postulate, both A and B must measure the speed of light to be c. Thus, B sees the beam travel at speed c at an angle to the vertical.

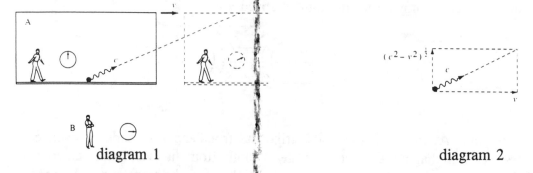

diagram 1   diagram 2

As shown in diagram 2, B uses the Pythagorean theorem and determines that the component of the light's speed in the vertical direction is $(c^2 - v^2)^{\frac{1}{2}}$. The time required for the light to travel from the floor to the ceiling, the vertical component of motion, is given by $\Delta t = D/(c^2 - v^2)^{\frac{1}{2}}$ and $D = (c^2 - v^2)^{\frac{1}{2}} \Delta t$ where Δt is the relativistic time.

Both observers measure D to be the same because there is no relative motion between their frames of reference in the vertical direction. Therefore, $c \Delta t_o = (c^2 - v^2)^{\frac{1}{2}} \Delta t$ and rearranging

gives $\Delta t = \Delta t_o/[1 - (v/c)^2]^{\frac{1}{2}}$. This equation is known as the time dilation equation.

Length Contraction

Not only would observer B measure time differently than observer A, but he would measure lengths differently as well. Objects in the spacecraft are at rest relative to observer A. Observer A measures the **proper length** (L_o) of objects in the spacecraft.

The spacecraft shown in the diagram on the previous page is moving in the x direction relative to observer B. Observer B will see the length of objects in the spacecraft shortened in the x direction but not in the y and z directions. The length of the object (L) as measured by observer B will differ from the length measured by observer A according to the following equation:

$$L = L_o[1 - (v/c)^2]^{\frac{1}{2}}$$

L is the **relativistic length** and is measured by the observer in motion relative to the object. The proper length L_o is the length measured by the observer at rest relative to the object.

Mass Increase

Special relativity leads to the prediction that the mass of an object will increase as its speed increases. If the object is moving at speed v relative to an observer, the observer will measure its mass to be m where

$$m = m_o/[1 - (v/c)^2]^{\frac{1}{2}}$$

m is the **relativistic mass** and is measured by an observer in motion relative to the object. The **proper mass** (m_o) is measured by an observer at rest relative to the object.

It should be noted that each of the equations discussed thus far, time dilation, length contraction, and mass increase, predict that there is an ultimate speed. As v approaches c, $[1 - (v/c)^2]^{\frac{1}{2}}$ approaches zero. Thus as v approaches c, the relativistic time and relativistic mass approach infinity while the relativistic length approaches zero. Since an infinite mass would require infinite energy to accelerate it, the principle of special relativity predicts that it is impossible for an object to travel at speeds equal to or greater than the speed of light.

Relativistic Addition of Velocities

Suppose observer A in the spacecraft shown in the diagram located at the top of page 26-5 throws an object in the positive x direction with a speed of u'. Observer B will not measure the relative speed of the object to be u where u = v + u'. The reason for this is because the addition of the two velocities might give a value larger than c. Instead, the speed of the object as measured by B is given by

$$u = (v + u')/(1 + vu'/c^2)$$

where u is the speed of the object as measured by observer B and v is the speed of observer A relative to observer B. u' is the speed of the object relative to observer A. If u' is in the same direction as v, then u' is positive. If u' is opposite from v, then u' is negative.

$E = mc^2$; Mass and Energy

When mechanical work is done on an object, the object's energy must change. However, according to the special theory, as the object's speed increases some of the energy is found in the increasing mass of the object instead of its speed and KE = ½ mv² no longer applies. The total energy that the object possesses is given by

$$E = mc^2$$

E is the object's total energy in joules or electron volts (eV) where 1.0 eV = 1.6 x 10⁻¹⁹ J. m is the object's relativistic mass in kg and $c^2 = (3.0 \times 10^8 \text{ m/s})^2 = 9.0 \times 10^{16} \text{ m}^2/\text{s}^2$.

The **total energy** of an object can be shown to be equal to the sum of the object's mechanical energy and **rest energy**. The object's rest energy is the energy which is the product of the rest mass and the speed of light squared, i.e., $E_o = m_o c^2$. For a moving object the total energy can be expressed as follows:

total energy = KE + rest energy

$$E = KE + E_o$$

The object's kinetic energy can be determined by rearranging the above formula:

$KE = E - E_o = mc^2 - m_oc^2 = (m - m_o) c^2$

According to the theory, even an object at rest has energy and it is possible to convert mass into energy and viceversa.

Although the relation $E = mc^2$ applies to all processes involving energy transfer, it is usually detected only in certain nuclear processes, e.g., nuclear fission and fusion. However, any situation where an object's energy changes should result in a change in its mass. For example, lifting a book and placing it on a table should result in an increase in the book's mass and heating a rod in a fire should result in an increase in its mass.

ANSWERS TO SELECTED TEXT QUESTIONS

QUESTION 5. If you were on a spaceship traveling at 0.5 c away from a star, at what speed would the starlight pass you?

ANSWER: Second postulate of special relativity: Light propagates through empty space with a definite speed c independent of the speed of the source or observer. Based on this postulate, you would measure the speed of the passing light to be c.

QUESTION 11. If you were traveling way from Earth at a speed of 0.5 c, would you notice a change in your heartbeat? Would your mass, height, or waistline change? What would observers on Earth using telescopes say about these things?

ANSWER: You are at rest with respect to objects in the spacecraft and of course you are at rest with respect to your own body. Because of this, you would not notice any change in your heartbeat, mass, height, or waistline. However, you are in motion relative to an Earthbound observer and

$[1 - (v/c)^2]^{1/2} = [1 - (0.5c/c)^2]^{1/2} = [1 - 0.25]^{1/2} = 0.87$

Using the relativity equations, it can be shown that according to the Earthbound observer, your heartbeat and waistline are reduced by the factor 0.87 while your mass has increased by $1/0.87 = 1.15$. Your height does not change because there is no relative motion between the Earthbound observer and your height.

QUESTION 22. It is not correct to say that "matter can neither be created nor destroyed." What must we say instead?

ANSWER: The formula $E = m c^2$ relates mass and energy and can be used to determine how much energy can be obtained from mass and vice versa. As stated in the text: "Mass and energy are interconvertible. The law of conservation of energy must include mass as a form of energy."

PROBLEM SOLVING SKILLS

For problems related to time dilation, length contraction, and mass increase:

1. Determine which observer is at rest relative to the objects to be measured. This observer measures the proper time, proper length, and proper mass. The observer who is moving relative to the objects measures the relativistic time, length, and mass.
2. Use the equations for time dilation, length contraction, and mass increase to solve the problem.

For problems related to the relativistic addition of velocities:

1. Determine the speed of the object as measured by one of the observers.
2. Determine the relative speed of the two observers.
3. Use the equation for relativistic addition of velocities to determine the speed of the object relative to the second observer.

For problems related to the total energy of an object:

1. Determine the object's mechanical energy in electron volts and joules.
2. Determine the object's rest energy ($E = m_0 c^2$) in electron volts and joules.
3. Determine the object's total energy ($E = m c^2$) in electron volts and joules.
4. If requested, use the total energy equation to determine the object's relativistic mass and the mass increase equation to determine the object's velocity.

PROGRAMMED PROBLEMS

PROBLEM 1. Observer A (diagram 1 on page 26-5) measures the passage of 1.0 hour in the spacecraft. Determine the amount of time that has passed according to observer B's clock if the spacecraft is moving at a) 0.80 c and b) 0.98 c.

a. 1	Solution: (Section 26-5)
Use the time dilation equation to determine the amount of time that passes on B's clock if $v = 0.8$ c.	Observer A measures the proper time (Δt_o) while B measures the relativistic time (Δt) and $v = 0.80$ c. Then
	$[1 - (v/c)^2]^{1/2} = [1 - (0.8c/c)^2]^{1/2} = (1 - 0.64)^{1/2} = 0.60$ and
	$\Delta t = \Delta t_o/(1 - (v/c)^2)^{1/2}$
	$\Delta t = (1.0 \text{ h})/(0.60) = 1.7$ h
b. 1	If $v = 0.98$ c, then
Use the time dilation equation to detemine the amount of time that passes on B's clock if $v = 0.98$ c.	$[1 - (v/c)^2]^{1/2} = [1 - (0.98 \text{ c}/c)^2]^{1/2} =$
	$(1 - 0.96)^{1/2} = 0.20$ and
	$\Delta t = \Delta t_o (1 - (v/c)^2)^{1/2}$
	$\Delta t = (1.0 \text{ h})/(0.20) = 5.0$ h

PROBLEM 2. Observer A (diagram 1 on page 26-5) measures the length of a piece of wood to be 1.0 meter and its mass to be 1.0 kg. The spacecraft is moving at 0.80 c relative to observer B. Determine the a) length of the wood and b) mass of the wood as observed by B.

a. 1	Solution: (Sections 26-6 and 26-8)
Determine which observer measures the proper length and mass and which observer measures the relativistic length and mass.	Observer A is at rest relative to the piece of wood; therefore, observer A measures the proper length and proper mass.
	$L_o = 1.0$ m and $m_o = 1.0$ kg
	The piece of wood is in motion relative to observer B; then B measures the relativistic length (L) and relativistic mass (m).
a. 2	In problem 1, it was determined that if $v = 0.80$ c, then $[1 - (v/c)^2]^{1/2} = 0.60$.
Determine the relativistic length of the wood.	$L = (1 - v/c)^2)^{1/2} L_o$
	$L = (0.60)(1.0 \text{ m}) = 0.60$ m

b. 1 Determine the relativistic mass of the wood.	$m = m_o /[1 - (v/c)^2]^{1/2}$ $m = (1.0 \text{ kg})/(0.60) = 1.7 \text{ kg}$

PROBLEM 3. Determine the a) relativistic mass and b) total energy of an object of rest mass 1.0 kg which is traveling at 0.98 c.

a. 1 Determine the relativistic mass.	Solution: (Sections 26-8 and 26-10) $m = m_o /[1 - (v/c)^2]^{1/2}$ $\quad = (1.0 \text{ kg})/[1 - (0.98c/c)^2]^{1/2}$ $m = (1.0 \text{ kg})/(0.20) = 5.0 \text{ kg}$
b. 1 Determine the total energy.	$E = m c^2 = (5.0 \text{ kg})(3.0 \times 10^8 \text{ m/s})^2$ $E = 4.5 \times 10^{17} \text{ J}$

PROBLEM 4. An electron is accelerated from rest through a potential difference of 1.0×10^6 volts. Determine the electron's a) kinetic energy, b) rest energy, c) total energy, and d) speed after passing through the potential difference.

a. 1 Determine the electron's kinetic energy in electron volts and joules after passing through the potential difference.	Solution: (Section 26-10) As the electron accelerates from rest through the potential difference, work is done on it and potential energy is converted to kinetic energy. $W = - \Delta PE = + \Delta KE$ where $\Delta PE = PE_f - PE_i$. $\Delta PE = 0 \text{ eV} - (1 \text{ electron})(1.0 \times 10^6 \text{ V})$ $\Delta PE = - 1.0 \times 10^6 \text{ eV}$ $- \Delta PE = \Delta KE, \quad \Delta KE = KE_f - KE_i \quad \text{and} \quad KE_i = 0 \text{ J}$ $-(-1.0 \times 10^6 \text{ eV}) = KE_f - 0$ $KE_f = 1.0 \times 10^6 \text{ eV}$ Express the electron's kinetic energy in joules. $KE_f = (1.0 \times 10^6 \text{ eV})(1.6 \times 10^{-19} \text{ J}/1.0 \text{ eV}) = 1.6 \times 10^{-13} \text{ J}$

b. 1 Determine the electron's rest energy in joules and MeV.	The rest energy (E_o) of an electron is given by $E_o = m_o\,c^2$ $\quad = (9.1 \times 10^{-31}\ \text{kg})(9.0 \times 10^{16}\ \text{m}^2/\text{s}^2)$ $E_o = 8.2 \times 10^{-14}\ \text{J}$ $1.0\ \text{MeV} = 1.0 \times 10^6\ \text{eV} = 1.6 \times 10^{-13}\ \text{J}$ Therefore, the rest energy of the electron may be expressed in MeV as follows: $E_o = (8.2 \times 10^{-14}\ \text{J})(1.0\ \text{MeV}/1.6 \times 10^{-13}\ \text{J})$ $E_o = 0.51\ \text{MeV}$
c. 1 Determine the electron's total energy in joules and MeV.	$E = KE + E_o$ $\quad = 1.6 \times 10^{-13}\ \text{J} + 8.2 \times 10^{-14}\ \text{J}$ $E = 2.4 \times 10^{-13}\ \text{J}$ or expressing the answer in MeV, $E = 1.0\ \text{MeV} + 0.51\ \text{MeV} = 1.5\ \text{MeV}$
d. 1 Determine the electron's relativistic mass.	$E = m\,c^2 \text{ and } m = E/c^2$ $m = (2.4 \times 10^{-13}\ \text{J})/(9.0 \times 10^{16}\ \text{m}^2/\text{s}^2)$ $m = 2.7 \times 10^{-30}\ \text{kg}$
d. 2 Use the relativistic mass equation to determine the electron's velocity.	$m = m_o/[1 - (v/c)^2]^{1/2} \text{ and } [1 - (v/c)^2]^{1/2} = m_o/m$ $[1 - (v/c)^2]^{1/2} = (9.1 \times 10^{-31}\ \text{kg})/(2.7 \times 10^{-30}\ \text{kg})$ $[1 - (v/c)^2]^{1/2} = 0.34$ Squaring both sides of the equation gives $1 - (v/c)^2 = 0.11 \text{ and}$ $- (v/c)^2 = -1 + 0.11 = -0.89$ $(v/c)^2 = 0.89$ $v/c = 0.94 \text{ and}$ $v = 0.94c = 2.8 \times 10^8\ \text{m/s}$

PROBLEM 5. The spacecraft shown in diagram 1 on page 26-5 is traveling at 0.90 c relative to observer B. Observer A throws an object in the same direction that the spacecraft is moving. The object's velocity relative to observer A is 0.80 c. Determine the velocity of the object as measured by observer B.

a. 1

Use the equation for the relativistic addition of velocities.

Solution: (Section 26-11)

$u = (v + u')/(1 + v\,u'/c^2)$

$= (0.90c + 0.80c)/(1 + (0.90\ c)(0.80\ c)/c^2)$

$u = 0.99\ c$

PRACTICE PROBLEMS

PROBLEM 1. Two twins are 20 years old when one of the twins sets out in a spaceship traveling at a constant speed. Both twins measure the time until the astronaut twin returns. The twin in the spaceship measures the passage of 10 years while the Earthbound twin measures the passage of 50 years. Using the theory of special relativity, determine the velocity of the spaceship.

ANS. v = 0.98 c

PROBLEM 2. Determine the speed and kinetic energy of a proton which has a relativistic mass of 2.03 x 10^{-27} kg. The rest mass of a proton is 1.67 x 10^{-27} kg.

ANS. v = 0.57 c and KE = 200 MeV

PROBLEM 3. The distance from the Earth to Alpha Centuri is approximately 4.3 light years. Determine the distance from the Earth to Alpha Centuri as measured by an astronaut traveling at 0.80 c toward Alpha Centuri.

ANS. 2.6 light years

PROBLEM 4. Assume that 1.0 kg of matter could be completely converted into electrical energy. Determine the value of this energy in dollars if electrical energy costs $0.08 per KWH.

ANS. $2.0 x 10^9

CHAPTER 27

EARLY QUANTUM THEORY AND MODELS OF THE ATOM

KEY TERMS AND PHRASES

e/m experiment showed that cathode rays consist of charged particles now known as electrons. As a result of this experiment, the ratio of the charge on an electron (e) to its mass (m) was determined.

Planck's quantum hypothesis predicts that the molecules in a heated object can vibrate only with discrete amounts of energy. Thus the energy of the vibrating atom is quantized.

Millikan oil drop experiment determined that the charge on a microscopically small oil drop is always a small whole-number multiple of 1.6×10^{-19} C. This value equals the charge on the electron. Once the value of the charge on an electron was determined, the accepted value for the mass of the electron was determined to be 9.1×10^{-31} kg.

blackbody radiation refers to the intensity of spectral radiation emitted by a "perfectly" radiating object.

photoelectric effect indicates that light has characteristics of particles. Light particles are called photons.

threshold frequency is the minimum frequency at which electrons are ejected from a surface.

work function (W_o) is the minimum energy required to break the electron free from the attractive forces which hold the electron to the surface of a metal.

pair production occurs when a high-energy photon known as a gamma ray traveling near the nucleus of an atom disappears and an electron and a positron may appear in its place.

positron has the same mass as an electron and carries the same magnitude of electric charge; however, the electron is negatively charged while the positron carries a positive charge.

wave-particle duality refers to the phenomena where both particles, such as electrons and protons, and light exhibit both the properties of waves and the properties of particles.

Compton effect shows that the interaction of a photon with an the electron can be viewed as a two-particle collision.

de Broglie wavelength is the wavelength of a particle of mass m traveling at speed v. The wavelength is given by $\lambda = h/(mv)$ where λ is the wavelength of the particle.

emission spectra are produced by a high voltage placed across the electrodes of a tube containing a gas under low pressure. The light produced can be separated into its component colors by a diffraction grating. Such analysis reveals a spectra of discrete lines and not a continuous spectrum.

ionization energy refers to the energy required to remove an electron from an atom.

SUMMARY OF MATHEMATICAL FORMULAS

Wien's law	$\lambda_p T = 2.90 \times 10^{-3}$ m K	the relationship between absolute temperature (T) and the peak wavelength (λ_p) in blackbody radiation
photon energy	$E = h f$	The energy (E) of a photon is related to the frequency (f) of the light.
photoelectric effect	$KE_{max} = hf - W_o$	The maximum kinetic energy (KE_{max}) of the emitted photoelectrons equals the difference between the energy of the incident photon (hf) and the work function (W_o) of the metal surface.
Compton effect	$\lambda' - \lambda = (h/mc)(1 - \cos \phi)$	A collision between a photon and an electron results in a change of wavelength for the photon.
de Broglie wavelength	$\lambda = h/mv$	λ represents the wavelength of a particle of mass m traveling at speed v.
Balmer equation	$1/\lambda = R\, Z^2(1/n'^2 - 1/n^2)$	An electron dropping from a higher energy level to a lower energy level emits a photon of wavelength λ.
ionization energy for an electron from a hydrogen-like atom	$E = (-13.6 \text{ eV})Z^2/n^2$	ionization energy (E) of an electron located in the nth level of the hydrogen-like atom
Bohr radius	$r = (0.53 \times 10^{-10} \text{ m})(n^2)$	radius (r) of the orbit of an electron in the hydrogen atom, where n = 1, 2, etc.
angular momentum of an electron	$L = m\, v\, r_n = n\, h/2\pi$	the angular momentum (L) of an electron orbiting the hydrogen nucleus

CONCEPT SUMMARY

The Electron

In 1897, J. J. Thomson performed the e/m experiment which showed that **cathode rays** consist of charged particles now known as electrons. As a result of his experiment, he was able to determine that the ratio of the charge on an electron (e) to its mass (m) is

$$e/m = v/Br$$

where v is the electron's velocity, B is the magnetic field strength which is directed perpendicular to the electron's path, and r is the radius of the circular path in which the particle travels.

B and r are readily measured while the velocity of the particle can be determined by using a device known as a **velocity selector**. In a velocity selector (see programmed problem 1), an electric field is arranged so that the electric force $F = q E$ balances the force exerted on the charged particle by the magnetic field $F = q v B \sin 90°$. As a result, $q E = q v B$ and $v = E/B$. The particle passes through the selector undeflected. Since $e/m = v/Br$ and $v = E/B$, then $e/m = E/B^2r$. Substituting experimental values for E, B, and r, Thomson obtained a value for the electron's charge to mass ratio close to the modern value of 1.76×10^{11} C/kg.

In 1909, R. A. Millikan was able to determine the charge on an electron in an experiment referred to as the **Millikan oil drop experiment**. Millikan found that the charge on a microscopically small oil drop is always a small whole-number multiple of 1.6×10^{-19} C. Once the charge on the electron became known it was possible to determine its mass. The accepted value for the mass of the electron is 9.1×10^{-31} kg.

Planck's Quantum Hypothesis

The intensity of spectral radiation emitted by a "perfectly" radiating object, known as **blackbody radiation,** when graphed as a function of wavelength, produces curves as shown at the top of the next page.

The wavelength at which peak intensity occurs decreases as the Kelvin temperature of the object increases. The relationship between the peak wavelength and the absolute temperature is given by **Wien's law,**

$$\lambda_p T = 2.90 \times 10^{-3} \text{ m K} \quad \text{where} \quad \lambda_p \text{ is the } \textbf{peak wavelength}$$

The peak wavelength is the wavelength of the emitted light at the point where the intensity of the emitted light is a maximum. T is the temperature of the blackbody in degrees Kelvin.

Max Planck produced a theory which agreed with the experimental data by assuming that the molecules in the heated object can vibrate only with discrete amounts of energy. Thus the energy of the vibrating atom is quantized.

The energy is related to the natural frequency of vibration of the molecules of the radiating object by the formula

$$E = n h f$$

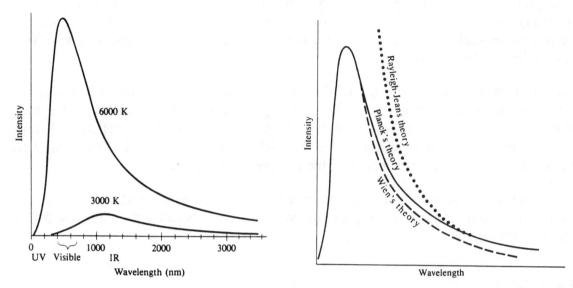

h is **Planck's constant**, h = 6.626 x 10^{-34} J s. E is the energy in joules, n is a whole number, n = 1, 2, 3, etc., while f is the object's natural frequency of vibration in hertz.

Photoelectric Effect

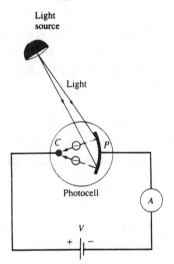

When light is incident on a metal surface, electrons are emitted. In the photocell shown in the diagram, the incident light causes the electrons to be emitted from the **plate** (P) and the difference in potential causes the electrons to travel to the **collector** (C).

The experimental results indicate that 1) the number of electrons emitted per second (electron current) increases as the light intensity increases, 2) the maximum kinetic energy of the electrons is not affected by the intensity of the light, and 3) below a certain frequency, called the **threshold frequency,** no electrons are ejected no matter how intense the light beam.

The wave theory agrees with the first result but cannot explain the second and third experimental results. In 1905, Albert Einstein proposed an extension of the quantum theory to explain the results of the **photoelectric effect**. His theory says that light is emitted in particles which are now called **photons**. Each photon has energy which is related to the frequency of the light according to the formula

$$E = h f$$

27—4

The experimental results of the photoelectric effect can be explained by the photon theory:

1) An electron in the metal making up the plate of the photocell absorbs the energy of the incident photon. If the photon energy is large enough, the electron escapes from the metal. As the intensity of the light beam increases, the number of photons increases and the number of electrons emitted each second increases. However, the intensity of the light does not affect the energy of the incident photons.

2) The maximum kinetic energy of the emitted electrons is related to the energy of the incident photon by the equation $hf = KE_{max} + W_o$ where hf is the energy of the incident photon in joules, KE_{max} is the kinetic energy of the most energetic electrons emitted from the metal in joules. The **work function** (W_o) is the minimum energy required to break the electron free from the attractive forces which hold the electron to the metal.

3) If the energy of the incident photon is below that of the work function, no electrons are emitted. The minimum frequency required to eject an electron is called the threshold frequency or "cut off" frequency.

Compton Effect

In 1922, Arthur Compton used the photon theory to explain why x-rays scattered from certain materials have different wavelengths than the incident x-rays. According to Compton, an incident photon transferred some of its energy to an electron in the material and was scattered with lower energy and therefore longer wavelength.

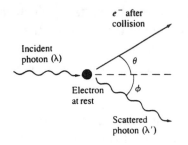

According to the theory, the incident photon carries both energy ($E = h f$) and momentum ($p = E/c = h f/c = h/\lambda$). The photon interaction with the electron can be considered to be a two-particle collision as shown in the diagram.

Applying the laws of conservation of energy and momentum to the collision, the theory correctly predicts a wavelength change given by the following equation:

$$\lambda' - \lambda = (h/mc)(1 - \cos \phi)$$

λ is the wavelength of the incident light, λ' is the wavelength of the scattered photon, m is the mass of the recoil electron, and ϕ is the angle between the direction of the incident photon and the scattered photon.

Pair Production

The equation $E = m c^2$ implies that it is possible to convert mass into energy and vice versa. One example of the conversion of energy to mass is **pair production**. In this process, a high-energy photon known as a gamma ray traveling near the nucleus of an atom may disappear and an electron and a positron may appear in its place. The electron and the positron have the same mass and carry the same magnitude of electric charge; however, while the electron is negatively charged the positron carries a positive charge. Thus, in addition to the laws of conservation of energy and momentum, the law of conservation of electric charge also holds true.

The minimum energy of a gamma ray required for the pair production of electron and positron can be shown to be about 1.02 Mev. If the energy of the gamma ray is above this amount, then excess energy is shared equally between the particles in the form of kinetic energy.

Wave Particle Duality; the Principle of Complementarity

Experiments such as the Young's interference experiment, polarization, and single slit diffraction indicate that light is a wave. The photoelectric effect and the Compton effect indicate that light is a particle. Light is a phenomena which exhibits both the properties of waves and the properties of particles. This is known as **wave-particle duality.**

Niels Bohr proposed the **principle of complementarity** which says that for any particular experiment involving light, we must either use the wave theory or the particle theory, but not both. The two aspects of light complement one another.

Wave Nature of Matter

Just as light exhibits properties of both particles and waves, particles such as electrons, protons, and neutrons also exhibit wave properties. Thus the wave-particle theory extends to matter as well as light. In 1923, Louis de Broglie suggested that the wavelength of a particle of mass m traveling at speed v is given by

$\lambda = h/(m\ v)$ where λ is the **de Broglie wavelength** of the particle.

In 1927, two Americans, Davisson and Germer, produced diffraction patterns by scattering electron beams from the surface of a metal crystal. The calculated wavelength of the electron waves agreed with de Broglie's prediction. It was later shown that protons, and neutrons as well as other particles exhibit wave properties as well as particle properties.

Rutherford's Model of the Atom

In 1909, Ernest Rutherford suggested to two of his students, Hans Geiger and Ernest Marsden, that they bombard a piece of thin gold foil with high-energy alpha particles.

The positively charged alpha particles are emitted spontaneously from a radioactive source. Because of their high energy, Rutherford expected that the alpha particles would pass through the gold foil without being significantly deflected from their original direction. However, as shown in the diagram at the top of the next page, Geiger and Marsden found that while most of the alpha particles did pass through undeflected, some were deflected by 30° or more and a few were deflected by 90° or more.

Based on observations, Rutherford concluded that most of the mass of the atom resided in a tiny region called the **nucleus**. The nucleus contains the positively charged protons while the electrons travel in orbits around the nucleus.

Rutherford's model had two problems: 1) since protons repel one another, the positively charged nucleus should not exist, and 2) the orbiting electrons are undergoing centripetal acceleration and, according to electromagnetic theory, accelerated electric charges give rise to electromagnetic waves. Thus the electrons should give up their energy in the form of a continuous spectrum as they spiral into the nucleus.

Atomic Spectra

Emission spectra are produced by a high voltage placed across the electrodes of a tube containing a gas under low pressure. The light produced can be separated into its component colors by a diffraction grating. Such analysis reveals a spectra of discrete lines and not a continuous spectrum.

In 1885, J. Balmer developed a mathematical equation which could be used to predict the wavelengths of the four visible lines in the hydrogen spectrum. Balmer's equation states

$$1/\lambda = R\ Z^2\ (1/n'^2 - 1/n^2)$$

where Z is the atomic number of the atom, λ is the wavelength of the spectral line in meters, and $R = 1.097 \times 10^7$ m^{-1} is is known as Rydberg's constant. n' is a whole number, 2 for lines in the visible spectrum, 3 for infrared, 1 for ultraviolet. n is a whole number; this number can be any number greater than n'. For example, for the visible spectrum n' = 2 while n = 3 (red light), n = 4 (blue light), n = 5 (violet light), and n = 6 (violet light).

Bohr's Model of the Hydrogen Atom

Based on the Rutherford model of the atom, line spectra, and the Balmer formula, Niels Bohr in 1913 proposed the following postulates for the hydrogen atom:

1. The electron travels in circular orbits about the positively charged nucleus. However, only certain orbits are allowed. The electron does not radiate energy when it is in one of these orbits, thus violating classical theory.

2. The allowed orbits have radii (r_n) where

$$r_n = (0.53 \times 10^{-10} \text{ m}) \, n^2 = (0.53 \text{ nm}) \, n^2 \quad \text{and} \quad n = 1, 2, 3, \text{ etc.}$$

If $n = 1$, then the electron is in its smallest orbit, which is known as its **ground state**.

3. The orbits have angular momentum (L) given by

$$L = m \, v \, r_n = n \, h/2\pi \qquad$$ where $n = 1, 2, 3$, etc. The angular momentum has only discrete values; thus it is **quantized**.

4. If an electron falls from one orbit, also known as **energy level,** to another, it loses energy in the form of a photon of light. The energy of the photon equals the difference between the energy of the orbits.

5. The energy level of a particular orbit is given by $E = -13.6\text{eV}/n^2$ where $n = 1, 2, 3$, etc. If $n = 1$, the electron is in its lowest energy level and it would take 13.6 eV to remove it from the atom (**ionization energy**).

6. A hydrogen atom can absorb only those photons of light which will cause the electron to jump from a lower level to a higher level. Thus the energy of the photon must equal the difference in the energy between the two levels. Therefore, a continuous spectrum passing through a cool gas will exhibit dark absorption lines at the same wavelengths as the emission lines.

The Bohr model proved to be successful for the hydrogen atom and for one electron ions; however, it did not work for multi-electron atoms.

De Broglie's Contribution to the Bohr Model

In 1923, de Broglie extended his idea of matter waves to the Bohr atom by stating that the electron wave must be a circular standing wave. The circumference ($2\pi \, r_n$) of the standing wave contains a whole number of wavelengths ($n\lambda$); thus $2\pi \, r_n = n\lambda$ where $n = 1, 2, 3$, etc.

$\lambda = 2\pi \, r_n/n$ but $\lambda = h/mv$. Then $2\pi \, r_n/n = h/mv$ and rearranging gives $m \, v \, r_n = n \, h/2\pi$.

This formula is Bohr's postulate concerning the quantization of the electron's angular momentum. de Broglie provided an explanation for the key postulate in Bohr's model of the hydrogen atom; that is, the idea that the angular momentum (L) of the electron is quantized, i.e., $L = m \, v \, r_n = n \, h/2\pi$

It is from this postulate that the equations for the discrete orbits and energy states are derived.

ANSWERS TO SELECTED TEXT QUESTIONS

QUESTION 7. If the threshold wavelength in the photoelectric effect increases when the emitting metal is changed, what can you say about the work functions of the two metals?

ANSWER: At the threshold wavelength, no electrons are ejected from the surface of the metal and $hf_o = W_o$. However, $\lambda_o = c/f_o$ and therefore, $W_o = hc/\lambda_o$. If the wavelength increases, the work function of the second metal must be less than the work function of the original metal.

QUESTION 10. If an X-ray photon is scattered by an electron, does its wavelength change? If so, does it increase or decrease?

ANSWER: The incident photon transfers some of its energy to the electron. The result is a scattered photon which has less energy and a different wavelength than the incident photon. Since $E = hf = hc/\lambda$, if the energy is less, the wavelength must increase.

QUESTION 16. If an electron and proton travel at the same speed, which has the shorter wavelength?

ANSWER: Based on de Broglie's hypothesis, the wavelength of a particle is given by $\lambda = h/mv$. The wavelength is inversely proportional to the particle's mass. The mass of the proton is approximately 1840 times greater than the mass of the electron. Therefore, the wavelength of the proton is approximately 1840 times less than the wavelength of the electron. The proton has the shorter wavelength.

PROBLEM SOLVING SKILLS

For problems involving the velocity selector:

1. The force exerted by the electric field on the particle is equal in magnitude to the force exerted by the magnetic field. The direction of the electric force may be determined by the conventions adopted in chapter 16 and the right-hand rule described in chapter 20.
2. Use the equation $v = E/B$ to solve for the unknown quantity.

For problems related to the photoelectric effect:

1. Construct a data table listing information both given and requested in the problem. For example, include the work function of the metal, threshold frequency and threshold wavelength for the metal, energy of the incident photons, wavelength of the incident photons, Planck's constant, and the speed of light.
2. Use the equation $E = KE_{max} + W_o$ where $W_o = h f_o$ to solve the problem.

For problems related to the Compton effect:

1. Construct a data table listing information both given and requested in the problem. For example, include the wavelength of the incident and scattered photons, the angle between the incident and scattered photons, and the energy of the recoil electron.
2. Use the equation $\lambda' - \lambda = (h/mc) (1 - \cos \phi)$ to solve for the unknown quantity.
3. Use the law of conservation of energy to solve for the energy of the recoil electron.

For problems related to pair production:

1. List the mass of the particle and its antiparticle.
2. Determine the minimum energy required for pair production.
3. Use the law of conservation of energy to determine the kinetic energy of the particles after they are produced.

For problems related to the de Broglie wavelength of a particle:

1. List the mass and velocity of the particle.
2. Use $\lambda = h/mv$ to solve for the de Broglie wavelength.

For problems related to emission spectra:

1. List information both given and requested in the problem.
2. Use the Balmer equation to solve for the wavelength of the light emitted when a electron transition occurs between two energy levels.
3. Use $f = c/\lambda$ to determine the frequency of the photon.
4. Use $E = hf$ to determine the energy of the photon.

PROGRAMMED PROBLEMS

PROBLEM 1. In a device known as a "velocity selector" a beam of protons is directed between the plates of a parallel plate capacitor as shown in the diagram. The electric field between the plates has a magnitude of 5000 N/C. Determine the magnitude and direction of the magnetic field required to allow protons traveling at 6.0×10^4 m/s to pass between the plates undeflected.

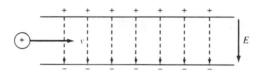

a. 1	Solution: (Section 27-1)
Locate the forces acting on the particle. Determine the net force acting on the particle.	In order for the beam to pass undeflected at constant speed, the net force on the particle must be zero. Based on the diagram, the electric force is directed downward. The magnetic force must be equal in magnitude to the electric force but directed upward in order that the net force equal zero.
	net F = 0
	$F_{elec} - F_{mag} = 0$

a. 2	$F_{elec} - F_{mag} = 0$
Determine the magnitude of the magnetic field required to allow the particle to pass undeflected.	$F_{elec} = F_{mag}$
	$q E = q v B \sin \theta$
	q cancels from both sides of the equation and solving for B gives
	$B = E/v \sin \theta$
	$= (5000 \text{ N/C})/(6.0 \times 10^4 \text{ m/s})(\sin 90°)$
	B = 0.083 N/C m/s = 0.083 T

a. 3	The magnetic force is directed upward. Using the right-hand rule, the fingers point toward the right of the page while the thumb is directed towards the top of the page. When you bend your fingers they point into the paper. Therefore, the magnetic field is directed directed into the paper.
Use the right-hand rule to determine the direction of the magnetic field.	

PROBLEM 2. a) Use Wien's law to determine the peak wavelength of light emitted by a blackbody whose temperature is 3000 K. b) Is this wavelength in the visible portion of the electromagnetic spectrum?

a. 1 Determine the peak wavelength.	Solution: (Section 27-2) $\lambda_p T = 2.90 \times 10^{-3}$ m K $\lambda_p (3000 \text{ K}) = 2.90 \times 10^{-3}$ m K $\lambda_p = 9.7 \times 10^{-7}$ m = 970 nm
b. 1 Is this wavelength the visible spectrum?	The visible spectrum extends from 400 to 700 nm. The peak wavelength is 970 nm; therefore, it is not in the visible range. 970 nm is located in the infrared portion of the electromagnetic spectrum.

PROBLEM 3. Show that the energy of the photon, E = h f, can be written as E = 1240 eV nm/λ.

a. 1 Express the formula for the energy of a photon in terms of joule meter divided by the wavelength.	Solution: (Section 27-3) $E = h f$ but $f = c/\lambda$ $E = hc/\lambda$ $\quad = (6.63 \times 10^{-34}$ J s$)(3.00 \times 10^8$ m/s$)/\lambda$ $E = 1.99 \times 10^{-25}$ J m/λ
a. 2 Use the conversion from joules to eV and from meters to nm to show that E = hf can be written as E = 1240 eV nm/λ.	$E = (1.99 \times 10^{-25}$ J m/$\lambda)(1.0$ eV/1.6×10^{-19} J$)$ $\quad = (1.243 \times 10^{-6}$ ev m$)(1.0$ nm/1.0×10^{-9} m$)/\lambda$ $E = 1240$ ev nm/λ

PROBLEM 4. The maximum kinetic energy of electrons emitted from a surface coated with sodium is 3.0 eV. If the work function is 2.28 eV, determine the a) energy of the incident photons and b) wavelength of the incident photons.

a. 1 Determine the energy of the incident photons.	Solution: (Section 27-3) $E = KE_{max} + W_o$ $E = 3.0$ eV + 2.28 eV = 5.3 eV

b. 1 Determine the frequency of the incident photons.	$E = h\,f$ and $f = E/h$ $f = (5.3\text{ eV})/(6.63 \times 10^{-34}\text{ J s}) \times (1.6 \times 10^{-19}\text{ J})/(1\text{ eV})$ $f = 1.27 \times 10^{15}\text{ Hz}$
b. 2 Determine the wavelength of the incident photons.	$\lambda = c/f = (3.0 \times 10^{8}\text{ m/s})/(1.27 \times 10^{15}\text{ Hz})$ $\lambda = 2.35 \times 10^{-7}\text{ m} = 235\text{ nm}$ alternate solution: $E = 1240\text{ eV nm}/\lambda$ and $\lambda = 1240\text{ eV nm}/E$ $\lambda = (1240\text{ eV nm})/(5.3\text{ eV}) = 235\text{ nm}$

PROBLEM 5. The threshold wavelength for incident photons to eject electrons from the surface of a particular metal is 600 nm. If light of wavelength 400 nm shines on the metal, determine the a) work function of the metal and b) energy of the most energetic electrons.

a. 1 Determine the work function of the metal.	Solution: (Section 27-3) At the threshold wavelength, the energy absorbed by the electron is used to break free from the attractive forces which hold it to the metal. As a result, it has no kinetic energy when it finally breaks free. Thus $E = KE_{max} + W_o$ but $KE_{max} = 0\text{ J}$ $hf = 0\text{ J} + W_o$ where $hf = hc/\lambda$ $W_o = (6.63 \times 10^{-34}\text{ J s})(3.0 \times 10^{8}\text{ m/s})/(6.00 \times 10^{-7}\text{ m})$ $W_o = 3.32 \times 10^{-19}\text{ J}$ or 2.07 eV alternate solution: $W_o = hf = hc/\lambda = 1240\text{ eV nm}/\lambda$ $W_o = (1240\text{ eV nm})/(600\text{ nm}) = 2.07\text{ eV}$
b. 1 Determine the energy of the most energetic electrons.	$E = KE_{max} + W_o$ where $E = h\,f = (1240\text{ eV})/(400\text{ nm}) = 3.10\text{ eV}$ $3.10\text{ eV} = KE_{max} + 2.07\text{ eV}$ $KE_{max} = 1.03\text{ eV}$

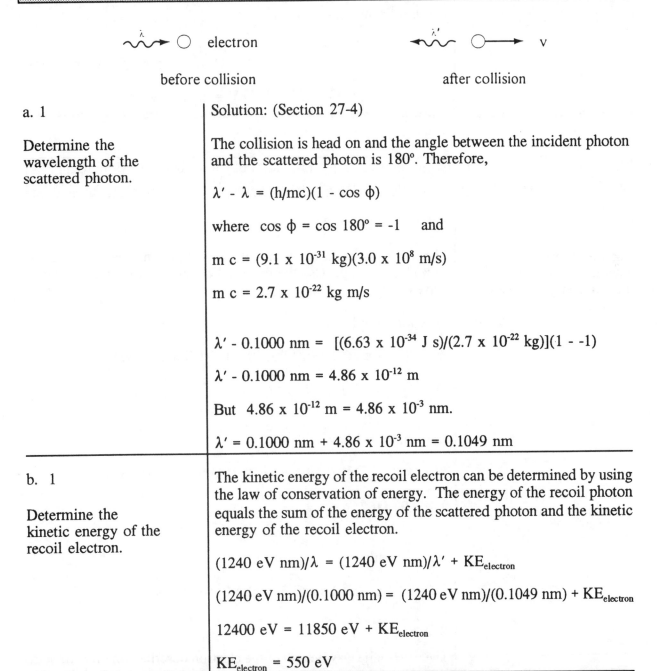

before collision   after collision

a. 1	Solution: (Section 27-4)
Determine the wavelength of the scattered photon.	The collision is head on and the angle between the incident photon and the scattered photon is 180°. Therefore, $$\lambda' - \lambda = (h/mc)(1 - \cos \phi)$$ where $\cos \phi = \cos 180° = -1$ and $$m\,c = (9.1 \times 10^{-31} \text{ kg})(3.0 \times 10^{8} \text{ m/s})$$ $$m\,c = 2.7 \times 10^{-22} \text{ kg m/s}$$ $$\lambda' - 0.1000 \text{ nm} = [(6.63 \times 10^{-34} \text{ J s})/(2.7 \times 10^{-22} \text{ kg})](1 - -1)$$ $$\lambda' - 0.1000 \text{ nm} = 4.86 \times 10^{-12} \text{ m}$$ But 4.86×10^{-12} m = 4.86×10^{-3} nm. $$\lambda' = 0.1000 \text{ nm} + 4.86 \times 10^{-3} \text{ nm} = 0.1049 \text{ nm}$$
b. 1	The kinetic energy of the recoil electron can be determined by using the law of conservation of energy. The energy of the recoil photon equals the sum of the energy of the scattered photon and the kinetic energy of the recoil electron.
Determine the kinetic energy of the recoil electron.	$$(1240 \text{ eV nm})/\lambda = (1240 \text{ eV nm})/\lambda' + KE_{electron}$$ $$(1240 \text{ eV nm})/(0.1000 \text{ nm}) = (1240 \text{ eV nm})/(0.1049 \text{ nm}) + KE_{electron}$$ $$12400 \text{ eV} = 11850 \text{ eV} + KE_{electron}$$ $$KE_{electron} = 550 \text{ eV}$$

a. 1

Use the law of
conservation of energy
to determine the minimum
energy of the photon.

Solution: (Section 27-4)

Using the law of conservation of energy, the minimum energy is
equal to the sum of the rest energies of the two particles.

$$E = m c^2 = m_{proton} c^2 + m_{antiproton} c^2$$

$$E = (1.67 \times 10^{-27} \text{ kg})(9.0 \times 10^{16} \text{ m}^2/\text{s}^2) +$$
$$(1.67 \times 10^{-27} \text{ kg})(9.0 \times 10^{16} \text{ m}^2/\text{s}^2)$$

$$E = 3.01 \times 10^{-10} \text{ J} = 1.88 \times 10^9 \text{ eV} = 1.88 \text{ BeV}$$

PROBLEM 8. Three particles, an electron, a proton, and a 0.500 kg ball are traveling at 5.0×10^6 m/s. Determine the de Broglie wavelength of each particle.

a. 1

Use the de Broglie
equation to determine
the wavelength of each
particle.

Solution: (Section 27-6)

The rest mass of an electron = 9.1×10^{-31} kg and that of a proton
is 1.67×10^{-27} kg. Use the de Broglie equation to solve for the
wavelength.

electron: $\lambda = h/mv$

$$\lambda = (6.63 \times 10^{-34} \text{ J})/[(9.1 \times 10^{-31} \text{ kg})(5.0 \times 10^6 \text{ m/s})]$$

$$\lambda = 1.5 \times 10^{-10} \text{ m} = 0.15 \text{ nm}$$

proton: $\lambda = h/mv$

$$\lambda = (6.63 \times 10^{-34} \text{ J s})/[(1.67 \times 10^{-27} \text{ kg})(5.0 \times 10^6 \text{ m/s})]$$

$$\lambda = 7.9 \times 10^{-14} \text{ m}$$

ball: $\lambda = h/mv$

$$\lambda = (6.63 \times 10^{-34} \text{ J s})/[(0.50 \text{ kg})(5.0 \times 10^6 \text{ m/s})]$$

$$\lambda = 2.7 \times 10^{-40} \text{ m}$$

The de Broglie wavelength of the electron described in this problem
corresponds to the distance between the layers of atoms in a crystal.
Therefore, it would be possible to detect the wave nature of the
electrons by using a diffraction experiment. The wavelength of the
proton and ball is such that the wave nature of these particles would
not be detected.

a. 1	Solution: (Section 27-8)
Use the law of conservation of energy to determine the closest approach of the alpha particle to the gold nucleus.	The collision is head on; therefore, all of the kinetic energy of the alpha particle will be converted into the form of electrical potential energy at the point of closest approach.
	Note: $q_\alpha = +2e$ $q_{gold} = +79e$
	$e = 1.6 \times 10^{-19}$ C $m_\alpha = 6.64 \times 10^{-27}$ kg
	$q_\alpha q_{gold} = (+2 \times 1.6 \times 10^{-19}$ C$)(+79 \times 1.6 \times 10^{-19}$ C$)$
	$q_\alpha\ q_{gold} = 4.04 \times 10^{-36}$ C^2
	initial kinetic energy = electrical potential energy at closest approach
	$\frac{1}{2}\ m\ v^2 = k\ q_\alpha\ q_{gold}/r$
	$\frac{1}{2}\ (6.64 \times 10^{-27}$ kg$)(2.00 \times 10^7$ m/s$)^2 =$ $(9 \times 10^9$ N m^2/C$^2)(4.04 \times 10^{-36}$ C$^2)/r$
	Solving for r gives
	$r = 2.74 \times 10^{-14}$ m

a. 1	Solution: (Sections 27-9 and 27-10)
Use the Balmer formula to determine the wavelength of the emitted photon.	$1/\lambda = R\ Z^2[(1/n'^2) - (1/n^2)]$
	$1/\lambda = (1.097 \times 10^7$ m$^{-1})(1^2)[(1/2^2) - (1/3^2)]$
	$1/\lambda = (1.097 \times 10^7$ m$^{-1})(0.25 - 0.11)$
	$1/\lambda = 1.52 \times 10^6$ m^{-1}
	$\lambda = 6.56 \times 10^{-7}$ m = 656 nm
b. 1	$c = f\ \lambda$
Determine the photon's frequency.	3.00×10^8 m/s $= f\ (6.56 \times 10^{-7}$ m$)$
	$f = 4.57 \times 10^{14}$ Hz

c. 1	$E = h\ f$
Determine the photon's energy in joules and eV.	$E = (6.63 \times 10^{-34}\ J\ s)(4.57 \times 10^{14}\ Hz)$
	$E = 3.03 \times 10^{-19}\ J = 1.89\ eV$
	alternate solution:
	The energy of any particular level in the hydrogen atom is given by the formula
	$E = -\ 13.6\ eV/n^2$
	The difference in the energy between the third and second levels equals the energy of the emitted photon.
	$\Delta E = E_2 - E_3 = (-\ 13.6\ eV/2^2) - (-\ 13.6\ eV/3^2)$
	$= (-\ 3.40\ eV) - (-\ 1.51\ eV)$
	$\Delta E = -\ 1.89\ eV$
	The electron loses 1.89 eV of energy as it drops from the third to the second level. The energy is in the form of a photon which has 1.89 eV of energy.

PROBLEM 11. Determine the energy required to ionize an electron which is in the ground state of an Li^{+2} ion.

a. 1	Solution: (Section 27-10)
Determine the energy required. Hint: n = 1 for the ground state n = infinity for for ionization.	The ionization energy is the energy required to completely remove the electron from the atom. The point at which it is completely free from the atom is referred to as the free state and at that point and E = 0. The ionization energy equals the difference in the energy between the ground state (n = 1) and the free state (n = ∞).
	The atomic number of lithium is 3; thus Z = 3. The energy of any level n in the lithium ion is given by
	$E = [(-13.6\ eV)\ Z^2]/n^2$
	$\Delta E = E_\infty - E_1$
	$= [(-13.6\ eV)(3^2)]/\infty^2] - [(-13.6\ eV)(3^2)]/1^2]$
	$\Delta E = (0\ eV) - (-122\ eV) = 122\ eV$

a. 1 Use the Balmer formula to solve for the wavelength.	Solution: (Section 27-10) The atomic number of Li is 3; therefore, Z = 3. $1/\lambda = (1.097 \times 10^7 \text{ m}^{-1})(3^2)[1/1^2 - 1/3^2]$ $1/\lambda = 8.78 \times 10^7 \text{ m}^{-1}$ $\lambda = 1.14 \times 10^{-8} \text{ m}$
b. 1 Solve for the frequency.	$c = f \lambda$ $3.00 \times 10^8 \text{ m/s} = f (1.14 \times 10^{-8} \text{ m})$ $f = 2.63 \times 10^{16} \text{ Hz}$
c. 1 Solve for the energy of the emitted photon.	The energy of the emitted photon equals the energy lost by the electron as it drops from the third level to the first level. $\Delta E = E_1 - E_3$ $\quad = [(-13.6 \text{ eV})(3^2)]/1^2] - [(-13.6 \text{ eV})(3^2)]/3^2]$ $\Delta E = (-122 \text{ eV}) - (-13.6 \text{ eV}) = -108 \text{ eV}$ The electron loses 108 eV of energy as it drops from n = 3 to n = 1. Therefore, the emitted photon has 108 eV of energy.

a. 1 Determine the radius of the orbit.	Solution: (Section 27-10) The radius of orbit is given by $r = (0.53 \times 10^{-10} \text{ m})(n^2)$ where n = 4 $r = (0.53 \times 10^{-10} \text{ m})(4^2)$ $r = 8.5 \times 10^{-10} \text{ m}$

a. 2

Determine the angular momentum of the electron.

$L = (n\ h)/(2\pi) = [(4)(6.63 \times 10^{-34}\ \text{J s})]/(2\pi)$

$L = 4.22 \times 10^{-34}\ \text{J s}$

PRACTICE PROBLEMS

PROBLEM 1. In a typical laboratory experiment involving the measurement of e/m of an electron, electrons are accelerated through a potential difference of 75 volts into a magnetic field of strength 0.0010 T. The magnetic field is directed perpendicular to the direction of motion of the electrons and the electrons are deflected into a circle of radius 0.030 m. Determine the experimental value of e/m.

Note: the accepted value of e/m is 1.76×10^{11} C/kg.

ANS. 1.67×10^{11} C/kg.

PROBLEM 2. Determine the de Broglie wavelength of a 100 eV proton.

ANS. 2.86×10^{-12} m

PROBLEM 3. Determine the work function of a metal if the threshold wavelength for incident photons is 546 nm.

ANS. 2.28 eV

PROBLEM 4. A line of wavelength 97.3 nm is observed in the spectra of hydrogen. Determine the transition required to produce this line. Hint: first determine the energy of the emitted photon and then use the energy level diagram for the hydrogen atom given in the text.

ANS. 4 to 1

CHAPTER 28

QUANTUM MECHANICS OF ATOMS

KEY WORDS AND PHRASES

quantum mechanics or **wave mechanics** unified the wave-particle duality into a single consistent theory.

Heisenberg uncertainty principle is an important result of quantum mechanics. This principle results from the wave-particle duality and an intrinsic limit in our ability to make accurate measurements. One form of the uncertainty principle states that it is impossible to know simultaneously both the precise position and momentum of a particle.

quantum numbers The state of an electron in the hydrogen atom is governed by four quantum numbers. The quantum numbers are n, ℓ, m_ℓ, and m_s. n is called the principle quantum number, where n = 1, 2, 3, 4, etc. ℓ is the orbital quantum number. m_ℓ is the magnetic quantum number. m_s is the spin quantum number.

Pauli exclusion principle is used to explain the arrangement of electrons in multi-electron atoms. This principle states that "no two electrons in an atom can occupy the same quantum state." Thus each electron has a unique set of quantum numbers: n, ℓ, m_ℓ, and m_s.

electronic configuration of the elements listed in the periodic table of the elements can be specified using the n and ℓ quantum numbers. Electrons with the same value of ℓ are in the same subshell within the main shell designated by the letter n. The subshells are designated by the letters s, p, d, f, etc.

X-rays exhibit properties of electromagnetic waves of very short wavelength. X-rays can be produced in two ways. One method is for high-energy electrons to knock an electron out of an inner energy level of certain atoms. When an electron drops from a higher level to a lower level an x-ray photon is emitted. The second method is bremsstrahlung or braking radiation. In this method, the electron is deflected as it passes near the nucleus of the atom.

SUMMARY OF MATHEMATICAL FORMULAS

Heisenberg uncertainty principle	$(\Delta x)(\Delta p) \geq h/2\pi$	It is impossible to know simultaneously both the precise position (Δx) and momentum of a particle (Δp). The product of uncertainty of the position (Δx) and the uncertainty of the momentum (Δp) must be greater than or equal to Plank's constant divided by 2π.
	$(\Delta E)(\Delta t) \geq h/2\pi$	Another form of the uncertainty principle states that the product of the uncertainty of energy (ΔE) and the uncertainty in time (Δt) must be greater than or equal to Plank's constant divided by 2π.
principle quantum numbers	n = 1, 2, 3, etc.	n is called the principle quantum number, where n = 1, 2, 3, 4, etc.
	$\ell \leq n - 1$	ℓ is the orbital quantum number. ℓ can take on integer values up to n - 1. For example, if n = 3, then ℓ can have the following values: 0, 1, 2.
	$-\ell \leq m_\ell \leq +\ell$	m_ℓ is the magnetic quantum number. It is related to the direction of the electron's angular momentum. m_ℓ is an integer and can have values from $-\ell$ to $+\ell$.
	m_s = +½ or -½	m_s is the spin quantum number. It is related to the spin angular momentum
electronic configuration of the elements	The order of filling is as follows: $1s^2$, $2s^2$, $2p^6$, $3s^2$, $3p^6$, $4s^2$, $3d^{10}$, $4p^6$, $5s^2$, $4d^{10}$, $5p^6$, $6s^2$, $4f^{14}$, $5d^{10}$, etc.	The s orbital holds up to 2 electrons, the p orbital up to 6 electrons, the d orbital up to 10 elecrons, and the f orbital up to 14 electrons.
cut-off wavelength of X-ray photons	$\lambda_o = (hc)/(eV)$	The cut-off wavelength (λ_o) is the shortest wavelength X-ray produced, e is the charge on the electron, and V is the accelerating voltage.

Moseley's formula	$\lambda = (1.22 \times 10^{-7} \text{ m})/(Z - 1)^2$	The wavelength (λ) of the K_α line is related to the atomic number (Z) of the atoms of the metal target.

CONCEPTS AND EQUATIONS

Quantum Mechanics

About 1925, Erwin Schrodinger and Werner Heisenberg produced a new theory, called wave mechanics or **quantum mechanics**, which unified the wave-particle duality into a single consistent theory.

Applied to the atom, quantum mechanics pictures the electron as spread out in space in the form of a cloud of negative charge. The shape and size of the electron cloud can be mathematically determined for a particular state of an atom. For the hydrogen atom, the shape of the electron cloud is spherically symmetric about the nucleus. The cloud model can be interpreted as an electron wave spread out in space or as a probability distribution for electrons as particles.

Quantum mechanics has been used to explain phenomena such as spectra emitted by complex atoms, the relative brightness of spectral lines, and even how atoms form molecules.

Quantum mechanics reduces to classical physics in instances where classical physics applies. Newtonian mechanics is a special case of quantum mechanics. Even Bohr's postulate on the quantization of angular momentum of the electron in the hydrogen atom can be shown to be a special case of the more general quantum mechanics.

Heisenberg Uncertainty Principle

Newtonian mechanics implies that if an object's position and momentum are known at a particular moment in time, and if the forces that are acting on it or will be acting on it are known, then its future position can be predicted. This idea is referred to as determinism.

However, an important result of quantum mechanics is the **uncertainty principle.** This principle results from the wave-particle duality and an intrinsic limit in our ability to make accurate measurements. One form of the uncertainty principle states that it is impossible to know simultaneously both the precise position and momentum of a particle. Expressed mathematically, the product of uncertainty of the position (Δx) and the uncertainty of the momentum (Δp) must be greater than or equal to Plank's constant divided by 2π, i.e., $(\Delta x)(\Delta p) \geq h/2\pi$.

Another form of the uncertainty principle states that the product of the uncertainty of energy (ΔE) and the uncertainty in time (Δt) must be greater than or equal to Plank's constant divided by 2π, i.e., $(\Delta E)(\Delta t) \geq h/2\pi$.

Thus, unlike Newtonian mechanics, quantum mechanics states that only approximate predictions are possible and that there is an inherent unpredictability in nature.

Quantum Mechanics of the Hydrogen Atom; Quantum Numbers

The state of an electron in the hydrogen atom is governed by four quantum numbers: n, ℓ, m_ℓ, m_s. The energy of a particular level in the hydrogen atom is related to n by the equation $E = (-13.6 \text{ eV})/n^2$. n is retained in quantum mechanics and is called the **principle quantum number**, where $n = 1, 2, 3, 4$, etc.

ℓ is the **orbital quantum number**. The angular momentum (L) is related to the orbital quantum number by the formula $L = [\ell(\ell + 1)]^{1/2} (h/2\pi)$. ℓ can take on integer values up to $n - 1$. For example, if $n = 3$, then ℓ can have the following values: 0, 1, 2.

m_ℓ is the **magnetic quantum number**. It is related to the direction of the electron's angular momentum. m_ℓ is an integer and can have values from $-\ell$ to $+\ell$. The component of the angular momentum in an assigned direction, usually along the z axis, is given by $L_z = m_\ell (h/2\pi)$.

m_s is the **spin quantum number**. This quantum number only has values of $+\frac{1}{2}$ or $-\frac{1}{2}$. The spin angular momentum in an assigned direction equals $m_s(h/2\pi)$. The word "spin" was originally given to this quantum number because it was thought to be associated with the electron spinning on its axis as it revolves around the nucleus. However, this is an oversimplification because the electron exhibits properties of waves as well as particles.

Pauli Exclusion Principle

In order to explain the arrangement of electrons in multi-electron atoms, the **Pauli exclusion principle** is used. This principle states that "no two electrons in an atom can occupy the same quantum state." Thus, each electron has a unique set of quantum numbers: n, ℓ, m_ℓ, and m_s. For example, helium has two electrons. Both electrons have $n = 1$, $\ell = 0$, and $m_\ell = 0$. However, one of the electrons has $m_s = +\frac{1}{2}$ and the other has $m_s = -\frac{1}{2}$. Thus each electron has a different set of quantum numbers.

Electronic Configuration of the Elements

The **electronic configuration** of the elements listed in the periodic table of the elements can be specified using the n and ℓ quantum numbers. Electrons with the same value of n are in the same shell. The shells are given letter symbols as shown in the table shown at the top of the next page. If $n = 1$, then the electrons are in the K shell.

As shown in the table, electrons with the same value of ℓ are in the same subshell within the main shell designated by the letter n. The subshells are designated by the letters s, p, d, f, etc.

The number of electrons in the subshell can be found by using the formula $2(2\ell + 1)$. Thus, if $\ell = 3$, then $2[2(3) + 1] = 14$ and the f subshell can hold 14 electrons.

The designation of an electron involves both n and ℓ plus a superscript which designates the number of electrons in the subshell. For example, if n = 2 and ℓ = 1 and there are three electrons in the orbital, then the designation would be $2p^3$.

value of n	symbol of subshell	value of ℓ	symbol of subshell	maximum number of electrons
1	K	0	s	2
2	L	1	p	6
3	M	2	d	10
4	N	3	f	14
.	.	4	g	18
.	.	5	h	32
.	.	.	.	.
		.	.	.
		.	.	.

The order of filling is as follows: $1s^2$, $2s^2$, $2p^6$, $3s^2$, $3p^6$, $4s^2$, $3d^{10}$, $4p^6$, $5s^2$, $4d^{10}$, $5p^6$, $6s^2$, $4f^{14}$, $5d^{10}$, etc. A simplified way of determining the order of filling is shown below:

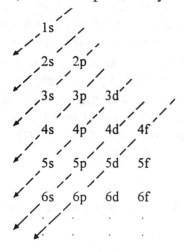

Write down the principle energy levels and their subshells and follow the diagonal lines. The diagonal lines follow the order of filling.

In filling the subshells, the lower energy subshells are filled first, as are the lower principle energy levels.

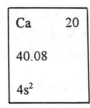

Each box in the periodic table contains the symbol of the element, its atomic number, its atomic mass and the ground state electronic configuration of the outermost electrons. For example, the symbol for calcium is Ca, 20 is the atomic number, 40.08 is the atomic mass and $4s^2$ is the electronic configuration of the outermost electrons.

X-rays and X-ray Production

X-rays were discovered in 1895 by Wilhelm Roentgen. The nature of x-rays was not determined until 1913 when it was shown that X-rays exhibit properties of electromagnetic waves of very short wavelength.

X-rays can be produced in two ways. One method is for high-energy electrons to knock an electron out of an inner energy level of certain atoms. When an electron drops from a higher level to a lower level an x-ray photon is emitted. The second method is a continuous spectrum called **bremsstrahlung** or **braking radiation**. In this method, the electron is deflected as it passes near the nucleus of the atom. During the resulting deceleration, energy in the form of an X-ray is produced.

An x-ray tube produces a spectrum of wavelengths. The shortest wavelength x-ray is the result of the electron losing all of its kinetic energy during the collision. In this case,

energy lost by electron = energy gained by x-ray photon

$e\ V = h\ f_o$ and because $f_o = c/\lambda_o$, then $\lambda_o = (hc/e)V$. f_o is the cut-off frequency. f_o is the highest frequency x-ray produced. λ_o is the cut-off wavelength and is the shortest wavelength X-ray produced, e is the charge on the electron and V is the accelerating voltage.

X-ray Diffraction

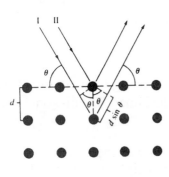

The wavelength of X-rays is comparable to the spacing of atoms in crystals such as sodium chloride (NaCl). If an x-ray beam is incident on the crystal, the subsequent reflection of the X-rays from the planes of atoms results in a diffraction pattern. Based on the diagram, constructive interference will occur if the extra distance, known as the path difference, that ray I travels before rejoining ray II equals a whole number of wavelengths. The path difference can be shown to be equal to $2\ d \sin \theta$ and therefore constructive interference occurs when $m\lambda = 2\ d \sin \theta$ (Bragg equation).

$m\lambda$ is a whole number of wavelengths and m = 1, 2, 3, etc., d is the distance between the layers of atoms, and θ is the angle of incidence of the X-ray beam with the surface of the crystal.

X-rays and Atomic Number

One method of producing X-rays is for high-energy electrons to knock an electron out of an inner shell of certain atoms. If the missing electron was in the K shell (n = 1) and is replaced by an electron which falls from the L shell (n = 2), then the X-ray is referred to as an K_α X-ray.

In 1914, Henry Moseley determined that the wavelength of the K_α x-ray line followed an empirical formula similar to Balmer's equation. Moseley's formula for the K_α line is as follows:

$1/\lambda = R(Z - 1)^2 (1/1^2 - 1/2^2)$ and upon rearranging this equation becomes

$\lambda = (1.22 \times 10^{-7} \text{ m})/(Z - 1)^2$

λ is the wavelength of the K_α X-ray and Z is the atomic number of the atoms of the metal target.

Moseley determined that a graph of $(1/\lambda)^{1/2}$ vs. Z produces a straight line. Because of this, the atomic number of a number of elements could be determined and his research led to the arrangement of the periodic table on the basis of atomic number. Moseley enlisted in the British army and was sent to Turkey. He was killed at Gallipoli on August 10, 1915, at the age of 27.

SELECTED TEXT QUESTIONS WITH ANSWERS

QUESTION 13. Which of the following electron configurations are forbidden?
a) $1s^2\ 2s^2\ 2p^6\ 3s^3$;   b) $1s^2\ 2s^2\ 2p^4\ 3s^2\ 4p^4$;   c) $1s^2\ 2s^2\ 2p^8\ 3s^1$

ANSWER: a) $3s^3$ is not allowed because the s orbital contains a maximum of two electrons; therefore, configuration a) is forbidden. b) The 2p orbital must fill before the 3s and 4p begin to fill. Since the 2p orbital contains only 4 electrons rather than the 6 required to fill it completely; therefore, configuration b) is forbidden. c) $2p^8$ is not allowed because the p orbital contains a maximum of six electrons; therefore, configuration c) is forbidden.

QUESTION 15. In what column of the periodic table would you expect to find the atom with each of the following configurations?
a) $1s^2\ 2s^2\ 2p^6\ 3s^2$;   b) $1s^2\ 2s^2\ 2p^6\ 3s^2\ 3p^6$;   c) $1s^2\ 2s^2\ 2p^6\ 3s^2\ 3p^6\ 4s^1$;   d) $1s^2\ 2s^2\ 2p^5$

ANSWER: The periodic table of the elements is given in the inside cover located in the back of the textbook. The table lists the outermost electrons (also known as valence electrons) for each atom. The outermost electrons for element a) is $3s^2$; therefore, the element is magnesium (Mg) which is in Group II or family IIA. Element b) is $3s^2\ 3p^6$, which is the element argon (Ar). Argon is in Group 0 or family VIIIA. Element c) is $4s^1$, which is the element potassium (K). Potassium is in Group I or family IA. Element d) is $2s^2\ 2p^5$, which is the element fluorine (F). Fluorine is in Group VII or family VIIA.

QUESTION 17. Why do chlorine and iodine exhibit similar properties?

ANSWER: Using the periodic table located in the text, it can be seen that the outer electron configuration of both elements consists of $s^2\ p^5$ electrons. Chlorine is $3s^2\ 3p^5$ while iodine is $5s^2\ 5p^5$. The physical and chemical properties of atoms are primarily determined by the outer electronic configuration, also known as the valence electrons. Therefore, both elements exhibit similar properties and are said to belong to the same chemical family, in this case the halogen family.

PROBLEM SOLVING SKILLS

For problems involving the uncertainty principle:

1. Determine the object's mass.
2. If necessary, determine the uncertainty in the object's speed.
3. Determine the uncertainty in the molecule's momentum.

4. Use the uncertainty principle to determine the uncertainty in the object's position.
5. If the uncertainty in position is given, use the uncertainty principle to determine the uncertainty in the object's momentum and velocity.
6. If the problem involves uncertainty in energy, then note either the uncertainty in the energy or time and solve the problem using the equation $(\Delta E)(\Delta t) \geq h/2\pi$.

For problems involving the possible quantum states of an atom.

1. Note the value of the principle quantum number (n) for the atom. For example, if an atom is in the third period of the periodic table, then the value of n = 3.
2. Determine the possible values of ℓ. Remember that ℓ can have positive values up to n - 1.
3. Determine the possible values of m_ℓ. Remember that m_ℓ can have values of $-\ell$ to $+\ell$.
4. m_s can have values of $+\frac{1}{2}$ or $-\frac{1}{2}$.
5. Use the Pauli exclusion principle to construct a table for the possible quantum states.

For problems involving the range of values of the angular momentum of an electron:

1. Note the principle quantum number (n) of the particular orbital in which the electron is located.
2. Determine the range of values of the orbital quantum number (ℓ).
3. Use the equation $L = [\ell(\ell + 1)]^{\frac{1}{2}} h/2\pi$ to determine the possible values of the angular momentum (L).

For problems involving the electronic configuration of an element:

1. Write down the order of filling of the s, p, d, and f subshells and the maximum number of electrons in each subshell. Hint: rather than attempting to memorize the order of filling, commit the table on page 28-5 to memory.
2. Take note of the atomic number of the element.
3. Write down the electronic configuration until the number of electrons in the atom is reached.
4. If the outer electron configuration is given, the element can be easily identified by using a periodic table. For example, if the outer configuration is $4s^2 4p^1$, then the element is located in period 4 and is a member of family IIIA. This element is gallium (Ga).

For problems involving x-ray spectra and atomic number:

1. If the accelerating voltage is given, then determine the electron's kinetic energy in electron volts ($W = \Delta KE = q \Delta V$).
2. The highest energy and shortest wavelength X-ray photon is produced if all of the electron's kinetic energy is used in producing the X-ray photon. Use the equation $q \Delta V = h c/\lambda$ to determine the wavelength of the X-ray.
3. If the wavelength of the shortest X-rays is given, then the equation $q \Delta V = h c/\lambda$ can be used to determine the voltage needed to give an electron the energy to produce the X-ray.
4. If the wavelength of the K_α X-ray is given, then the formula $\lambda = (1.22 \times 10^{-7} m)/(Z - 1)^2$ is used to determine the atomic number (Z) of the target element.

PROGRAMMED PROBLEMS

An air molecule travels at 1150 miles per hour (515 m/s) at 0°C. Suppose the uncertainty in an experimental measurement of its speed is ±5.0%. Compute the minimum uncertainty in its position.

a. 1 Determine the uncertainty in the molecule's speed in m/s.	Solution: (Section 28-3) The uncertainty in its speed is 5.0%; thus $v = 5.0\% \times 515$ m/s $= 25.8$ m/s
a. 2 Determine the mass of a nitrogen molecule.	Air is mainly nitrogen and the molecular weight of diatomic nitrogen is 28 grams/mole. $m = (28$ g/mole$)(1$ mole$/6.02 \times 10^{23}$ molecules$)$ $m = 4.7 \times 10^{-23}$ g $= 4.7 \times 10^{-26}$ kg
a. 3 Determine the uncertainty in the molecule's momentum.	$\Delta p = m\, \Delta v$ $\quad = (4.7 \times 10^{-26}$ kg$)(25.8$ m/s$)$ $\Delta p = 1.2 \times 10^{-24}$ kg m/s
a. 4 Determine the minimum uncertainty in the molecule's position.	Using the Heisenberg uncertainty principle, $(\Delta x)(\Delta p) \geq h/2\pi$ $(\Delta x)(1.2 \times 10^{-24}$ kg m/s$) \geq (6.63 \times 10\text{-}34$ J s$)/(2\pi)$ $(\Delta x) \geq 8.80 \times 10^{-11}$ m.

List the possible quantum states for $_{10}$Ne.

a. 1 Use the Pauli exclusion principle to list the possible quantum states.	Solution: (Sections 28-6 and 28-7) The Pauli exclusion principle states, "no two electrons in an atom can occupy the same quantum state." Each atom must have a unique set of quantum numbers: n, ℓ, m_ℓ, and m_s. Based on the exclusion principle (see top of next page) it is possible to construct a table of possible quantum states for $_{10}$Ne.

sub-shell	n	ℓ	m	m_s	sub-shell	n	ℓ	m	m_s
1s	1	0	0	$+\frac{1}{2}$	2p	2	1	-1	$-\frac{1}{2}$
1s	1	0	0	$-\frac{1}{2}$	2p	2	1	0	$+\frac{1}{2}$
2s	2	0	0	$+\frac{1}{2}$	2p	2	1	0	$-\frac{1}{2}$
2s	2	0	0	$-\frac{1}{2}$	2p	2	1	1	$+\frac{1}{2}$
2p	2	1	-1	$+\frac{1}{2}$	2p	2	1	1	$-\frac{1}{2}$

PROBLEM 3. Determine the values of m_ℓ that are allowed for the a) 1s subshell and b) 3d subshell.

a. 1

Determine the values for the 1s subshell.

Solution: (Section 28-6)

For the 1s subshell, n = 1, and since $\ell \le n - 1$, then $\ell = 0$. m_ℓ can have values from $-\ell$ to $+\ell$, but because $\ell = 0$, $m_\ell = 0$.

b. 1

Determine the values for the 3d subshell.

For the 3d subshell, n = 3; therefore, $\ell = 1$ or 2, and since m can have values from $-\ell$ to $+\ell$, m_ℓ can have the values -2, -1, 0, +1, +2.

PROBLEM 4. What is the range of values of the angular momentum of an electron in the n = 4 state of the hydrogen atom.

a. 1

Determine the possible values of the orbital quantum number. Then determine the range of values of the angular momentum.

Solution: (Section 28-6)

n = 4 and ℓ can take on integer values from 0 to n - 1; then ℓ can be 0, 1, 2, 3. The range of values for the angular momentum (L) can be determined by substituting the possible values of ℓ into the following equation:

$L = [\ell(\ell + 1)]^{\frac{1}{2}} h/2\pi$

$\ell = 0$, $L = [0(0 + 1)]^{\frac{1}{2}} h/2\pi = 0$

$\ell = 1$, $L = [1(1 + 1)]^{\frac{1}{2}} (6.63 \times 10^{-34} \text{ J s})/2\pi = 1.49 \times 10^{-34} \text{ J s}$

$\ell = 2$, $L = [2(2 + 1)]^{\frac{1}{2}} (6.63 \times 10^{-34} \text{ J s})/2\pi = 2.58 \times 10^{-34} \text{ J s}$

$\ell = 3$, $L = [3(3 + 1)]^{\frac{1}{2}} (6.63 \times 10^{-34} \text{ J s})/2\pi = 3.66 \times 10^{-34} \text{ J s}$

a. 1

Write the
electronic
configuration
for each element.

Solution: (Section 28-8)

The order of filling of the subshells is

$1s^2$ $2s^2$ $2p^6$ $3s^2$ $3p^6$ $4s^2$ $3d^{10}$, etc.

Sodium (Na) has 11 electrons, and its electronic configuration is

$1s^2$ $2s^2$ $2p^6$ $3s^1$

Using the same method, the other elements can be shown to have the following configurations:

$_{17}$Cl $1s^2$ $2s^2$ $2p^6$ $3s^2$ $3p^5$

$_{20}$Ca $1s^2$ $2s^2$ $2p^6$ $3s^2$ $3p^6$ $4s^2$

$_{21}$SC $1s^2$ $2s^2$ $2p^6$ $3s^2$ $3p^6$ $4s^2$ $3d^1$

---

a. 1

Use the periodic table at
thethe back of the text to
determine the symbol for
each element.

Solution: (Section 28-8)

The periodic table in the textbook specifies the configuration of outermost electrons and any other nonfilled subshells. Using the periodic table as a guide, it can be seen that arsenic (As) has an outer configuration of $4s^2$ $4p^3$.

In the same manner, the symbol of each of the other elements can now be determined.

$3d^7$ $4s^2$; the element is cobalt (Co).

$5s^1$; the element is rubidium (Rb).

---

a. 1	Solution: (Section 28-9)
Determine the minimum accelerating voltage necessary for an electron to produce an X-ray photon of wavelength 0.1540 nm.	Assume that all of the electron's energy is converted into the energy of the X-ray photon upon collision with the copper target.

kinetic energy   =   energy of the
of the electron   X-ray photon

$q V = h c/\lambda = 1240$ eV nm$/\lambda$

(1 electron) $V = (1240$ eV nm$)/(0.1540$ nm$)$

$V = 8052$ volts

PROBLEM 8. The K_α X-ray of copper has an energy of 8000 eV. Determine the a) wavelength of this X-ray, and b) atomic number of Cu.

---

a. 1	Solution: (Section 28-9)
Determine the wavelength of this X-ray.	$E = h f = h c/\lambda = 1240$ eV nm$/\lambda$
	8000 eV $= 1240$ eV nm$/\lambda$; $\lambda = 1240$ eV nm$/8000$ eV
	$\lambda = 0.16$ nm $= 1.6 \times 10^{-10}$ m

b. 1	The wavelength of a K_α X-ray is related to the atomic number of the target metal by the following equation:
Use Moseley's formula to determine the atomic number of the copper target.	$\lambda = (1.22 \times 10^{-7}$ m$)/(Z - 1)^2$
	and rearranging gives
	$(Z - 1)^2 = (1.22 \times 10^{-7}$ m$)/\lambda$
	$(Z - 1)^2 = (1.22 \times 10^{-7}$ m$)/(1.6 \times 10^{-10}$ m$)$
	$(Z - 1)^2 = 760$ and $Z - 1 = 28$
	$Z = 29$. The atomic number of copper is 29.

PRACTICE PROBLEMS

PROBLEM 1. Write the electronic configuration for each of the following:

a) $_8$O   b) $_{18}$Ar   c) $_{32}$Ge   d) $_{40}$Zr

ANS. a) $_8$O $1s^2\ 2s^2\ 2p^4$

b) $_{18}$Ar $1s^2\ 2s^2\ 2p^6\ 3s^2\ 3p^6$

c) $_{32}$Ge $1s^2\ 2s^2\ 2p^6\ 3s^2\ 3p^6\ 4s^2\ 3d^{10}\ 4p^2$

d) $_{40}$Zr $1s^2\ 2s^2\ 2p^6\ 3s^2\ 3p^6\ 4s^2\ 4p^6\ 3d^{10}\ 5s^2\ 4d^2$

PROBLEM 2. Determine the values of m_ℓ that are allowed for the a) 4p subshell, and b) 5f subshell.

ANS. a) -1, 0, +1,   b) -3, -2, -1, 0, +1, +2, +3

PROBLEM 3. Write the symbols for the elements whose outer electron configurations are as follows: a) $3d^{10} 4s^1$ b) $5p^2$

ANS. a) Cu, b) Sn

PROBLEM 4. The K_α X-ray produced by an unknown metal target has a wavelength of 0.847 nm. a) Determine the atomic number of the element and b) identify the element.

ANS. a) 13, b) aluminum

CHAPTER 29

MOLECULES AND SOLIDS

KEY TERMS AND PHRASES

chemical bonds refer to the forces that hold the atoms of a molecule together.

pure covalent bond is the type of chemical bond in which the electrons are shared equally.

polar covalent bond refers to the type of bond in which the electrons which form the bond are not shared equally. One end of the molecule is charged positively while the other end is charged negatively; the molecule is called a **polar molecule**.

ionic bond is formed when one or more electrons are transferred from one atom to another. The bond formed is based on the electrostatic attraction of the negatively charged ion for the positively charged ion.

bond energy or **binding energy** refers to the energy required to break a chemical bond which holds the atoms of a molecule together. The bonds which hold atoms of a molecule together are called "**strong**" bonds.

"**weak**" **bond** or **Van der Waals bond** usually refers to electrostatic attraction between molecules. An example of a weak bond is between two dipoles and such a bond is often called a **dipole-dipole bond**.

activation energy refers to the energy which must be added in order to force atoms together to form a molecule.

molecular spectra or **band spectra** is exhibited by molecules. This is because molecules have additional energy levels due to the vibration of the atoms of the molecule with respect to each other and the rotational energy of the molecule.

n-type semiconductors are semiconductors where electrons (negative charge) carry the current.

p-type semiconductors are semiconductors where positive holes "appear" to carry the current.

semiconductor diode is produced when a p-type and an n-type semiconductor are joined. This combination is called a **pn junction diode.**

diode is forward biased if voltage from a battery connected to the diode is large enough to overcome the internal potential difference. The result is a current flow through the diode.

diode is reversed biased if voltage from a battery connected to the diode causes the holes and electrons to be separated, which means that the negative charge tends to be separated from the positive charge. The result is a diode that is essentially nonconducting.

half-wave rectifier allows current from an ac source to flow only in one direction. A pn junction diode acts as a half-wave rectifier. It can be used to change an ac current to a dc current.

SUMMARY OF MATHEMATICAL FORMULAS

potential energy between two point charges	$PE = (1/4\pi\epsilon_o)q_1\,q_2/r$ $1/4\pi\epsilon_o = k$ where $k = 9 \times 10^9$ N m^2/C^2	The potential energy between two point charges is related to the product of the charges ($q_1\,q_2$) and is inversely proportional to the distance r between their centers. If the charges are both positive or both negative, then the PE is positive and decreases with increasing r. If the charges are of opposite sign, then PE is negative and increases with increasing r.
molecular rotational energy	$E_{rot} = L(L + 1)\,\hbar^2/2I$ where $L = 0, 1, 2$, etc.	Molecular rotational energy (E_{rot}) depends on the quantum number (L), the moment of inertia (I), and $\hbar$, where $\hbar = h/2\pi$.
molecular vibrational energy	$E_{vib} = (v + \frac{1}{2})\,h\,f$ where $v = 0, 1, 2$, etc.	Energy levels for vibrational motion depend on the frequency of vibration (f) of the molecule and the vibrational quantum number (v).

CONCEPT SUMMARY

Chemical Bonds

The force that holds the atoms of a molecule together is referred to as a **bond**. The following is a summary of the two main types of chemical bonds: covalent and ionic.

Covalent Bond

In the case of diatomic molecules, such as H_2, O_2, and N_2, the outermost electrons are shared equally by both atoms. The type of bond in which the electrons are shared is called a **covalent bond**.

The cloud model from quantum mechanics is useful in attempting to explain chemical bonding. A simple molecule to consider is hydrogen. When two hydrogen atoms are at a distance, the electron clouds repel and the positively charged nuclei repel. There is no unbalanced force between the atoms.

As the atoms approach, the positively charged nucleus of one atom attracts the electron cloud of the other atom and the shapes of the electron clouds become distorted. The nuclear charge is concentrated and as a result the attraction of one nucleus for the electron cloud of the other atom is greater than the repulsion between the clouds.

As the electron clouds overlap, the overlapping regions cause the repulsion between the clouds to be further reduced. However, as the atoms come closer the repulsion between the positively charged nuclei increases until the forces balance. The distance between the nuclei when the balance point is reached is called the bond length.

Polar Covalent Bond

The chemical bond formed by molecules such as H_2, O_2, and N_2, is a pure covalent bond. This is because the electrons which form the chemical bond are shared equally by the atoms which form the molecule. When the atoms involved are from different elements, then the electrons which form the bond are not shared equally and the bond is not a pure covalent bond.

The water molecule contains two atoms of hydrogen and one atom of oxygen. Hydrogen has one proton in its nucleus while oxygen has eight. The result is that oxygen's large nuclear charge tends to pull the electron cloud of the hydrogen toward it so that the region near the oxygen atom is negatively charged while the region near each hydrogen is positively charged. Because one end of the molecule is charged positively while the other end is charged negatively, the molecule is called a **polar molecule**. This type of a bond is known as **polar covalent**.

Ionic Bond

An **ionic bond** is formed when one or more electrons are transferred from one atom to another. The bond formed is based on the electrostatic attraction of the negatively charged ion for the positively charged ion.

An example of a compound exhibiting ionic bonding is sodium chloride, NaCl. The nucleus of the sodium atom contains 11 protons while that of chlorine contains 17 protons. The nuclear charge of the chlorine exerts a greater force on the outer electron of the sodium ion than does the sodium nucleus. The result is the transfer of the outer electron of the sodium ion to the chlorine atom. The sodium ion (Na^+) exerts a force of electrostatic attraction on the chlorine (Cl^-). Because the force is between two ions, the bond is called an ionic bond.

Strong vs. Weak Bonds

Energy is required in order to break a chemical bond which holds the atoms of a molecule together. The energy required to break a bond is called the **bond energy** or **binding energy**. The binding energy for covalent and ionic bonds is usually in the range of 2 eV to 5 eV. The bonds which hold atoms of a molecule together are called "**strong**" bonds.

The term "**weak**" **bond** or **Van der Waals bond** usually refers to electrostatic attraction between molecules. An example of a weak bond is between two dipoles and such a bond is often called a **dipole-dipole bond**. When one of the atoms in a dipole-dipole bond is hydrogen, then the bond is usually referred to as a hydrogen bond. Another type of weak bond is a dipole-induced dipole bond. This type of bond results when a polar molecule with a permanent dipole moment induces a dipole moment in an electrically balanced, nonpolar molecule.

The strength of a weak bond is in the range of 0.04 to 0.3 eV. In a biological cell the average kinetic energy of molecules is in the same range. A weak bond can be broken during molecular collisions and therefore weak bonds are not permanent. Strong bonds are almost never broken by molecular collisions and are therefore relatively permanent. They can be broken by chemical action in a biological cell with the aid of an enzyme.

Potential Energy Diagrams for Molecules

As discussed in chapter 17, the potential energy between two point charges separated by a distance r is given by

$$PE = (1/4\pi\epsilon_o)q_1 q_2/r \quad \text{where} \quad 1/4\pi\epsilon_o = k \quad \text{and} \quad k = 9 \times 10^9 \text{ N m}^2/\text{C}^2$$

If the charges are both positive or both negative, then the PE is positive and decreases with increasing r as shown in figure A. If the charges are of opposite sign, then PE is negative and increases with increasing r as shown in figure B.

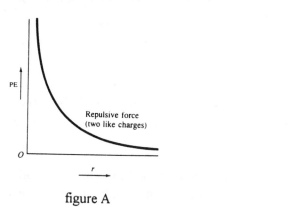

figure A

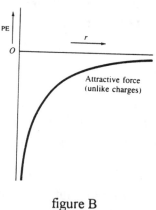

figure B

Covalent Bond

The potential energy function of a covalent bond, e.g., H_2, is shown in figure C. As discussed on page 29-3, as the two atoms approach, they tend to attract and share their valence electrons. The value of the PE decreases to a minimum value at a certain optimum distance between their nuclei (r_o). This distance is known as the bond length. However, if the distance between the nuclei becomes less than r_o, then the nuclei repel and the PE increases. In figure

C, r_0 is the approximate point of greatest stability for the molecule and the approximate point of lowest energy. The energy at this point is called the **binding energy**. The binding energy is the amount of energy required to separate the two atoms to infinity, at infinity PE = 0.

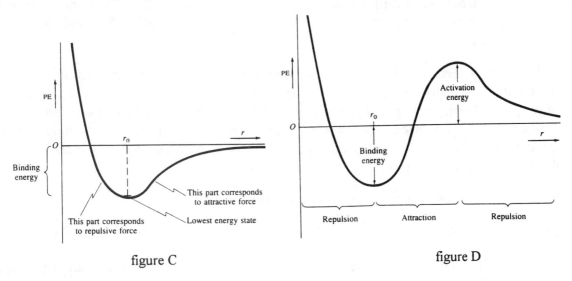

<table>
<tr><td>figure C</td><td>figure D</td></tr>
</table>

Activation Energy

For many molecules, the force between the atoms as they approach is repulsive. In order for the atoms to form a molecule, additional energy called the **activation energy** must be added to force them together. Figure D represents the PE curve for a situation where the activation energy must be considered.

Ionic Bond

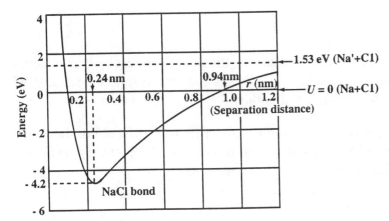

figure E

Figure E is the PE diagram for a typical ionic bond, in this case NaCl. As discussed on page 29-3, the sodium atom tends to lose its 3s electron to the chlorine atom. The result is a force of attraction between Na^+ and Cl^-. As shown in the diagram, the equilibrium distance between the ions is 0.24 nm and at this point the PE is a minimum and corresponds to a bonding energy of 4.2 eV.

Molecular Spectra

As stated in the text, "When atoms combine to form molecules, the energy levels of the outer electrons are altered because they now interact with each other. Additional energy levels also become possible because the atoms can vibrate with respect to each other, and the atom as a whole can rotate."

"The energy levels for both vibrational and rotational motion are quantized, and are very close together (typically 10^{-1} to 10^{-3} eV apart). Each atomic energy level thus becomes a set of closely spaced levels corresponding to the vibrational and rotational motions. Transitions from one energy level to another appear as many very closely spaced lines." The simple line spectra associated with the Bohr model of the hydrogen atom is not observed in molecules. Instead, molecules exhibit band spectra.

The quantized rotational energy levels are given by

$E_{rot} = L(L + 1) \hbar^2/2I$ where $L = 0, 1, 2$, etc.

where I is the moment of inertia and $\hbar = h/2\pi$. Transitions from one rotational energy level to another are subject to the selection rule $\Delta L = \pm 1$.

The energy levels for vibrational motion are given by

$E_{vib} = (\nu + \frac{1}{2}) h f$

where ν is the vibrational quantum number, $\nu = 0, 1, 2$, etc. and f is the classical frequency of vibration of the molecule. Transitions from one vibrational energy level to another are subject to the selection rule $\Delta \nu = \pm 1$.

Bonding in Solids

In chapter 16, the ability of a solid to conduct an electrical current resulted in it being classified as a conductor, semiconductor, or insulator. This classification can now be discussed in terms of what is referred to as the **band theory of solids**.

As stated in the text, "If a large number of atoms come together to form a solid, then each of the original atomic levels becomes a band" [as shown in figure F]. "The energy levels are so close together in each band that they seem essentially continuous."

figure F

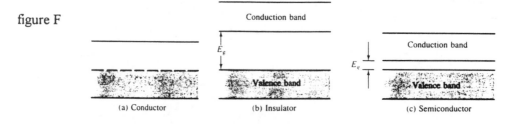

"In a good conductor, e.g., a metal, the highest energy band (valence band) containing electrons is only partially filled. As a result, many electrons are relatively free to move throughout the volume of the metal and the metal can carry an electric current."

"In a good insulator, the highest band (valence band) is completely filled with electrons. The next highest energy band, called the conduction band, is separated from the valence band by a large energy gap (E_g) of 5 to 10 eV." "At room temperature, molecular kinetic energy available due to collisions is only about 0.04 eV, so almost no electrons can jump from the valence to the conduction band." "When a potential difference is applied across the material, there are no available states accessible to the electrons, and no current flows. Hence the material is a good insulator."

"The bands for a pure semiconductor, such as silicon or germanium, are like those for an insulator, except that the unfilled conduction band is separated from the filled valence band by a much smaller energy gap (E_g), typically on the order of 1 eV. At room temperature, there will be a few electrons that can acquire enough energy to reach the conduction band and so a very small current can flow when a voltage is applied. At higher temperatures, more electrons will have enough energy to jump the gap. This effect can often more than offset the effects of more frequent collisions due to increased disorder at higher temperature, so that the resistivity of semiconductors can decrease with temperature."

Semiconductors and Doping

The electronic configuration of the valence electrons of silicon is $3s^2\ 3p^2$ and for germanium $4s^2\ 4p^2$, which means that each element has four outer electrons and is a relatively poor conductor of electricity. Silicon and germanium are examples of semiconductors.

However, if a small amount of an impurity such as arsenic ($4s^2\ 4p^3$) is introduced into the crystal structure of germanium, then arsenic's fifth electron is not bound and is free to move about. As a result, the "**doped**" semiconductor becomes highly conducting. An arsenic-doped germanium crystal is called an **n-type** semiconductor because electrons (negative charge) carry the current.

If a small amount of gallium ($4s^2\ 4p^1$) is added to the germanium crystal, then an empty place or **hole** is introduced because gallium has only three outer electrons. An electron from the germanium atom can jump into this hole but as a result the hole moves to a new location. Because most of the atoms of the crystal are germanium, this new location is invariably next to a germanium atom which is now positively charged because it has lost an electron. An electron can jump from another germanium atom to fill the previous hole and thus the hole can move through the crystal. The flow of electricity in this instance is called a hole current. A germanium crystal doped with gallium is called a **p-type** semiconductor since it is the positive holes which "appear" to carry the current.

Semiconductor Diodes

A semiconductor **diode** is produced when a p-type and an n-type semiconductor are joined.

This combination is called a **pn junction diode.** At the junction, a few electrons from the n-type diffuse into holes in the p-type and an internal difference in potential develops between the sections.

If a battery is connected as shown in figure G, the diode is said to be **forward biased.** If the battery voltage is large enough to overcome the internal potential difference, a current will flow through the diode.

figure G	figure H

If the battery is connected as shown in figure H, the diode is **reversed biased.** This causes the holes and electrons to be separated, which means that the negative charge tends to be separated from the positive charge. The result is a diode that is essentially non-conducting. A graph of current versus voltage for a typical diode is shown in Fig. 29-30 of the text. As a reference, figure 29-30 is shown below.

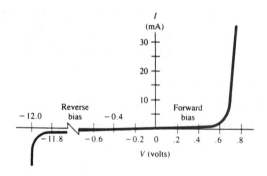

Since a pn junction diode allows current to flow only in one direction, it will allow current from an ac source to flow only in one direction through the circuit. Therefore, it can be used to change an ac current to a dc current. This is called rectification and in the simple circuit shown below, the diode is acting as a half-wave rectifier.

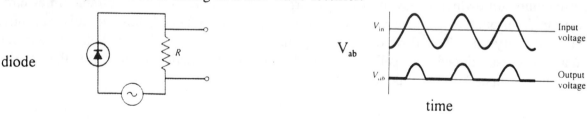

diode

ac source
simple rectifier circuit

output voltage across R as a
function of time

It should be noted that the symbol for a semiconductor diode is —▷|— . The diode allows current to flow in the direction of the arrow but not in the reverse direction.

SELECTED TEXT QUESTIONS WITH ANSWERS

QUESTION 3. Does the H_2 molecule have a permanent dipole moment? Does O_2? Does H_2O? Explain.

ANSWER: Both H_2 and O_2 have a pure covalent bond. The electrons which form the bond between the atoms are equally shared by the atoms; therefore, neither molecule has a permanent dipole moment.

H_2O consists of two hydrogen atoms and one oxygen atom. As described on page 29-3, oxygen's large nuclear charge tends to pull the electron cloud of each hydrogen toward it with the result that the region near the oxygen atom is negatively charged while the region near each hydrogen atom is positively charged. Overall, the water molecule is electrically neutral. However, one end of the molecule is charged positively while the other end is charged negatively and the molecule does have a permanent dipole moment.

QUESTION 5. The energy of a molecule can be divided into four categories. What are they?

ANSWER: The categories are electrostatic potential energy, translational kinetic energy, rotational kinetic energy, and vibrational kinetic energy.

QUESTION 7. A silicon semiconductor is doped with phosphorous. Will these atoms be donors or acceptors? What type of semiconductor will this be?

ANSWER: The outer electronic configuration of phosphorous is $2s^2 \, 3p^3$. If a small amount of phosphorous is introduced into the crystal structure of silicon, then the fifth electron of the phosphorous atom is not bound and is free to move through the solid. Phosphorous is a donor atom and a semiconductor consisting of silicon doped with phosphorous is an n-type.

QUESTION 11. Compare the resistance of a pn junction diode when connected in forward bias to when connected in reverse bias.

ANSWER: A diode is forward biased when connected to a battery as shown in figure G. A battery voltage large enough to overcome the internal potential difference can cause a current to flow through the diode and the resistance to current flow is small.

A diode is reverse biased when connected to a battery as shown in figure H. This causes the holes and electrons to be separated. The result is that the diode is essentially nonconducting and the electrical resistance to current flow is very large.

PROBLEM SOLVING SKILLS

For problems related to electrostatic potential energy:

1. Determine the charge on each ion and the distance between the ions.
2. Use $PE = (1/4\pi\epsilon_o)q_1q_2/r$ to solve for the potential energy.

For problems related to rotational energy states:

1. Determine the difference in energy between the rotational energy states.
2. Use the energy difference to determine the moment of inertia of the molecule.
3. Use the moment of inertia to determine the bond length.

For problems related to the energy gap between the valence band and conduction band for a semiconductor or insulator:

1. If the energy for the electron to travel from the valence band to the conduction band is provided by a photon, determine the energy of the photon. The energy of a photon can be determined by using $E = h\,f$ or $E = h\,c/\lambda$.
2. The energy difference (energy gap) between the valence band and conduction band is equal to the lowest frequency (or longest wavelength) photon capable of causing the electron's transition.

For problems related to a series circuit containing a voltage source, a diode, and an additional resistor:

1. Use a figure giving the current-voltage characteristics of the diode to determine the approximate voltage drop across the diode.
2. Use Ohm's law to determine the voltage drop across the series resistor.
3. Use Kirchhoff's voltage rule to determine the voltage of the source.

PROGRAMMED PROBLEMS

PROBLEM 1. a) Determine the electrostatic potential energy of the Rb^+ and Cl^- ions in an RbCl molecule which are at a stable separation of 0.329 nm. Assume each ion carries a charge of +1.0e. Express your answer in both joules and electron volts. b) What is the magnitude of the electrostatic attractive force acting between the ions?

---

a. 1

Calculate the magnitude of the electrostatic potential energy in joules.

Solution: (Section 29-2)

$PE = (1/4\pi\epsilon_o)q_1q_2/r$

But $q_1 q_2 = (1.6 \times 10^{-19} C)(-1.6 \times 10^{-19} C) = -2.6 \times 10^{-38} C^2$.

$PE = (9 \times 10^9 N m^2/kg^2)(2.6 \times 10^{-38} C^2)/(0.329 \times 10^{-9} m)$

$PE = -7.0 \times 10^{-19} J$

---

a. 2

Determine the energy in eV.

$PE = (-7.0 \times 10^{-19} J)(1.0 eV/1.6 \times 10^{-19} J)$

$PE = -4.37 eV$

---

b. 1

Determine the magnitude of the electrostatic force attraction between the two ions.

$F = (1/4\pi\epsilon_o)q_1 q_2/r^2$

$= (9 \times 10^9 N m^2/kg^2)(-2.6 \times 10^{-38} C^2)/(0.329 \times 10^{-9} m)^2$

$F = -2.12 \times 10^{-9} N$

PROBLEM 2. It is determined that a photon of wavelength 0.00698 m causes the O_2 molecule to make a transition from the lowest rotational energy state (L = 0) to the first excited state (L = 1). Determine the a) energy of the photon, b) difference in energy between the rotational energy states, c) moment of inertia of the O_2 molecule, and d) bond length of the molecule. Note: the mass of an oxygen atom is 2.67×10^{-26} kg.

---

a. 1

Determine the energy of the photon.

Solution: (Section 29-4)

$E = h f$ but $f = c/\lambda$ Therefore,

$E = hc/\lambda$

$E = (6.63 \times 10^{-34} J s)(3 \times 10^8 m/s)(0.00698 m) = 2.85 \times 10^{-23} J$

---

b. 1 Determine the difference in energy between the states where $L = 0$ and $L = 1$.	The photon caused the transition from $L = 0$ to $L = 1$; therefore, the energy difference between the two states equals the energy of the photon. $\Delta E = E_1 - E_0 = 2.85 \times 10^{-23}$ J
c. 1 Determine the moment of inertia of the O_2 molecule.	The formula for the quantized rotational energy level is $E_{rot} = L(L + 1)\, \hbar^2/2I$ If $L = 0$, $E_0 = 0(0 + 1)\, \hbar^2/2I = 0$ J. If $L = 1$, $E_1 = 1(1 + 1)\, \hbar^2/2I = 2\, \hbar^2/2I = \hbar^2/I$ $\Delta E = E_1 - E_0 = \hbar^2/I - 0$ J $= \hbar^2/I$ 2.85×10^{-23} J $= \hbar^2/I$ and $I = \hbar^2/(2.85 \times 10^{-23}$ J$)$ $I = [(6.63 \times 10^{-34}$ J s$)/2\pi]^2/(2.85 \times 10^{-23}$ J$)$ $I = 3.91 \times 10^{-46}$ J s$^2 = 3.91 \times 10^{-46}$ kg m^2
d. 1 Determine the bond length of the molecule.	Both oxygen atoms have the same mass and each atom can be considered to be a point mass. The center of mass of the molecule lies midway between the two atoms. The moment of inertia about the center of mass is given by $I = m_1 r_1^2 + m_2 r_2^2$ But $m_1 = m_2 = m$ and $r_1 = r_2 = r$. Therefore, $I = 2\,m\,r^2$ 3.91×10^{-46} kg m$^2 = 2\,(2.67 \times 10^{-26}$ kg$)\,r^2$ $r^2 = 7.3 \times 10^{-21}$ m^2 $r = 8.6 \times 10^{-11}$ m The bond length is the distance between the oxygen atoms. bond length $= 2r = 1.71 \times 10^{-10}$ m

a. 1	Solution: (Section 29-6)
Determine the energy gap in joules.	E_g is the minimum energy required to cause an electron to jump from the valence band to the conduction band. The energy of the wavelength radiation must equal E_g.
	$E_g = E_{photon} = hc/\lambda$
	$\quad = (6.63 \times 10^{-34} \text{ J s})(3 \times 10^8 \text{ m/s})/(1.85 \times 10^{-6} \text{ m})$
	$E_g = 1.1 \times 10^{-19}$ J
a. 2	$E_g = (1.1 \times 10^{-19} \text{ J})(1.0 \text{ eV}/1.6 \times 10^{-19} \text{ J})$
Determine the energy in eV.	$E_g = 0.69$ eV

a. 1	Solution: (Section 29-6)
Express the energy in joules.	$E_g = (0.72 \text{ eV})(1.6 \times 10^{-19} \text{ J}/1.0 \text{ eV})$
	$E_g = 1.15 \times 10^{-19}$ J
a. 2	$E_g = E_{photon} = hf$
Determine the frequency of the photon.	$1.15 \times 10^{-19} \text{ J} = (6.63 \times 10^{-34} \text{ J s}) \, f$
	$f = 1.74 \times 10^{14}$ Hz
b. 1	$c = f\lambda$
Determine the wavelength of the photon.	$3.0 \times 10^8 \text{ m/s} = (1.74 \times 10^{14} \text{ Hz}) \, \lambda$
	$\lambda = 1.72 \times 10^{-6}$ m

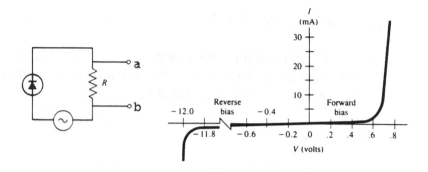

a. 1

Use the diagram to to determine the approximate voltage drop across the diode.

Solution: (Section 29-8)

Using the figure it can be determined that a voltage of approximately 0.75 volt is needed to produce a 20 mA current.

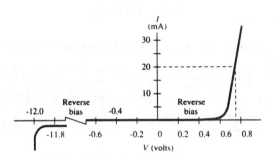

- -

a. 2

Use Ohm's law to determine the voltage drop across the resistor.

The circuit is a series circuit; therefore, the diode current equals the current flowing through the 300 ohm resistor.

$V_R = I R = (20 \text{ mA})(300 \text{ } \Omega)$

$V_R = (20 \times 10^{-3} \text{ A})(300 \text{ } \Omega) = 6.0 \text{ volts}$

- -

a. 3

Use Kirchhoff's voltage rule to determine the battery voltage.

The resistor is in series with the diode and the voltage source,

$V_{battery} = V_{diode} + V_R$

$= 0.75 \text{ V} + 6.0 \text{ volts}$

$V_{battery} = 6.75 \text{ V}$

PRACTICE PROBLEMS

PROBLEM 1. The electrostatic potential energy between two ions is 5.3 eV. The charge on the positive ion is +1.0e and on the negative ion -1.0e. Determine separation between the ions.

ANS. 0.271 nm

PROBLEM 2. The bond length between the carbon and oxygen atoms in a carbon monoxide molecule is 0.113 nm. The mass of the carbon atom is 2.00×10^{-26} kg and the mass of the oxygen atom is 2.67×10^{-26} kg. Determine the a) location of the center of mass as measured from the carbon atom, b) moment of inertia of the molecule about its center of mass, and c) wavelength of the longest wavelength photon which could cause a transition from the L = 0 rotational energy state to the L = 1 state.

ANS. a) 0.0646 nm from the carbon atom, b) 1.46×10^{-46} kg m^2, c) 2.60×10^{-3} m

PROBLEM 3. The energy gap for particular element is 10.0 eV. Determine the a) lowest frequency photon which will cause an electron to jump from the valence band to the conduction band, and b) wavelength of the photon found in part a. c) What portion of the electromagnetic spectrum contains photons of this wavelength? d) Would this particular element be considered an insulator, conductor, or semiconductor?

ANS. a) 2.41×10^{15} Hz, b) 1.24×10^{-7} m or 124 nm,   c) ultraviolet, d) insulator

PROBLEM 4. Solve programmed problem 5 if the resistance of the series resistor is 100 Ω and the current through the circuit is 5.0 mA.

ANS. approximately 1.15 volt

CHAPTER 30

NUCLEAR PHYSICS AND RADIOACTIVITY

KEY TERMS AND PHRASES

nucleus of an atom contains protons and neutrons. These particles are called **nucleons**.

1 unified atomic mass unit (u) = 1.660×10^{-27} kg.

atomic number is the number of protons contained in the nucleus.

atomic mass number is the total number of protons and neutrons in the nucleus.

isotopes are atoms which have the same number of protons but different number of neutrons in the nucleus.

binding energy is the energy required to break apart a nucleus into its constituent protons and neutrons.

strong nuclear force refers to attractive force which holds the nucleus together. The strong nuclear force acts between all nucleons, protons and neutrons alike. This force is much greater than the force of electrostatic repulsion which exists between the protons. The strong nuclear force is a short-range force. It acts between nucleons if they are less than 10^{-15} m apart, but is essentially zero if the separation distance is greater than 10^{-15} m.

weak nuclear force is much weaker than the strong nuclear force and appears in a type of radioactive decay called **beta decay**.

radioactive decay results from the instability of certain nuclei. There are three different radiations produced by radioactive decay: **alpha, beta,** and **gamma**.

parent nucleus refers to the nucleus before radioactive decay.

daughter nucleus refers to the nucleus after radioactive decay.

alpha decay occurs when a helium nucleus is spontaneously emitted from the nucleus. An alpha particle (α) consists of two protons and two neutrons but no electrons.

disintegration energy (Q) is the total energy released during alpha decay.

beta decay occurs when an electron (e⁻) and an antineutrino ($\overline{v}$) are spontaneously emitted from the nucleus. Beta decay (β-) is observed in nuclei which have a high ratio of neutrons to protons.

electron capture occurs when the nucleus absorbs one of the inner orbital electrons in the atom to form a neutron.

transmutation is the changing of one element into a new element. Transmutation is the result of alpha or beta decay.

gamma decay occurs when a nucleus in an excited state drops to a lower energy state. In the process a photon called a gamma ray (γ) is emitted.

law of conservation of nucleon number states that the total number of nucleons before decay equals the total number of nucleons after decay.

half-life of an isotope is the time required for half of the radioactive nuclei present in the sample to decay.

radioactive decay series is a series of successive decays which starts with one parent isotope and proceeds through a number of daughter isotopes. The series ends when a stable, nonradioactive isotope is produced.

radioactive dating refers to a method of estimating the age of an object based on the object's half-life and the amount of the isotope present in the sample being analyzed.

SUMMARY OF MATHEMATICAL FORMULAS

atomic mass number	$^{A}_{Z}X$ $A = Z + N$	The nucleus of a chemical element is designated by $^{A}_{Z}X$. X is the chemical symbol of the element, Z is the atomic number, which is the number of protons contained in the nucleus, A is the atomic mass number, and N is the number of neutrons in the nucleus.
atomic radius	$r \approx (1.2 \times 10^{-15} \text{ m})(A^{\frac{1}{3}})$	The approximate atomic radius (r) of the nucleus increases with the mass number (A).

binding energy	$E = (\Delta m) c^2$ $E = (u)(931.5 \text{ MeV/u})$	The energy equivalent of the mass difference (Δm) is the binding energy (E) of the nucleus. The binding energy is determined by multiplying the mass difference, expressed in u, by the conversion factor 931.5 MeV/u.
alpha decay	$N(A, Z) \rightarrow N(A\text{-}4, Z\text{-}2) + {}^{4}_{2}\text{He}$	An alpha decay particle consists of a helium nucleus. $N(A, Z)$ is the parent nucleus (N) with atomic number (Z) and mass number (A). $N(A\text{-}4, Z\text{-}2)$ is the daughter nucleus and and ${}^{4}_{2}\text{He}$ is the alpha particle.
disintegration energy	$Q = (M_p - M_d - m_\alpha) c^2$	The total energy released during alpha decay is called the disintegra tion energy (Q). M_p is the mass of the parent nucleus, M_d is the mass of the daughter nucleus, and m_α is the mass of the alpha particle.
beta decay	$N(A, Z) \rightarrow N(A, Z+1) + {}^{0}_{-1}e + \bar{\nu}$	Beta decay occurs when an electron (e⁻) and an antineutrino ($\bar{\nu}$) are spontaneously emitted from the nucleus. The electron carries one negative charge and as a result the atomic number of the daughter nucleus is one greater than the atomic number of the parent nucleus. The antineutrino carries no electric charge and has zero rest mass.
electron capture	$N(A, Z) + {}^{0}_{-1}e \rightarrow N(A, Z\text{-}1) + \nu$	Electron capture occurs when the nucleus absorbs one of the inner orbital electrons in the atom to form a neutron. When electron capture occurs, a neutrino (ν) is spontaneously emitted from the nucleus.

gamma decay	$N^*(A, Z) \rightarrow N(A, Z) + \gamma$	Gamma decay occurs when a nucleus in an excited state drops to a lower energy state. In the process a photon called a gamma ray (γ) is emitted. Since the gamma ray is a photon, it carries no electric charge and has no rest mass. No change in nucleon number or atomic number occurs due to a gamma decay. $N^*(A, Z)$ is the nucleus in an excited state.
radioactive decay rate	$\Delta N = - \lambda N \Delta t$ or $\Delta N/\Delta t = - \lambda N$	The number of radioactive decays (ΔN) that occur in a short time interval (Δt) is proportional to the length of the time interval and the total number (N) of radioactive nuclei present. λ is the decay constant which is different for different isotopes. $\Delta N/\Delta t$ is the rate of decay or the activity of the isotope. The negative sign indicates that the number of radioactive nuclei present is decreasing.
law of radioactive decay	$N = N_o e^{-\lambda t}$	Based on the law of radioactive decay, the number of radioactive nuclei present (N) after time (t) depends on the number of radio-active nuclei present in the original sample (N_o) and the decay constant (λ) for the particular isotope.
half-life	$T_{1/2} = - 0.693/\lambda$	The half-life ($T_{1/2}$) of an isotope is the time required for half of the radioactive nuclei present in the sample to decay. The half life is related to the decay constant (λ).

CONCEPT SUMMARY

Structure of the Nucleus

The nucleus of an atom contains protons and neutrons. These particles are called **nucleons**. The neutron is electrically neutral, while the proton carries a single positive charge of magnitude 1.6×10^{-19} C. The mass of each particle is $m_p = 1.673 \times 10^{-27}$ kg $= 1.0073$ u; $m_n = 1.675 \times 10^{-27}$ kg $= 1.0087$ u where 1 u $= 1.660 \times 10^{-27}$ kg $= 1$ unified atomic mass unit.

The nucleus of a chemical element is designated by $^A_Z X$. X is the chemical symbol of the element, e.g., H for hydrogen, Ca for calcium. Z is the **atomic number**, which is the number of protons contained in the nucleus. A is the **atomic mass number**. The atomic mass number is the total number of protons and neutrons in the nucleus.

Recent experiments indicate that the nuclei have a roughly spherical shape. The radius (r) of the nucleus increases with the mass number (A), and the approximate radius is given by

$$r \approx (1.2 \times 10^{-15} \text{ m})(A^{\frac{1}{3}})$$

Isotopes

Atoms which have the same atomic number but different mass numbers are called **isotopes**. Isotopes of the same element have the same 1) atomic number, 2) number of electrons, 3) electronic configuration, and 4) chemical properties. Isotopes differ in the number of neutrons in the nucleus. For example, the two isotopes of lithium are Li-6 and Li-7. Each isotope contains three protons; however, one contains three neutrons while the other contains four neutrons.

Binding Energy and Nuclear Forces

The total mass of the nucleus is always less than the sum of the masses of the protons and neutrons of which it is composed. The energy equivalent of the mass difference (Δm) is the **binding energy** of the nucleus and can be determined by multiplying the mass difference, expressed in u, by the conversion factor 931.5 MeV/u.

The average binding energy per nucleon is defined as the total binding energy divided by the mass number (A). Figure 30-1 in the text shows that the average binding energy per nucleon increases until A = 15, levels off at about 8.8 MeV per nucleon until A = 60, and then slowly decreases.

The nucleus is held together by an attractive force which acts between all nucleons, protons and neutrons alike. This force is called the **strong nuclear force** and is much greater than the force of electrostatic repulsion which exists between the protons. The strong nuclear force is a short-range force. It acts between nucleons if they are less than 10^{-15} m apart, but is essentially zero if the separation distance is greater than 10^{-15} m.

A second type of nuclear force is the **weak nuclear force**. This force is much weaker than the strong nuclear force and appears in a type of radioactive decay called **beta decay**.

Radioactive Decay Mode

Certain nuclei are unstable and undergo **radioactive decay**. There are three different radiations produced by radioactive decay: **alpha, beta**, and **Gamma**.

Alpha Decay

An **alpha decay** particle consists of a helium nucleus. Thus, an alpha particle (α) consists of two protons and two neutrons but no electrons. An alpha decay can be represented by the following equation:

$$N(A, Z) \rightarrow N(A\text{-}4, Z\text{-}2) + {}^4_2He$$

$N(A, Z)$ is the **parent nucleus (N)** with atomic number (Z) and mass number (A).

$N(A\text{-}4, Z\text{-}2)$ is the **daughter nucleus** and ${}^4_2He$ is the alpha particle.

When alpha decay occurs, the atomic number of the element decreases by two while the mass number decreases by four. Alpha decay occurs in large nuclei in which the strong nuclear force is not strong enough to hold the nucleus together. It occurs when the mass of the parent nucleus is greater than the mass of the daughter plus the mass of the alpha particle, i.e., $M_p > M_d + \alpha$. The mass difference appears in the form of kinetic energy of the daughter nucleus and the alpha particle (but mainly in the alpha particle).

The total energy released during alpha decay is called the disintegration energy Q where $Q = (M_p - M_d - m_\alpha) c^2$

Beta Decay

Beta⁻ decay occurs when an electron (e⁻) and an antineutrino ($\bar{v}$) are spontaneously emitted from the nucleus. Beta decay (β-) is observed in nuclei which have a high ratio of neutrons to protons. The electron is not a nucleon; therefore, there is no change in the mass number of the daughter nucleus. However, the electron carries one negative charge and as a result the atomic number of the daughter nucleus is one greater than the atomic number of the parent nucleus. Beta⁻ decay is represented by the following equation:

$$N(A, Z) \rightarrow N(A, Z+1) + {}^0_{-1}e + \bar{v}$$

The **antineutrino** ($\bar{v}$) carries no electric charge and has zero rest mass.

Certain unstable isotopes have a low neutron to proton ratio. In such a situation, a positron and a neutrino (v) may be emitted from the nucleus. The charge on a positron (${}^0_{+1}e$) is opposite that of the electron but the mass of both particles is the same. The positron is represented as e⁺ or β+ and is the antiparticle to the electron. As in the case of the antineutrino, the neutrino carries no electric charge and has zero rest mass. The equation for positron decay is as follows:

$$N(A, Z) \rightarrow N(A, Z\text{-}1) + {}^0_{+1}e + v$$

Electron Capture

Electron capture occurs when the nucleus absorbs one of the inner orbital electrons in the atom to form a neutron. When electron capture occurs, a neutrino is spontaneously emitted from the nucleus. Electron capture may occur if the neutron-proton ratio is low. Electron capture is represented by the following equation:

$$N(A, Z) + {}_{-1}^{0}e \rightarrow N(A, Z-1) + \nu$$

Transmutation

A new element is formed when alpha or beta decay occurs. The changing of one element (parent nucleus) into a new element (daughter nucleus) is called **transmutation**.

Gamma Decay

Gamma decay occurs when a nucleus in an excited state drops to a lower energy state. In the process a photon called a gamma ray (γ) is emitted. The process is analogous to photon emission when an orbital electron drops from a higher energy level to a lower energy level.

The nucleus may be in the excited state due to a violent collision with another particle or a previous radioactive decay leaves the nucleus in an excited state.

Since the gamma ray is a photon, it carries no electric charge and has no rest mass. Therefore, no change in nucleon number or atomic number occurs due to a gamma decay. A gamma decay can be represented as follows:

$$N^*(A, Z) \rightarrow N(A, Z) + \gamma$$

where $N^*(A, Z)$ is the nucleus in an excited state.

Conservation Laws

In addition to conservation of energy, linear momentum, angular momentum, and electric charge, a radioactive decay also obeys the law of **conservation of nucleon number**. This law states that the total number of nucleons before decay equals the total number of nucleons after decay.

Half-life and Rate of Decay

The number of radioactive decays (ΔN) that occur in a short time interval (Δt) is proportional to the length of the time interval and the total number (N) of radioactive nuclei present.

$$\Delta N = - \lambda N \Delta t \quad \text{or} \quad \Delta N/\Delta t = - \lambda N$$

λ is the decay constant, which is different for different isotopes. $\Delta N / \Delta t$ is the rate of decay or the activity of the isotope. The negative sign in the equation indicates that the number of radioactive nuclei present is decreasing.

Based on the law of radioactive decay, the number of radioactive nuclei present is given by the equation

$$N = N_o \, e^{-\lambda t}$$

N_o is the number of radioactive nuclei present in the original sample, i.e., the number at $t = 0$ s. N is the number of radioactive nuclei present at time t and $e = 2.718$.

The rate of decay of any isotope is usually given by its **half-life**. The half-life ($T_{1/2}$) of an isotope is the time required for half of the radioactive nuclei present in the sample to decay. The relationship between the half-life and the decay constant is given by

$$T_{1/2} = - 0.693 / \lambda$$

The graph shown below indicates the number of undecayed nuclei (parent nuclei) present as a function of time where the time is expressed in terms of half-lives. This type of curve is known as an **exponential decay curve**.

Note: in this diagram, the time is given in terms of the half-life of an isotope of carbon, carbon-14.

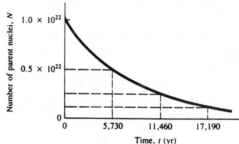

Decay Series

A radioactive decay often results in a daughter nucleus that is also radioactive. A **radioactive decay series** is a series of successive decays which starts with one parent isotope and proceeds through a number of daughter isotopes. The series ends when a stable, nonradioactive isotope is produced.

Radioactive Dating

If the half-life of a radioactive isotope is known, an estimate of the age of an object oftentimes can be made. For example, the ratio of carbon-14 to carbon-12 in a living object is relatively constant. However, a living object stops absorbing carbon-14 when it dies. Therefore, by knowing the half-life of carbon-14 (5700 years) and the object's $^{14}C/^{12}C$ ratio, an estimate of the object's age can be made.

Geologists determine the amount of ^{238}U remaining in a rock relative to the amount of daughter nuclei present to estimate the passage of time since the rock solidified from molten material.

SELECTED TEXT QUESTIONS WITH ANSWERS

QUESTION 4. Why are the atomic masses of many elements (see the periodic table) not close to whole numbers?

ANSWER: Isotopes of the same element contain the same number of protons but different number of neutrons. The weighted average of the percent abundance of the isotopes is used to determine the atomic mass. For most elements this weighted average is not close to a whole number. For example, in appendix C of the text, chlorine is listed as having 17 protons but one isotope (^{35}Cl) has 18 neutrons while ^{37}Cl has 20 neutrons. The percent abundance of ^{35}Cl is 75.77% while ^{37}Cl is 24.23%. The atomic mass is determined as follows:
(0.7577)(35) + (0.2423)37 = 35.45

QUESTION 11. What element is formed by the radioactive decay of

a) $^{24}_{11}Na(\beta-)$; b) $^{22}_{11}Na(\beta+)$; c) $^{210}_{84}Po(\alpha)$.

ANSWER:

PART a) $^{24}_{11}Na(\beta-)$ is an example of $\beta-$ decay. The general equation for this type of decay is

$$N(A, Z) \rightarrow N(A, Z+1) + {}^{0}_{-1}e + \nu \quad \text{therefore,}$$

$$^{24}_{11}Na \rightarrow {}^{24}_{12}Mg + {}^{0}_{-1}e + \bar{\nu}$$

PART b) $^{22}_{11}Na(\beta+)$ is an example of beta^{+} decay. The general equation for this type of decay is

$$N(A, Z) \rightarrow N(A, Z-1) + {}^{0}_{+1}e + \nu \quad \text{therefore,}$$

$$^{22}_{11}Na \rightarrow {}^{22}_{10}Ne + {}^{0}_{+1}e + \nu$$

PART c) $^{210}_{84}Po(\alpha)$ is an example of alpha decay. The general equation for this type of decay is

$$N(A, Z) \rightarrow N(A-4, Z-2) + {}^{4}_{2}He \quad \text{therefore,}$$

$$^{210}_{84}Po \rightarrow {}^{206}_{82}Pb + {}^{4}_{2}He$$

QUESTION 20. An isotope has a half-life of one month. After two months, will a given sample of this isotope have completely decayed? If not, how much remains?

ANSWER: After two months the sample will not have completely decayed, in fact, ¼ of the original amount remains. The half-life of the isotope is one month, which means after one month half of the original sample remains. After another month has passed, half of the sample which remained after one month remains. If the initial amount is N, then after one month ½N remains. After two months, ½(½N) = ¼N is present.

PROBLEM SOLVING SKILLS

For problems involving nuclear density:

1. Use the equation $r \approx (1.2 \times 10^{-15} \text{ m})(A^{1/3})$ to determine the approximate radius of the nucleus.
2. Use the equation $V = (4/3) \pi r^3$ to determine the approximate volume of the nucleus.
3. Convert the mass of the nucleus from unified atomic mass units to kg. Note: the mass of the nucleus expressed in atomic mass units can be determined by noting the mass number.
4. Use the equation $\rho = m/V$ to determine the approximate density.

For problems related to the binding energy of the nucleus:

1. Use a table of nuclear masses to determine the total mass of the constituent particles which make up the nucleus. Note also the exact atomic mass of the nucleus.
2. Determine the difference between the mass of the nucleus and the mass of the constituent particles.
3. Use the equation $\Delta E = (\Delta m) c^2$ to determine the binding energy.

For questions related to a particular type of nuclear decay when the parent nucleus is known:

1. Take note whether the decay is alpha, beta, or gamma.
2. Write down the general equation for the particular type of decay.
3. Use the conservation laws to determine the atomic number and mass number of the daughter nucleus.
4. Use a periodic table to identify the daughter element.

For problems related to the calculation of the disintegration energy of an alpha decay:

1. Use a table of nuclear masses to determine the mass of the parent nucleus, daughter nucleus, and alpha particle in atomic mass units.
2. Determine the difference between the mass of the parent nucleus and the mass of the daughter nucleus + alpha particle. Express this difference in atomic mass units.
3. Determine the disintegration energy by multiplying the mass difference expressed in atomic mass units by 931.5 MeV/u.
4. If the problem asks for the kinetic energy of the daughter nucleus and alpha particle:
   a. Use the law of conservation of momentum to determine the ratio of the velocity of the alpha particle to that of the daughter nucleus.
   b. Determine the ratio of the kinetic energy of the alpha particle to the daughter nucleus.
   c. Knowing the total kinetic energy of the daughter nucleus and alpha particle, and the ratio of the two kinetic energies, algebraically solve for the kinetic energy of each particle.

For problems related to radioactive decay:

1. Determine the number of radioactive nuclei in the sample.
2. Use $\Delta N / \Delta t = -\lambda N$ to determine the activity of the sample.
3. If the problem involves carbon dating, then determine the ratio of decays per second in the artifact to the number of decays per second in the sample.
4. Knowing the half-life of C-14, determine the age of the artifact.

For problems related to half-life, decay constant, and the law of radioactive decay:

1. If either the half-life or decay constant is given, use $T = 0.693/\lambda$ to determine either the half life or decay constant.
2. Use the law of radioactive decay $N = N_0 e^{-\lambda t}$ to determine the number of nuclei remaining after a certain time has passed.

PROGRAMMED PROBLEMS

PROBLEM 1. Determine the approximate density of the nucleus of an ^{27}Al atom in kg/m³.

a. 1	Solution: (Section 30-1)
Determine the radius of the nucleus.	The atomic mass number (A) of an aluminum atom is 27. The approximate radius of the nucleus may be determined as follows: $$r \approx (1.2 \times 10^{-15} \text{ m})(A^{\frac{1}{3}})$$ $$r \approx (1.2 \times 10^{-15} \text{ m})(27^{\frac{1}{3}}) \approx 3.6 \times 10^{-15} \text{ m}$$
a. 2 Determine the approximate volume of the nucleus.	$$V = (4/3) \pi r^3$$ $$\approx (4/3)(3.14)(3.6 \times 10^{-15} \text{ m})^3$$ $$V \approx 2.0 \times 10^{-43} \text{ m}^3$$
a. 3 Determine the mass of the nucleus in kg.	$$m = (27 \text{ u})(1.66 \times 10^{-27} \text{ kg/u})$$ $$m = 4.48 \times 10^{-26} \text{ kg}$$
a. 4 Determine the approximate density of the nucleus.	$$\rho = m/V \approx (4.48 \times 10^{-26} \text{ kg})/(2.0 \times 10^{-43} \text{ m}^3)$$ $$\rho \approx 2.2 \times 10^{17} \text{ kg/m}^3$$

PROBLEM 2. The exact atomic mass of a ^{65}Cu atom is 64.92779 u. Determine the total mass of its constituent particles and calculate the binding energy of the atom. The mass of an electron, proton, and neutron are 0.00055 u, 1.00728 u, and 1.00867 u, respectively.

a. 1	Solution: (Section 30-2)
Determine the total mass of the constituent particles.	The atomic number is 29; therefore, the Cu atom has 29 electrons and 29 protons. The mass number is 65; therefore, Cu has 65 - 29 = 36 neutrons.

electrons	29 x 0.00055 u	=	0.01595 u
protons	29 x 1.00728 u	=	29.21112 u
neutrons	36 x 1.00867 u	=	36.31212 u
	total mass of nucleons		65.53919 u

a. 2	mass of nucleons	65.53919 u
Determine the difference in mass between Cu-65 and its constituent particles.	mass of Cu	64.92779 u
	mass difference	0.61140 u

| a. 3 | $E = (\Delta m)c^2$ |
| Determine the binding energy. | $E = 0.61140$ u x 931.5 MeV/u = 569.5 MeV |

PROBLEM 3. a) Use figure 30-1 from the text to determine the binding energy per nucleon for Cu-65. b) Use this value to determine the total binding energy for the Cu-65 nucleus.

| a. 1 | Solution: (Section 30-2) |
| Determine the binding energy per nucleon. | It is not possible to determine an exact value from figure 30-1. An approximate value for the binding energy per nucleon for Cu-65 is 8.8 MeV per nucleon. |

| b. 1 | 8.8 MeV/nucleon x 65 nucleons = 570 MeV |
| Determine the total binding energy. | |

PROBLEM 4. Cite the type of reaction and then complete the following.

a) $^{239}_{92}U \rightarrow X + ^{0}_{-1}e + \bar{v}_e$ b) $^{105}_{48}Cd + ^{0}_{-1}e \rightarrow X + v_e$

c) $^{22}_{88}Ra \rightarrow X + ^{4}_{2}He$ d) $^{15}_{8}O \rightarrow X + ^{0}_{+1}e + v_e$

| a. 1 | Solution: (Sections 30-4, 30-5, and 30-7) |
| Cite the type of reaction, then use the law of conservation of charge and nucleon number to solve for X. | This reaction is a beta decay. The product nucleus must have an atomic number of 92 - (-1) = 93 in order to satisfy the law of conservation of charge. It must have a mass number of 239 in order to conserve nucleon number. Using the periodic table, it can be determined that X is neptunium; thus

 X is $^{239}_{93}Np$ |

| b. 1 | This is an example of electron capture. The atomic number of the product nucleus is 48 + (-1) = 47. The mass number is 105. Using the periodic table, the unknown nucleus is silver; thus |
| Cite the type of reaction and repeat step a. 1. | X is $^{105}_{47}Ag$ |

c. 1	This is an example of alpha decay. The atomic number of the product nucleus is 88 - 2 = 86. The mass number is 226 - 4 = 222. The unknown nucleus is radon; thus
Cite the type of reaction and repeat step a. 1.	X is $^{222}_{86}$Ra

d. 1	This is an example of positron emission. The atomic number of the product nucleus is 8 - (+1) = 7. The mass number is 15 - 0 = 15. The element is nitrogen; thus
Cite the type of reaction and repeat step a. 1.	X is $^{15}_{7}$N

PROBLEM 5. a) Determine the disintegration energy released when the following reaction occurs

$$^{214}_{84}\text{Po} \quad \rightarrow \quad ^{210}_{82}\text{Pb} \quad + \quad ^{4}_{2}\text{He}$$

b) Determine the kinetic energy of each of the products.

a. 1	Solution: (Section 30-4)		
Determine the mass before and after the decay.	before		after
	$^{214}_{84}$Po 213.9952 u	$^{210}_{82}$Pb	209.9842 u
		$^{4}_{2}$He	4.0026 u
	total mass of products		213.9868 u

a. 2	mass difference = 213.9952 u - 213.9868 u
Determine the mass difference.	mass difference = 0.0084 u

a. 3	(0.0084 u)(931.5 MeV/u) = 7.82 Mev
Determine the energy released.	

b. 1 Use the law of conservation of momentum to express the final velocity of the alpha particle in terms of the final velocity of the lead nucleus. Note: assume that the the initial momentum of the system was zero.	In any reaction, both energy and momentum must be conserved. Assuming that the polonium nucleus was initially at rest, then the initial momentum was zero and the total momentum after the decay must also be zero. This means that the alpha particle's momentum must be equal to but opposite that of the lead nucleus. initial momentum = final momentum $0 = m_\alpha \, v_\alpha + m_{Pb} \, v_{Pb}$ $m_\alpha \, v_\alpha = - m_{Pb} \, v_{Pb}$ and $v_\alpha = -(m_{Pb}/m_\alpha) \, v_{Pb}$ $v_\alpha = -(209.9842 \text{ u}/4.0026 \text{ u}) \, v_{Pb}$ $v_\alpha = - 52.46 \, v_{Pb}$
b. 2 Determine the ratio of the KE of the lead nucleus to the KE of the alpha particle.	$KE_{Pb}/KE_\alpha = (\frac{1}{2} m_{Pb} \, v_{Pb}{}^2)/(\frac{1}{2} m_\alpha \, v_\alpha{}^2)$ but $v_\alpha = - 52.46 \, v_{Pb}$ $KE_{Pb}/KE_\alpha = [\frac{1}{2} (209.9842 \text{ u})(v_{Pb})^2]/[\frac{1}{2} (4.0026 \text{ u})(-52.46 \, v_{Pb})^2]$ Upon simplifying, $KE_{PB}/KE_\alpha = 0.01906$
b. 3 Determine the kinetic energy of the alpha particle and the daughter nucleus.	The total kinetic energy of the daughter nucleus and the alpha particle is 7.82 MeV. $KE_\alpha + KE_{Pb} = 7.82$ MeV but $KE_{Pb} = 0.01906 \, KE_\alpha$ and substituting gives $KE_\alpha + 0.01906 \, KE_\alpha = 7.82$ MeV $1.0196 \, KE_\alpha = 7.82$ MeV $KE_\alpha = 7.67$ MeV $KE_{Pb} = 7.82$ MeV - 7.67 MeV = 0.15 MeV

PROBLEM 6. a) A sample of wood from a living tree is found to contain 20 grams of pure carbon. If carbon contains 1 part in 10^{12} of carbon-14, determine the activity of the sample. b) A wood artifact from an archeological site is found to contain 20 grams of pure carbon and produces 57 decays/min. Determine the age of the artifact. Note: the half-life of carbon-14 is 5700 years and the decay constant is 3.8×10^{-12}/s.

a. 1	Solution: (Sections 30-8, 30-9, and 30-11)
Determine the number of carbon atoms in the sample.	20 g x 6.02 x 10^{23} atoms/12 g = 1.0 x 10^{24} atoms
a. 2	(1.0 x 10^{24} atoms)(1 carbon-14 nucleus/10^{12} atoms)
Determine the number of carbon-14 nuclei in the sample.	$\qquad$ = 1.0 x 10^{12} carbon-14 nuclei
a. 3	- $\Delta N/\Delta t = \lambda \ \Delta N$
Determine the activity in the sample.	$\qquad$ = (3.8 x 10^{-12}/sec)(1.0 x 10^{12} nuclei) - $\Delta N/\Delta t$ = 3.8 decays/sec
b. 1	(57 decays/min)(1 min/60 s) = 0.95 decays/s
Determine the number of decays per second.	
b. 2	(0.95 decays/sec)/(3.8 decays/sec) = ¼
Determine the ratio of the decays/sec in the artifact to the decays/sec in the sample.	
b. 3	The rate of activity in the artifact is ¼ that of the sample from a living tree. Therefore, two half-lives have passed since the artifact was made. Since the half-life of carbon-14 is 5700 years, the age of the artifact is 2 x 5700 year = 11,400 yrs.
Determine the age of the artifact.	

PROBLEM 7. A 2.00 microgram sample of pure ^{49}Cr is to be used in a laboratory experiment. The half-life of the isotope is 42.0 min. Determine the a) decay constant of the isotope, and b) number of original nuclei remaining after 2.80 hours.

a. 1	Solution: (Section 30-8)
Determine the half-life in seconds.	(42.0 min)(60 s/1 min) = 2520 s

a. 2 Determine the decay constant for this element.	$\lambda = 0.693/T_{1/2}$ $= 0.693/2520$ s $\lambda = 2.75 \times 10^{-4}/s$
b. 1 Determine the number of half-lives that have passed in 2.8 h.	(2.8 hours)(60 min/1 hour) = 168 min (168 min)(1 half-life/42.0 min) = 4 half-lives
b. 2 Determine the number of moles of the original sample that is present after 4 half-lives.	The number of moles of the original present after 4 half-lives should be $(1/2)^4$ moles of the original; thus $(1/2)^4(2.00 \times 10^{-6}$ g)(1 mole/49 g) = 2.55×10^{-9} moles
b. 3 Determine the number of nuclei of Cr-49 present after 2.8 h.	N = $(2.55 \times 10^{-9}$ moles)$(6.02 \times 10^{23}/$mole) N = 1.54×10^{15} nuclei

ALTERNATE SOLUTION: Use the law of radioactive decay to solve part b.

b. 1 Determine the number of moles of Cr-49 in the original sample.	$(2.0 \times 10^{-6}$ g)(1 mole/49 g) = 4.1×10^{-8} moles
b. 2 Determine the number of nuclei present in the original sample.	N_o = $(4.1 \times 10^{-8}$ moles)$(6.02 \times 10^{23}/$mole) N_o = 2.46×10^{16} nuclei
b. 3 Use the law of radio-active decay to solve for the number of Cr-49 nuclei remaining after 2.8 h.	$N = N_o\, e^{-\lambda t}$ where $\lambda = 2.75 \times 10^{-4}/s$, t = 2.8 h = 1.01×10^4 s, and $\lambda t = (2.75 \times 10^{-4}/s)(1.01 \times 10^4$ s) = 2.77 N = $(2.46 \times 10^{16}$ nuclei) $e^{-2.77}$ = $(2.46 \times 10^{16}$ nuclei)(0.063) N = 1.54×10^{15} nuclei

PRACTICE PROBLEMS

PROBLEM 1. a) Use the data given in appendix C of the text to determine the total binding energy of the nucleus of ^{235}U. b) Determine the binding energy per nucleon. c) Use figure 29-1 as an alternate means of determining the binding energy per nucleon.

ANS. a) 1790 MeV, b) 7.62 MeV per nucleon, c) approximately 7.6 MeV per nucleon.

PROBLEM 2. Complete the following reactions.

a) $^{214}_{82}Pb \rightarrow {}^{214}_{83}Bi + X + \bar{\nu}$ c) $^{228}_{90}Th \rightarrow X + {}^{4}_{2}He$

b) $^{30}_{15}P \rightarrow {}^{30}_{14}Si + X + \nu$ d) $^{13}_{7}N \rightarrow X + {}^{0}_{+1}e + \nu$

ANS. a) $^{0}_{-1}e$, b) $^{0}_{+1}e$, c) $^{224}_{88}Ra$, d) $^{13}_{6}C$

PROBLEM 3. a) Determine the energy released by the following reaction.

$$^{231}_{91}Pa \rightarrow \ ^{227}_{89}Ac \ + \ ^{4}_{2}He$$

b) Determine the kinetic energy of each of products.

ANS. a) 5.15 MeV, b) KE_α = 5.06 MeV, KE_{AC} = 0.089 MeV

PROBLEM 4. A sample of pure strontium-90 (^{90}Sr) contains 90 grams and has a decay constant of 7.63 x 10^{-10} decays/sec. Determine the a) half-life of the isotope, b) initial activity of the sample, and c) number of original nuclei remaining after 86.4 years have passed.

ANS. a) 9.08 x 10^8 s or 28.8 years, b) 4.59 x 10^{14}/s, c) 7.53 x 10^{22} nuclei

CHAPTER 31

NUCLEAR ENERGY; EFFECTS AND USES OF RADIATION

KEY TERMS AND PHRASES

nuclear reaction occurs when two nuclei collide and two (or more) nuclei are produced. In the process, the nucleus of one element is changed into a different element, and therefore a transmutation of elements has occurred.

exothermic or **exoergic reaction** occurs when energy is released in the reaction.

endothermic or **endoergic reaction** occurs when the total kinetic energy of the projectile particle and the target particle must be greater than or equal to a minimum amount for the reaction to occur. This minimum energy is called the **threshold energy**.

nuclear fission is the process in which a nucleus is split into approximately equal parts along with the release of neutrons and a large amount of energy.

nuclear chain reaction occurs when the neutrons released in a typical fission cause other nuclei to fission. .

critical mass is the minimum amount of fissionable material required for a self-sustaining chain reaction to occur.

nuclear reactor is a device where the chain reaction is controlled and the energy is released gradually.

atomic bomb is a device in which the chain reaction is uncontrolled and the energy release occurs in a few moments of time.

nuclear fusion is the process by which small nuclei combine to form heavier nuclei.

thermonuclear, or **hydrogen,** bomb is a device in which uncontrolled fusion reactions release enormous amount of energy in a few moments of time.

dosimetry refers to measuring the quantity, or dose, of radiation that passes through a material.

1 **rad** is the amount of radiation which deposits 10^{-2} J of energy per kg of absorbing material.

rem (rad equivalent man) is a measure of the biological damage caused by radiation.

relative biological effectiveness (RBE) of a certain type of radiation is defined as the number of rad of x or γ radiation that produces the same biological damage as 1 rad of the given radiation.

Geiger counter is a gas-filled device with electrodes that detects electrons and ions produced by emissions of radioactive nuclei.

scintillation counter makes use of a solid, liquid, or gas which emits light when struck by ionizing radiation.

semiconductor detector makes use of a pn junction diode. The device detects ionizing radiation by the electrons and holes produced in the semiconductor.

bubble chambers detect the paths of elementary particles by photographing the vapor tracks left by the particles as they pass through a superheated liquid.

cloud chambers detect ionizing radiation by photographing the condensation track left in a supercooled gas.

SUMMARY OF MATHEMATICAL FORMULAS

| nuclear reaction | $x + X \rightarrow y + Y$

 or written in short form as

 $X(x, y)Y$. | A nuclear reaction occurs when two nuclei collide and two (or more) nuclei are produced. In the process, the nucleus of one element is changed into a different element, and therefore a transmutation of elements has occurred.

 x is the bombarding particle, sometimes called the bullet. X is the nucleus hit by the bullet and usually referred to as the target nucleus. y is a small particle or possibly a photon and is called the product particle. Y is the recoil nucleus. |

reaction energy or Q-value	$Q = (M_a + M_X - M_b - M_Y)c^2$ $Q = (\Delta m)c^2$ or $Q = KE_b + KE_Y - KE_a - KE_X$	The reaction energy (or Q-value) is related to the mass defect (Δm) by Einstein's equation. a is a projectile particle or small nucleus that strikes nucleus X, producing nucleus Y and particle b. Q can also be written in terms of the change in kinetic energy.
nuclear fission	A typical fission reaction is $_0^1 n + _{92}^{235}U \rightarrow _{38}^{90}Sr + _{54}^{136}Xe + 10\, _0^1 n$	Nuclear fission is the process in which a nucleus is split into approximately equal parts along with the release of neutrons and a large amount of energy. Fission occurs only in the nuclei of certain elements. These elements have a large nucleon number, e.g., U-235. The nucleus may fission if struck by a slow-moving neutron.
dosimetry curie	1 Ci = 3.70×10^{10} disintegrations per second 1 μCi = 10^{-6} Ci.	Dosimetry refers to measuring the quantity, or dose, of radiation that passes through a material. The activity of a radioactive isotope is measured in terms of curies (Ci) or microcuries (μCi).
dosimetry roentgen	One roentgen (1 R) is the amount of radiation that will produce 2.1×10^9 ion pairs/cm^3 of air at STP. Today the roentgen is defined as that amount of X or γ radiation that deposits 0.878×10^{-2} J of energy per kg of air.	The roentgen (R) is a unit of dosage that was previously defined in terms of the amount of ionized air produced by the radiation.

| dosimetry

rad

radiation equivalent man (rem) and relative biological effectiveness (RBE) | 1 rad is the amount of radiation which deposits 10^{-2} J of energy per kg of absorbing material.

rem = rad x RBE | The roentgen has been replaced by the rad.

The rem (rad equivalent man) is the unit which refers to the biological damage caused by radiation. RBE is the relative biological effectiveness of a certain type of radiation. It is defined as the number of rad of x or γ radiation that produces the same biological damage as 1 rad of the given radiation. The RBE of x and γ rays is 1.0, β particles 1.0, α particles 10-20, slow neutrons 3-5, and fast neutrons and protons 10. |

CONCEPT SUMMARY

Nuclear Reactions and Transmutation of Elements

A **nuclear reaction** occurs when two nuclei collide and two (or more) nuclei are produced. In the process, the nucleus of one element is changed into a different element, and therefore a transmutation of elements has occurred.

A nuclear reaction can be written as an equation as follows:

$x + X \rightarrow y + Y$ which can be written in short form as $X(x, y)Y$

x is the bombarding particle, sometimes called the bullet. X is the nucleus hit by the bullet and usually referred to as the target nucleus. y is a small particle or possibly a photon and is called the product particle. Y is the recoil nucleus.

For example, in the reaction $^{1}_{1}H + ^{23}_{11}Na \rightarrow ^{20}_{10}Ne + ^{4}_{2}He$

x is $^{1}_{1}H$, X is $^{23}_{11}Na$, y is $^{4}_{2}He$ and Y is $^{20}_{10}Ne$. The reaction can be written in short form as $^{23}_{11}Na(p, \alpha)^{20}_{10}Ne$. The proton $^{1}_{1}H$ is written as p, while the alpha particle $^{4}_{2}He$ is written as α.

Reaction Energy or Q Value

In any nuclear reaction, all the conservation laws hold, and therefore it is possible to determine if a particular reaction can occur. If, in a reaction, the rest mass of the products is less than the rest mass of the reactants, it is possible for the reaction to occur and energy be released to account for the missing mass. However, in some reactions, the rest mass of the products may be greater than that of the reactants. In such a case, the reaction will not occur unless the bombarding particle has sufficient kinetic energy. The reaction energy (or Q value) can be determined as follows:

$$Q = (M_a + M_X - M_b - M_Y) c^2$$

where a is a projectile particle or small nucleus that strikes nucleus X, producing nucleus Y and particle b. Q can also be written in terms of the change in kinetic energy as follows:

$$Q = KE_b + KE_Y - KE_a - KE_X$$

If $Q > 0$, then the reaction is exothermic or exoergic and energy is released in the reaction. If $Q < 0$, then the reaction is endothermic or endoergic and the total kinetic energy of the projectile particle and the target particle must be greater than or equal to a minimum amount for the reaction to occur. This minimum energy is called the **threshold energy**. The threshold energy, when added to Q, results in an energy great enough to allow the final products to have velocities which obey both the law of conservation of momentum as well as the law of conservation of energy.

Nuclear Fission

Nuclear fission is the process in which a nucleus is split into approximately equal parts along with the release of neutrons and a large amount of energy. Fission occurs only in the nuclei of certain elements. These elements have a large nucleon number, for example, $^{235}_{92}U$ and $^{239}_{94}Pu$. Both of these nuclei may fission if struck by a slow-moving neutron. A typical fission reaction is

$$^1_0n + ^{235}_{92}U \rightarrow ^{90}_{38}Sr + ^{136}_{54}Xe + 10\,^1_0n$$

It can be shown that in this reaction, the sum of the rest masses of the reactants is greater than the sum of the rest masses of the products. The missing mass is converted to energy, $Q > 0$, and the amount of energy released can be found by using

$$E = (\Delta m)c^2$$

Nuclear Chain Reaction

The neutrons released in a typical fission can be used to cause other nuclei to fission. If enough U-235 or Pu-239 is available, it is possible to create a self-sustaining **chain reaction**. The minimum amount of fissionable material required for a self-sustaining chain reaction is known as the critical mass. The critical mass depends on a number of factors, including: the moderator used to slow the neutrons since the probability of causing U-235 or Pu-239 to fission increases with slow-moving neutrons, the type of fuel used, e.g., U-235 or Pu-239, and whether or not the fuel is enriched. Only 0.7% of naturally occurring uranium is the fissionable U-235 and in order to have a self-sustaining chain reaction the percentage of U-235 must be increased.

In a **nuclear reactor**, the chain reaction is controlled and the energy is released gradually.

In an "atomic bomb," the chain reaction is uncontrolled and the energy release occurs in a few moments of time.

Nuclear Fusion

Nuclear fusion is the process by which small nuclei combine to form heavier nuclei. In the reaction, large amounts of energy are released. The force of electrostatic repulsion keeps the nuclei from combining, and therefore the fusion of light nuclei does not occur spontaneously. In order for fusion to happen, it is necessary for the nuclei to collide while traveling at high velocity. Such velocities do occur if the nuclei are part of a hot gas in which temperature approximates 100 million degrees Kelvin. Temperatures of this magnitude are found in the interior of stars, and thus fusion accounts for the enormous energy released by the Sun and other stellar objects.

Uncontrolled fusion reactions have been achieved in the form of the thermonuclear, or hydrogen, bomb. This is because the temperature required for the fusion is achieved by detonating a fission or atomic bomb. The fission bomb creates the necessary temperatures and pressures for the fusion process to occur. However, the difficulty of achieving and sustaining the very high temperatures needed for fusion has thus far prevented the construction of a controlled fusion reactor.

Dosimetry

Dosimetry refers to measuring the quantity, or dose, of radiation that passes through a material. There are several units used for this purpose. The activity of a radioactive isotope is measured in terms of curies (Ci) or microcuries (μCi) where 1 Ci = 3.70 x 10^{10} disintegrations per second and 1 μCi = 10^{-6} Ci.

The **roentgen (R)** is a unit of dosage that was previously defined in terms of the amount of ionized air produced by the radiation. One roentgen (1 R) is the amount of radiation that will produce 2.1 x 10^9 ion pairs/cm^3 of air at STP. Today the roentgen is defined as that amount of x or γ radiation that deposits 0.878 x 10^{-2} J of energy per kg of air. The roentgen has been replaced by the **rad** where 1 rad is the amount of radiation which deposits 10^{-2} J of energy per kg of absorbing material.

Units have been developed which refer to the biological damage caused by radiation. The **rem** (rad equivalent man) is such a unit and can be used for all types of radiation.

rem = rad x RBE

RBE is the relative biological effectiveness of a certain type of radiation. It is defined as the number of rad of x or γ radiation that produces the same biological damage as 1 rad of the given radiation. The RBE of x and γ rays is 1.0, β particles 1.0, α particles 10-20, slow neutrons 3-5, and fast neutrons and protons 10. While we are all subject to about 0.13 rem/year of background radiation from our environment, large doses of radiation can cause radiation sickness leading to death. However, the length of time over which the body receives the dose as well as the size of the dose are important.

Detection of Radiation

A number of ways have been developed to measure radiation dose. A film badge is worn by people who work around sources of radiation. When developed, the darkness of the film is a measure of the radiation dose received. The **Geiger counter, scintillation counter,** and **semiconductor detector** all detect the presence and level of activity of radiation. The paths of elementary particles can be photographed using devices such as **bubble chambers, cloud chambers**, and **spark chambers**.

SELECTED TEXT QUESTIONS WITH ANSWERS

QUESTION 1(d). Fill in the missing particles or nuclei:
$^{197}_{79}$Au(α, d) where d stands for a deuterium nucleus.

ANSWER: The equation of the reaction in the long form is

$$^{197}_{79}\text{Au} + {}^{4}_{2}\text{He} \rightarrow {}^{A}_{Z}\text{X} + {}^{2}_{1}\text{H}$$

conservation of nucleon number	conservation of charge
$197 + 4 = A + 2$	$79 + 2 = Z + 1$
$201 = A + 2$	$81 = Z + 1$
$A = 199$	$Z = 80$

Therefore, $A = 199$ and $Z = 80$. Using the periodic table, it is found that the atomic number of mercury is 80. Therefore, the resulting nuclide is $^{199}_{80}$Hg.

QUESTION 3. When $^{22}_{11}$Na is bombarded by deuterons ($^{2}_{1}$H) an α particle is emitted. What is the resulting nuclide?

ANSWER: Writing the equation of the reaction in the short form gives $^{22}_{11}$Na(d, α)X. Writing the reaction in the long form gives

$$^{22}_{11}\text{Na} + {}^{2}_{1}\text{H} \rightarrow {}^{A}_{Z}\text{X} + {}^{4}_{2}\text{He}$$

conservation of nucleon number	conservation of charge
$22 + 2 = A + 4$	$11 + 1 = Z + 2$
$24 = A + 4$	$12 = Z + 2$
$A = 20$	$Z = 10$

Therefore, A = 20 and Z = 10. Using the periodic table, it is found that the atomic number of neon is 10. Therefore, the resulting nuclide is

$^{20}_{10}$Ne.

QUESTION 4. Why are neutrons such good projectiles for producing nuclear reactions?

ANSWER: Neutrons, which are electrically neutral, are not repelled by the positive charges located in the nucleus of the atom. As a result they can approach the nucleus to the point where the strong nuclear force plays a dominant role.

PROBLEM SOLVING SKILLS

For problems which involve determination of a missing particle in a nuclear reaction:

1. If the reaction is given in the short form, re-write the reaction in the long form.
2. Apply the law of conservation of charge to determine the atomic number (Z) of the nuclide.
3. Apply the law of conservation of nucleon number to determine the mass number of the nuclide.
4. Use the periodic table to identify the unknown nuclide.

For problems involving the calculation of the Q-value of a nuclear reaction:

1. Determine the total mass of the reactants and the total mass of the products.
2. Determine the mass difference between the sum of the reactants and the sum of the products.
3. If the mass difference is expressed in kg, use $E = (\Delta m)c^2$ to determine the Q-value in joules. The Q-value may be expressed in MeV by using the conversion factor 1 MeV = 1.6 x 10^{-13}J.
4. If the mass difference is expressed in atomic mass units (u), then the Q-value may be expressed in MeV by multiplying the mass difference by 931.5 MeV/u.
5. If the Q-value is positive, then the reaction is exothermic or exoergic and can occur spontaneously. If the Q-value is negative, then the reaction is endothermic or endoergic and energy must be added for the reaction to occur.

If the problem involves dosimetry:

1. Determine the number of decays which occur per second.
2. Determine the amount of energy absorbed per kg of mass per second.
3. Convert the energy absorbed per kg of mass per second to rads per second.
4. If required, determine the number of rads absorbed per year.
5. If the RBE value is given, then the yearly dosage in rem may be determined.

PROGRAMMED PROBLEMS

PROBLEM 1. Fill in the missing particle or nucleus.

a) $^{24}_{12}Mg(d, \alpha)?$,   b) $^{12}_{6}C(d, n)?$,   c) $^{35}_{17}Cl(?, \alpha)^{32}_{16}S$,   d) $^{200}_{80}Hg(p, ?)^{197}_{79}Au$

a. 1 Write the reaction in the long form.	Solution: (Section 31-1) d represents a deuteron ^2_1H, while α represents an alpha particle ^4_2He. $^{24}_{12}Mg + ^2_1H \rightarrow ? + ^4_2He$
a. 2 Apply the law of conservation of charge to determine the atomic number.	$12 + 1 = ? + 2$; thus the atomic number of the unknown nucleus is 11 and the element is sodium (Na).
a. 3 Apply the law of conservation of nucleon number and use the periodic table to identify the unknown particle.	$24 + 2 = ? + 4$; thus the mass number is 22. The unknown nucleus is $^{22}_{11}Na$.
b. 1 Write the reaction in the long form.	The symbol n refers to a neutron ^1_0n. $^{12}_6C + ^2_1H \rightarrow ? + ^1_0n$
b. 2 Use conservation of charge.	$6 + 1 = ? + 0$; therefore, the atomic number of the nucleus is 7 and the element is nitrogen (N).
b. 3 Use conservation of nucleon number to identify the unknown nuclide.	$12 + 2 = ? + 1$ and the mass number of the nucleus is 13. The unknown nucleus is $^{13}_7N$.
c. 1 Write the reaction in the long form.	$^{35}_{17}Cl + ? \rightarrow ^{32}_{16}S + ^4_2He$

c. 2 Apply conservation of charge and determine the atomic number of the particle.	$17 + ? = 16 + 2$; therefore, the atomic number of the bombarding particle is 1. The element is hydrogen (H).
c. 3 Apply conservation of nucleon number and identify the nuclide.	$35 + ? = 32 + 4$; therefore, the mass number is 1 and the bombarding particle is a proton $_1^1H$.
d. 1 Write the reaction in the long form.	$_{80}^{200}Hg + _1^1H \rightarrow _{79}^{197}Au + ?$
d. 2 Apply conservation of charge and determine the atomic number of the particle.	$80 + 1 = 79 + ?$; therefore, the atomic number of the product particle is 2 and the element is helium (He).
d. 3 Apply conservation of nucleon number and identify the nuclide.	$200 + 1 = 197 + ?$; therefore, the mass number of the product particle is 4. The product particle is an alpha particle.

PROBLEM 2. Determine the Q-value of the following reaction. State whether the reaction is exothermic or endothermic.

$$_2^3He + _0^1n \rightarrow _1^2H + _1^2H$$

a. 1
 	
Determine the total mass of the reactants and the total mass of the products.
	Solution: (Section 31-1)
	
	    reactants products
	
	$_2^3He$ 3.01603 u $_1^2H$ 2.01410 u
	
	$_0^1n$ 1.00867 u $_1^2H$ 2.01410 u
	
	     4.02470 u 4.02820 u

31-10

a. 2	mass difference = 4.02470 u - 4.02820 u
Determine the mass difference between the products and the the reactants.	mass difference = - 0.00350 u

The products have greater mass than the reactants; therefore, the reaction is endothermic. Energy must be added for the reaction to happen. |
| a. 3 | $Q = (M_a + M_X - M_b - M_Y) c^2$ |
| Determine the value of Q in joules and Mev. | $Q = (\Delta m) c^2$

$\Delta m = (- 0.00350 \text{ u})(1.66 \times 10^{-27} \text{ kg/u}) = - 5.81 \times 10^{-30}$ kg

$Q = (- 5.81 \times 10^{-30} \text{ kg})(3 \times 10^8 \text{ m/s})^2 = - 5.229 \times 10^{-13}$ J

$Q = (- 5.229 \times 10^{-13} \text{ J})(1 \text{ MeV}/1.6 \times 10^{-13} \text{ J}) = - 3.26$ MeV

Alternate solution:

$Q = (- 0.00350 \text{ u})(931.5 \text{ MeV/u}) = - 3.26$ MeV |

PROBLEM 3. Determine the energy released by the following fusion reaction:

$$^{235}_{92}U + ^{1}_{0}n \rightarrow ^{138}_{56}Ba + ^{93}_{41}Nb + 5\,^{1}_{0}n + 5\,^{0}_{-1}e$$

a. 1	Solution: (Section 31-2)
Determine the total mass of the reactants and the total mass of the products.	reactants products

$^{235}_{92}U$ 235.0439 u $^{138}_{56}Ba$ 137.9050 u

$^{1}_{0}n$ $\underline{1.0087 \text{ u}}$ $^{93}_{41}Nb$ 92.9060 u

  236.0526 u $5\,^{1}_{0}n$ 5.0435 u

       $5\,^{0}_{-1}e$ $\underline{0.0028 \text{ u}}$

         235.8573 u |
| a. 2 | mass difference = 236.0526 u - 235.8573 u |
| Determine the mass difference. | mass difference = 0.1953 u |
| a. 3

Determine the energy released in the reaction. | The reactants have a greater mass than the products; therefore, the reaction can occur spontaneously.

$(0.1953 \text{ u})(931.5 \text{ MeV/u}) = 182$ MeV |

PROBLEM 4. a) Determine the energy released when the following fusion reaction occurs:

$$^3_1H + {}^3_1H \rightarrow {}^4_2He + 2{}^1_0n$$

b) Determine the amount of energy in joules produced per u of tritium consumed in the reaction.
c) Determine the amount of energy in joules produced per kg of tritium consumed in the reaction.

a. 1	Solution: (Section 31-3)
Determine the total mass of the reactants and the total mass of the products.	reactants products
	^3_1H 3.016049 u ^4_2He 4.002613 u
	^3H 3.016049 u $2\,{}^1_0n$ 2.017330 u
	6.032098 u 6.019933 u
a. 2	mass difference = 6.03098 u - 6.019933 u
Determine the mass difference and the energy released per reaction.	mass difference = 0.012165 u
	(0.012165 u)(931.5 MeV/u) = 11.3 MeV
b. 1	(11.3 MeV)(1.6 x 10^{-13} J/MeV) = 1.81 x 10^{-12} J
Express the energy released per reaction in terms of joules.	
b. 2	(1.81 x 10^{-12} J)/(6.032098 u) = 3.00 x 10^{-13} J/u
Determine the energy released per atomic mass unit of tritium consumed in the reaction.	
c. 1	1 u = 1.66 x 10^{-27} kg
Determine the energy released per kg of tritium consumed in the reaction.	(3.00 x 10^{-13} J/u)(1 u/1.66 x 10^{-27} kg) = 1.81 x 10^{14} J/kg

PROBLEM 5. Suppose the reaction discussed in problem 4 could be used to produce electrical power in a nuclear power plant of the future. Determine the amount of tritium that would be consumed each day if the output from the plant is to be 100,000 watts and the plant is 33.3% efficient.

a. 1	Solution: (Section 31-3)
Determine the input power required to generate 100,000 watts of output power.	efficiency = output/input x 100% 33.3% = 100,000 watts/input x 100% input = (100,000 watts)(100%)/(33.3%) input = 300,000 watts
a. 2 Determine the energy in joules of input energy during a 24 hour period.	Note: 24 hours = 86,400 second and 1 watt = 1 J/s. 300,000 J/s x 86400 s = 2.59×10^{10} J
a. 3 Determine the amount of tritium consumed. Hint: see part c of problem 4.	In part c of problem 4 it was determined that 1.80×10^{14} joules of energy is produced by each kg of tritium consumed. m = $(2.59 \times 10^{10}\text{J})/(1.81 \times 10^{14}$ J/kg) m = 1.43×10^{-4} kg = 0.143 g

PROBLEM 6. At one time it was possible to purchase a watch that contained the radium isotope Ra-226. This isotope emits 4.76 MeV alpha particles which struck a special material in the paint used for the dial of the watch. This material gave off visible light when struck by the alpha particles. Assume that the dial contained 5.00 microcuries of the isotope and that the wearer had a mass of 70.0 kg and wore the watch 24 hours per day. Determine the a) number of decays per second, and b) yearly dosage in rem to the wearer if 10% of the disintegrations interact with the person's body and deposit all of their energy. Assume an RBE of 10 for the alpha particles. c) The maximum dosage recommended by the U.S. government is 5.0 rem/year. Does the answer to part b exceed this recommendation?

a. 1	Solution: (Section 31-5)
Determine the number of decays per second.	1.0 Ci = 3.7×10^{10} decays/s; thus $(5.00 \times 10^{-6}$ Ci)$(3.7 \times 10^{10}$ decays/s)/1 Ci) = 1.85×10^5 decays/s

b. 1 Determine the amount of energy absorbed by the person's body each second in MeV.	Only 10% (i.e., 0.10) of the disintegrations deposit their energy in the person's body. $(0.10)(1.85 \times 10^5 \text{ decays/s}) = (1.85 \times 10^4 \text{ decays/s})$ $(1.85 \times 10^4 \text{ decays/s})(4.76 \text{ MeV/decay}) = 8.81 \times 10^4 \text{ MeV}$
b. 2 Express this energy in joules.	$(8.81 \times 10^4 \text{ MeV})(1.6 \times 10^{-13} \text{ J/MeV}) = 1.41 \times 10^{-8} \text{ J/s}$
b. 3 Determine the amount of energy absorbed per kg of mass by the person's body.	$(1.41 \times 10^{-8} \text{ J/s})/(70 \text{ kg}) = 2.01 \times 10^{-10} \text{J/kg s}$
b. 4 Convert the answer in step 3 to rad/s.	Note: 1 rad $= 10^{-2}$ J/kg $(2.01 \times 10^{-10} \text{ J/kg s})(1 \text{ rad}/10^{-2} \text{ J/kg}) = 2.01 \times 10^{-8} \text{ rad/s}$
b. 5 Determine the number of rads absorbed by the body in 1 year.	1 year $= 3.15 \times 10^7$ seconds $(2.01 \times 10^{-8} \text{ rad/s})(3.15 \times 10^7 \text{ s/year}) = 0.634 \text{ rad/year}$
b. 6 Determine the dosage in rem/year.	dosage in rem = rad x RBE $= (0.634 \text{ rad/year})(10)$ dosage in rem = 6.34 rem/year
c. 1 Does the dosage exceed recommended limit?	Since the maximum recommended dosage for a person is 5.0 rem/year, the 6.34 rem/year due to the watch is in excess of the recommended limit set by the government.

PRACTICE PROBLEMS

PROBLEM 1. Fill in the missing particle or nucleus.

a) $_3^6Li(d, \alpha)$?   b) $_{92}^{238}U(n, \beta)$?   c) $_4^8Be(\alpha, \gamma)$?   d) $_5^{10}B(n, \alpha)$?

ANS. a) $_2^4He$ b) $_{93}^{239}Np$ c) $_6^{12}C$ d) $_3^7Li$

PROBLEM 2. Determine the Q-value of the following reaction and state whether the reaction is exothermic or endothermic.

$$_8^{16}O + _1^2H \rightarrow _7^{14}N + _2^4He$$

ANS. - 3.11 MeV, endothermic

PROBLEM 3. A typical fission reaction is

$$^{235}_{92}U + ^{1}_{0}n \rightarrow ^{90}_{38}Sr + ^{136}_{54}Xe + 10\ ^{1}_{0}n$$

Determine the energy released during this reaction. Assume that the kinetic energy of the incident neutron is negligible. Use appendix C in the text in order to determine the atomic mass of each isotope.

ANS. 140 MeV

PROBLEM 4. One step in the "carbon cycle," a fusion reaction that occurs in stars, is

$$^{12}_{6}C + ^{1}_{1}H \rightarrow ^{13}_{6}C + ^{0}_{+1}e + \nu$$

Determine the energy released during this reaction.

ANS. 3.65 MeV

CHAPTER 32

ELEMENTARY PARTICLES

KEY TERMS AND PHRASES

Van de Graaff generator electrostatically accelerates charged particles through potential differences as high as 30 MV along a straight line.

cyclotron accelerates protons or electrons through a potential difference into a magnetic field directed perpendicular to their path. As their speed increases, the radius of the circle in which they travel increases until they reach the outer edge of the apparatus and either strike a target within the cyclotron or leave the cyclotron and strike an external target.

cyclotron frequency is the number of revolutions per second made by the charged particle.

synchrotron or synchrocyclotron was developed as a way to compensate for the mass increase with increasing speed. In this type of accelerator, the radius of the particle's path is held constant while the magnetic field increases as the particle's mass increases.

linear accelerator, or LINAC, accelerates a charged particle in steps along a straight line.

colliding beam accelerator causes two beams of charged particles moving in opposite directions to collide head on. The kinetic energy of the colliding particles is available for causing reactions or creating new particles.

electromagnetic force is explained by the wave-particle duality. It is possible to suggest that the electromagnetic force between charged particles is the result of an electromagnetic field produced by one particle that affects the second particle (wave theory) or by an exchange of photons (γ particles) between the charged particles (particle theory).

strong nuclear force is postulated to be caused by pi meson or pion.

weak nuclear force accounts for beta decay (β). The weak nuclear force is presumed to be due to particles known as W^+, W^-, and Z°.

graviton is postulated to be responsible for the gravitational force.

positron is the antiparticle of the electron. The positron has the same mass as an electron but has the opposite charge. When the positron encounters an electron, the two annihilate each other and a gamma ray is produced.

antiparticle of the proton is the antiproton and the antiparticle of the neutron is the antineutron. A few particles, e.g., the photon and the neutral pion (π_0), have no antiparticle.

leptons (light particles) include the electron, muon, and two types of neutrino: the electron neutrino (ν_e) and muon neutrino (ν_μ) as well as their antiparticles.

hadrons are composed of two subgroups, the **baryons** (heavy particles) and the **mesons** (intermediate particles). The baryons include the proton and all heavier particles through the omega particle plus their antiparticles. The mesons include the pion, kaon, and eta particles plus their antiparticles.

conservation laws for reaction and decay processes to occur include energy (including mass), linear momentum, charge, angular momentum (including spin), baryon family number, electron-lepton family number, and muon-lepton family number.

resonances are super short lived particles which decay via the strong interaction. They do not travel far enough in a bubble chamber or a spark chamber to be detected and their existence is inferred because of decay products which can be detected.

strangeness and the **principle of conservation of strangeness** provide an explanation for the observation of certain reactions and also why other reactions do not occur. They also provide an explanation for the "long" lifetimes, 10^{-10} s to 10^{-8} s, for certain particles which interact via the strong interaction.

quarks are composed of four particles which are named up (u), down (d), strange (s), and charm and carry a fractional charge, either ⅓ or ⅔ the charge on the electron. The quarks have antiparticles called antiquarks. Two more quarks called top (t) or truth and bottom (b) or beauty have been postulated.

color and flavor are properties associated with quarks. Each of the flavors has three colors; red, green, and blue, while the antiquarks are colored antired, antigreen and antiblue.

gluons are particles which transmit the strong force. According to the theory there are eight gluons. They are massless and six of the gluons have color charge.

electroweak theory proposes that the weak force and the electromagnetic force are two different manifestations of a single, more fundamental, electroweak interaction.

grand unified theory suggests that at very short distances (10^{-31} m) and very high energy, the electromagnetic, weak, and strong forces are different aspects of a single underlying force. In this theory the fundamental difference between quarks and leptons disappears. Attempts are presently being made to incorporate the four forces found in nature - gravity, electromagnetic, weak, and strong forces - into a single theory.

SUMMARY OF MATHEMATICAL FORMULAS

cyclotron frequency	$f = 1/T$ $f = (qB)/(2\pi m)$	The cyclotron frequency (f) is the number of revolutions per second made by the charged particle.
type of force	relative strength	field particle
strong nuclear	1	mesons/gluons, very short distance ($\approx 10^{-15}$ m)
electromagnetic	10^{-2}	photon
weak nuclear	10^{-9}	W± and Z°, extremely short range ($\approx 10^{-15}$ m)
gravitational	10^{-38}	graviton

CONCEPT SUMMARY

Particle Accelerators

As discussed in chapter 25, the sharpness or resolution of the details of an image is limited by the wavelength of the incident radiation. The de Broglie wavelength of particles such as electrons or protons is given by $\lambda = h/mv$, and particles which have greater momentum have a shorter wavelength. There are a number of different types of particle accelerators available to produce the high energy, and therefore short wavelength particles needed to investigate nuclear structure.

The **Van de Graaff generator** electrostatically accelerates charged particles through potential differences as high as 30 MV along a straight line. The **cyclotron** accelerates protons or electrons through a potential difference into a magnetic field directed perpendicular to their path. The charged particles move within two D-shaped cavities, called **dees,** and each time they move into the space between the dees a potential difference is applied which increases their speed. As their speed increases, the radius of the circle in which they travel increases until they reach the outer edge of the dee and either strike a target within the cyclotron or leave the cyclotron and strike an external target.

The **cyclotron frequency** is given by $f = 1/T = (qB)/(2\pi m)$.

f is the cyclotron frequency, which is the number of revolutions per second made by the charged particle. T is the period in seconds of the particle's motion. q is the charge in coulombs on the particle. B is the strength in tesla of the magnetic field which is directed perpendicular to the particle's path and m is the mass of the particle in kg.

The cyclotron frequency determines the frequency of the applied voltage needed to accelerate the charged particle. The frequency does not depend on the radius of the circle in which the particle is traveling but it does depend on the particle's mass. As the speed increases, the mass of the particle increases according to Einstein's formula $m = m_o/(1- v^2/c^2)^{1/2}$. In order to continue to accelerate the particle, the cyclotron frequency must be reduced as the mass of the particle increases. A device known as a **synchrocyclotron** or **synchrotron** was developed for this purpose.

The synchrotron was developed as an alternative way to compensate for the mass increase with increasing speed. In this type of accelerator, the radius of the particle's path is held constant while the magnetic field increases as the particle's mass increases. The radius of the synchrotron may exceed 1.0 km and accelerate protons to energies exceeding 500 GeV.

In both the cyclotron and the synchrotron the charged particle travels in a circle and undergoes centripetal acceleration. The accelerated charges radiate electromagnetic energy and considerable energy is lost through radiation. This effect is called synchrotron radiation.

In a **linear accelerator,** or **LINAC,** a charged particle is accelerated in steps along a straight line. Linear accelerators have been constructed which have accelerated electrons to over 20 GeV. No magnetic fields are used in a linear accelerator.

In a **colliding beam accelerator,** two beams of charged particles moving in opposite directions collide head on. If the particles have the same mass and energy, then the momentum before and after is zero. If the kinetic energy after impact is zero, then all of the initial kinetic energy of the colliding particles is available for causing reactions or creating new particles.

The Four Forces in Nature

Because of the wave-particle duality, it is possible to suggest that the electromagnetic force between charged particles is the result of an electromagnetic field produced by one particle that affects the second particle (wave theory) or by an exchange of photons (γ particles) between the charged particles (particle theory).

In 1935, Hidecki Yukawa postulated the existence of a particle that produces the strong nuclear force that holds the atomic nucleus together. This particle, now known as the pi meson or pion and represented by the symbol π, was discovered in 1947. The particle can be charged positively ($\pi+$) or negatively (π-) or be uncharged (πo). The mass of the $\pi+$ or π- particle is 140 MeV/c^2 or 273 times the rest mass of the electron while the mass of the πo particle is 135 MeV/c^2, which is 264 times the rest mass of the electron.

The **weak nuclear force** is used to account for beta decay (β). The weak nuclear force is presumed to be due to particles known as W^+, W^-, and Z^o. In 1983, the discovery of the W and Z particles was announced by a group of scientists led by Carlo Rubbia. The group used the high-energy accelerator at CERN.

The gravitational force is believed to be due to a particle known as the **graviton.** The graviton has not yet been observed. The electromagnetic force and the gravitational force are known as "long-range" forces, decreasing as the square of the distance between interacting particles. The strong nuclear force and the weak nuclear force are very short range forces, limited to distances of approximately 10^{-15} m. This distance is the approximate size of the atomic nucleus. The following table lists each type of force, relative strength of the particular force as compared to the strong nuclear force, and the name of the particle credited with producing the force.

type	relative Strength	field particle
strong nuclear	1	mesons/gluons
electromagnetic	10^{-2}	photon
weak nuclear	10^{-9}	W± and Z° particle
gravitational	10^{-38}	graviton

Particles and Antiparticles

The first **antiparticle**, the **positron**, was discovered in 1932. The positron has the same mass as an electron but has the opposite charge. When the positron encounters an electron, the two annihilate one another and a gamma ray is produced. The energy of the gamma ray is equal to the mass equivalent of the electron and positron plus any kinetic energy the two possessed at the time of interaction.

It is now known that most particles have antiparticles, e.g., proton (p)-antiproton ($\bar{p}$),

neutron (n)-antineutron ($\bar{n}$). A few particles, e.g., the photon and the neutral pion (π°), have no antiparticle.

Particle Classification

Table 32-2 in the text lists a number of particles according to their family group and the way they interact. Photons interact only through the electromagnetic force, while **leptons** (light particles) interact only through the weak nuclear force. The leptons include the electron, **muon**, and two types of **neutrino**: the electron neutrino (v_e) and muon neutrino (v_μ) as well as their antiparticles.

The **hadrons** interact through the strong nuclear force. The hadrons are composed of two subgroups, the **baryons** (heavy particles) and the **mesons** (intermediate particles). The baryons include the proton and all heavier particles through the omega particle plus their antiparticles. The mesons include the pion, kaon, and eta particles plus their antiparticles.

Particle Interactions and Conservation Laws

In order for reaction and decay processes to occur, seven quantities must be conserved: energy (including mass), linear momentum, charge, angular momentum (including spin), baryon family number, electron-lepton family number, and muon-lepton family number.

Energy is conserved if the sum of the rest energies and total kinetic energy of the reactants equals the sum of the rest energies and total kinetic energy of the products. Linear momentum must be conserved. In the case of a decay of a nucleus at rest, linear momentum can be satisfied if the decay products travel in different directions such that the total final momentum equals zero.

The law of conservation of charge is satisfied if the sum of the charges on the reactants equals the sum of the charges on the products.

Angular momentum is satisfied if the sum of the spin angular momentum quantum numbers of the reactants equals that of the products. The spin angular momentum of each particle is a multiple of $h/2\pi$ or $\hbar$. For example, the spin angular momentum of an electron equals $\pm\frac{1}{2}(h/2\pi)$, where the + or - is used since the spin can be up (+) or down (-). The spin angular momentum quantum number for each particle listed in table 32-2 is as follows:

category	particle name	spin quantum number
leptons	electron neutrino	½
	muon neutrino	½
	electron	½
	muon	½
photons	photon	1
hadrons		
mesons	pion	0
	kaon	0
	eta	0
baryons	proton	½
	neutron	½
	lambda	½
	sigma	½
	xi	½
	omega	3/2

The law of conservation of baryon number states that in every interaction the total baryon number remains unchanged. The baryon number for a baryon is B = + 1 and for an antibaryon B = - 1. All non-baryons have a baryon number B = 0.

The law of conservation of lepton number must be applied separately to each lepton family. In the electron-lepton family the electron and electron neutrino (v_e) are L_e = + 1, while the

positron and electron antineutrino ($\overline{v}_e$) are assigned L_e = - 1. The electron-lepton family number must be conserved in order for a reaction to occur.

The law of conservation of muon-lepton family number states that the muon-lepton number (L_μ) must be conserved for a process to occur. The μ- and v_μ particles have numbers L_μ = +1 while for μ+ and μ- the number is L_μ = -1. All other particles have a lepton number equal to zero.

Resonances

In addition to the particles listed in table 32-2, there are a great many particles which decay in 10^{-23} seconds or less. Such super short lived particles are known as **resonances** and decay via the strong interaction. They do not travel far enough in a bubble chamber or a spark chamber to be detected and their existence is inferred because of decay products which can be detected.

Strange Particles

The production and decay of certain particles led to the introduction of a new quantum number called **strangeness** (S). Particles are assigned a number called the strangeness number (see table 32-2). The use of strangeness and the principle of conservation of strangeness provides an explanation for the observation of certain reactions and also why other reactions do not occur. It also provides an explanation for the "long" lifetimes, 10^{-10} s to 10^{-8} s, for certain particles which interact via the strong interaction. Strangeness is conserved in the strong interaction but not in the weak interaction. Thus conservation of strangeness is an example of a "partially conserved" quantity.

Quarks

The four known leptons (e-, μ-, v_e, v_μ) seem to be truly elementary particles. There is no evidence that they have internal structure; also they have no measurable size and do not decay.

Experiments indicate that hadrons, mesons, and baryons do have internal structure, have a definite size (10^{-15} m in diameter) and, with the exception of the proton, do decay. It was proposed in 1963 that the hadrons are combinations of three constituent particles known as **quarks**. The quarks are given names up (u), down (d), and strange (s) and have a fractional charge, either ⅓ or ⅔ of the charge on the electron. The quarks have antiparticles called **antiquarks** and the properties of both quarks and antiquarks are listed in table 32-3. All of the hadrons known in 1963 could be constructed from a combination of quarks and antiquarks.

The up and down quarks were named because of their spin directions while the strange quark was named because it is associated with the concept of strangeness. In 1964, the existence of a fourth quark, called charmed (c) was postulated. The c quark has a charm number C = +1 while the $\bar{c}$ antiquark is C = - 1. In 1974, a particle called the J/ψ or ψ was found that did not fit the three-quark scheme but could be accounted for on the basis of a four-quark theory.

In recent years the existence of two more quarks called top (t) or truth and bottom (b) or beauty has been postulated. In 1977, a new meson was detected which is believed to be a combination of a beauty quark and its antiquark $(\bar{b}\bar{b})$.

Quantum Electrodynamics

An extension of the quark theory suggests that quarks have a property called **color** with the distinction between the different quarks called **flavor**. Each of the flavors has three colors; red, green, and blue, while the antiquarks are colored antired, antigreen, and antiblue.

The concept of color is used to explain the force which binds the quarks together in a hadron. Each quark carries a **color charge** and the strong force between the quarks is called the **color force**. This theory of a strong force is called **quantum chromodynamics**, or QCD, in order to indicate that the theory refers to the force between color charges.

The particles which transmit the strong force are called **gluons**. According to the theory there are eight gluons. They are massless and six of the gluons have color charge.

Grand Unified Theories

In the 1960s, a theory called the **electroweak theory** was proposed in which the weak force and the electromagnetic force are viewed as two different manifestations of a single, more fundamental, electroweak interaction. Attempts have been made to incorporate the electroweak force and the strong (color) force into a theory known as the **grand unified theory**. One such theory suggests that at very short distances (10^{-31} m) and very high energy, the electromagnetic, weak, and strong forces are different aspects of a single underlying force. In this theory the fundamental difference between quarks and leptons disappears.

Attempts are presently being made to incorporate the four forces found in nature - gravity, electromagnetic, weak, and strong forces - into a single theory.

SELECTED TEXT QUESTIONS WITH ANSWERS

QUESTION 1. What limits the maximum energy attainable for protons in an ordinary cyclotron? How is this overcome in a synchrotron?

ANSWER: The cyclotron frequency is related to the proton's mass by the equation

$f = (2 \pi m)/q B$. At high speeds, usually taken to be velocities in excess of 0.1c, the relativistic mass rather than the rest mass must be used in the equation. In order to compensate for the increasing mass, the cyclotron frequency is reduced as the mass increases.

An alternative way to continue to add energy to the proton is to use a synchrotron rather than a cyclotron. In a synchrotron, the radius is kept constant while the magnetic field strength is increased at the same rate that the mass changes. As a result, the energy can be increased well beyond that attainable using a cyclotron. The radius of the synchrotron may exceed 1.0 km and the energy of the protons may exceed 500 GeV.

QUESTION 4. What would an "antiatom," made up of antiparticles to constituents of normal atoms, consist of? What might happen if antimatter, made of such antiatoms, came in contact with our normal world of matter?

ANSWER: The antiatom would contain the antiparticles to those found in the normal atom. A normal atom contains electrons, protons, and neutrons. Therefore, the antiatom would contain positrons, antiprotons, and antineutrons. The conservation laws which apply to atoms would also apply to antiatoms. It is known that antimatter-matter interactions, such as electron-positron interactions, result in the annihilation of the particles involved with the production of their energy equivalent.

QUESTION 8. Which of the four interactions (strong, electromagnetic, weak, gravitational) does an electron take part in?

ANSWER: The electron possesses electric charge and therefore interacts with other charged particles via the electromagnetic force and photon-photon exchange. It possesses mass and therefore interacts via the gravitational force with other particles by exchanging gravitons. The electron is a lepton and therefore exhibits weak interaction by exchanging W and Z particles. The electron does not contain "color" constituents and therefore does not take part in interactions via the strong force.

PROBLEM SOLVING SKILLS

For problems involving high-energy projectiles:

1. The kinetic energy of particles in the 1.0 GeV range is approximately equal to mc^2. Express the energy of the particle in joules.
2. The de Broglie wavelength of such a particle (or the energy if the wavelength is given) can be found by using $\lambda = h/mv \approx hc/mc^2$ where $mc^2 = E$, the particle's energy.

For problems involving the cyclotron:

1. Determine the particle's velocity. If the velocity is above 0.1c, then equations used for special relativity must be applied.

2. Derive an equation for the radius of the orbit, the cyclotron frequency, and the period of the motion.
3. Use the equations derived in step 2 to solve the problem.

For problems involving particle interactions:

1. Apply each of the conservation laws.
2. Each of the conservation laws must hold for the interaction to be allowed.

For problems involving the width of a resonance:

1. Use the uncertainty principle, i.e., $(\Delta E)(\Delta t) = h/2\pi$, to determine the width of the resonance in joules.
2. Express the width of the resonance in MeV.
3. Express the width of the resonance in terms of the rest mass of an electron.

PROGRAMMED PROBLEMS

PROBLEM 1. Determine the kinetic energy in GeV of a proton whose de Broglie wavelength is 1.00 fm.

a. 1

Use the de Broglie formula to determine the energy. Note: the energy of the proton is in the order of GeV. Assume that the proton's velocity is approximately that of the speed of light.

Solution: (Section 32-1)

$E = h c/\lambda$ where $\lambda = 1.00$ fm $= 1.00 \times 10^{-15}$ m

$\quad = [(6.63 \times 10^{-34} \text{ J s})(3.0 \times 10^{8} \text{ m/s})]/(1.00 \times 10^{-15} \text{ m})$

$E = 1.99 \times 10^{-10}$ J

$E = (1.99 \times 10^{-10} \text{ J})(1.00 \text{ eV}/1.6 \times 10^{-19} \text{ J}) = 1.24 \times 10^{9}$ eV

$\quad = (1.24 \times 10^{9} \text{ eV})(1.00 \text{ GeV}/10^{9} \text{ eV})$

$E = 1.24$ GeV

PROBLEM 2. A cyclotron is used to accelerate protons to an energy of 30 MeV. The strength of the magnetic field is 1.00 T. Determine the a) radius of the proton's orbit when the proton's energy reaches 30 MeV, b) cyclotron frequency, and c) period of the proton's motion.

a. 1

Determine the proton's velocity when its energy reaches 30 MeV.

Solution: Section (32-2)

$KE = \frac{1}{2} m v^2$

$(30 \text{ MeV})(1.6 \times 10^{-13} \text{ J/MeV}) = (1.67 \times 10^{-27} \text{ kg}) v^2$

$v^2 = 3.36 \times 10^{13}$ m^2/s^2 and solving for v gives

$v = 7.58 \times 10^{7}$ m/s

At this speed, the proton's mass is no longer equal to its rest mass; however, the relativistic mass is approximately equal to the rest mass ($m = 1.03 \text{ m}_o$) and the rest mass will be used.

a. 2

Solve for the radius of the proton's orbit.

$F = q v B \sin \theta$ but $\theta = 90°$ and $\sin 90° = 1$

The motion is circular; therefore, $F = m v^2/r$, then

$q v B = m v^2/r$ and solving for r gives

$r = (m v)/(q B)$

$r = [(1.67 \times 10^{-27} \text{ kg})(7.58 \times 10^{7} \text{ m/s})]/[(1.6 \times 10^{-19} \text{ C})(1.00 \text{ T})]$

$r = 0.191$ m or 19.1 cm

b. 1 Determine the cyclotron frequency.	For a particle traveling in a circle at constant speed (v), $v = (2 \pi r)/T = 2 \pi r f$ From step a. 2, $r = (m\ v)/(q\ B)$; therefore, $r = (m\ 2 \pi r f)/(q\ B)$ Solving for f gives $f = (2 \pi m)/(q\ B)$ $f = [(2 \pi \times 1.67 \times 10^{-27}\ kg)]/[(1.6 \times 10^{-19}\ C \times 1.00\ T)]$ $f = 1.53 \times 10^7\ Hz$
c. 1 Determine the period of the proton's motion.	The period is inversely related to the frequency; therefore, $T = 1/f = 1/(1.53 \times 10^7\ Hz)$ $T = 6.55 \times 10^{-8}\ s$

PROBLEM 3. A head-on collision occurs between two protons having equal kinetic energy. Determine the minimum kinetic energy of each proton in order to produce the following reaction:

$$p\ +\ p\ \rightarrow\ n\ +\ p\ +\ \pi^+$$

a. 1 Determine the total mass of the reactants and the total mass of the products. Note: in table 32-2 of the text, the rest mass is expressed in MeV, therefore, use MeV in solving the problem.	Solution: (Section 32-3) <table><tr><th>reactants</th><th>products</th></tr><tr><td>p 938.3 MeV</td><td>n 939.6 MeV</td></tr><tr><td>p 938.3 MeV</td><td>p 938.3 MeV</td></tr><tr><td></td><td>π^+ 139.6 MeV</td></tr><tr><td>――――― 1876.6 MeV total mass of reactants (in MeV)</td><td>――――― 2017.5 MeV total mass of products (in MeV)</td></tr></table>
a. 2 Determine the mass difference (expressed in MeV) between the products and the reactants.	mass difference = 2017.5 MeV - 1876.6 MeV mass difference = 140.9 MeV

a. 3	The mass of the products is greater than the mass of the reactants; therefore, the kinetic energy of the colliding protons must provide the energy necessary for the reaction to occur. Since the protons have equal kinetic energy, then each proton must have
Determine the kinetic energy of each of the colliding protons.	140.9 MeV/2 = 70.5 MeV of kinetic energy

PROBLEM 4. Use the conservation laws to determine if the following reactions may occur. In each case demonstrate which conservation laws are satisfied and which are violated. In part a assume that the reactants have sufficient kinetic energy for the reaction to occur. In parts b and c assume that the nucleus which decays is at rest.

a) $p + p \rightarrow p + \bar{n} + n$

b) $\kappa \rightarrow \mu^- + \bar{\nu}_\mu$

c) $e \rightarrow \nu_e + \gamma$

a. 1	Solution: (Section 32-5 and 32-6)
Apply the law of conservation of charge.	$p + p \rightarrow p + \bar{n} + n$
	$+1 + +1 \rightarrow +1 + 0 + 0$ (not allowed)
	The left side has a net charge of +2 while the right side has a net charge of +1. The law of conservation of charge is violated.
a. 2	$\pm\frac{1}{2} + \pm\frac{1}{2} \rightarrow \pm\frac{1}{2} + \pm\frac{1}{2} + \pm\frac{1}{2}$ (not allowed)
Apply the law of conservation of angular momentum.	By inspection, there is no way of combining the possible values so that the left side will equal the right side of the equation. The law of conservation of angular momentum is violated.
a. 3	$1 + 1 \rightarrow 1 + (-1) + 1$ (not allowed)
Apply the law of conservation of baryon number.	The left side has a net result of +2 while the right side has a net value of +1. The law of conservation of baryon family number is violated.
a. 4	$0 + 0 \rightarrow 0 + 0 + 0$
Apply the law of conservation of lepton number.	No leptons are involved in this process. The law is not applicable.

a. 5 Apply the laws of conservation of energy and conservation of of momentum.	The law of conservation of energy would be satisfied if the reactants have sufficient kinetic energies. Also, it is possible for the reaction to satisfy the law of conservation of momentum. However, for a reaction to be allowed, every law must be satisfied. Therefore, this reaction is not allowed.
b. 1 Apply the law of conservation of charge.	$\kappa^- \rightarrow \mu^- + \bar{\nu}$ charge $-1 \rightarrow -1 + 0$ (allowed)
b. 2 Apply the law of conservation of angular momentum.	spin $0 \rightarrow \pm\frac{1}{2} + \pm\frac{1}{2}$ (allowed)
b. 3 Apply the law of conservation of baryon number.	baryon number $0 \rightarrow 0 + 0$ (not applicable)
b. 4 Apply the law of conservation of lepton number.	muon-lepton number $0 \rightarrow +1 + (-1)$ (allowed)
b. 5 Apply the laws of conservation of energy and conservation of momentum.	energy 493.8 MeV $\rightarrow$ 105.7 MeV + 0 (allowed) The missing energy can be accounted for in the kinetic energy of the product particles. momentum (allowed) The product particles could be moving such that they have equal but opposite momenta. All of the conservation laws are satisfied; therefore, the reaction is allowed.
c. 1 Apply the law of conservation of charge.	$e^- \rightarrow \nu_e + \gamma$ charge $-1 \rightarrow 0 + 0$ (not allowed)

c. 2	spin $\pm\frac{1}{2}$ $\rightarrow$ $\pm\frac{1}{2}$ $+$ 1 (allowed)
Apply the law of conservation of angular momentum.	

c. 3	baryon 0 $\rightarrow$ 0 $+$ 0 (not applicable) number
Apply the law of conservation of baryon number.	

c. 4	electron-lepton 1 $\rightarrow$ 1 $+$ 0 (allowed) number
Apply the law of conservation of lepton number.	

c. 5	energy 0.51 MeV $\rightarrow$ 0 MeV + 0 MeV (allowed)
Apply the laws of conservation of energy and conservation of momentum.	The product particles have no rest energy. However, the missing energy can be accounted for in the kinetic energy of each product particle. momentum (allowed) The product particles could be moving such that they have equal but opposite momenta. The law of conservation of charge is not satisfied; therefore, the reaction cannot occur.

PROBLEM 5. a) The estimated lifetime of a certain resonance is 8.6×10^{-23} s. Estimate the width of the resonance in joules and MeV. b) Express the width of the resonance in terms of the mass of an electron.

a. 1	Solution: (Section 32-7)
Use the uncertainty principle to determine the width of the resonance in joules.	If the particle's lifetime is uncertain by an amount Δt, then its rest energy will be uncertain by an amount given by $\Delta E = (h/2\pi)/\Delta t$ $= (6.63 \times 10^{-34} \text{ J s}/2\pi)/(8.6 \times 10^{-23} \text{ s})$ $\Delta E = 1.23 \times 10^{-12}$ J

a. 2	$\Delta E = (1.23 \times 10^{-12} \text{ J})(1 \text{ MeV}/1.6 \times 10^{-13} \text{ J})$
Express the answer in MeV.	$\Delta E = 7.7 \text{ Mev}$
b. 1	The rest energy of an electron is 0.51 Mev. Therefore, the width of this particular resonance is approximately
Express the width of the resonance in terms of the rest mass of an electron.	7.7 Mev/0.51 Mev = 15 times the rest mass of an electron

PRACTICE PROBLEMS

PROBLEM 1. A cyclotron is used to accelerate deuterons to an energy of 33 MeV. The magnetic field used to deflect the deuterons into a circular path has a strength of 3.00 T. Determine the a) cyclotron frequency and b) radius of the deuteron's orbit when it reaches 33 MeV.

ANS. a) 2.3×10^7 Hz, b) 5.62×10^7 m/s

PROBLEM 2. A 2.0 GeV photon produces a neutron-antineutron pair. Determine the kinetic energy of each of the product particles.

ANS. each particle has 61 MeV

PROBLEM 3. Determine the maximum kinetic energy of a gamma ray emitted during the following decay:

$$\Sigma_0 \rightarrow \Lambda_0 + \gamma$$

ANS. 76 MeV

PROBLEM 4. Use the conservation laws to determine if the following reactions may occur. In each case demonstrate which conservation laws are satisfied and which are violated. In parts a and b assume that the nucleus which decays is at rest. In parts c and d assume that the reactants have sufficient kinetic energy for the reaction to occur.

a) $n \rightarrow \pi^+ + \pi^-$ b) $p + p \rightarrow p + \Lambda_0 + \Sigma^+$

c) $\mu^+ \rightarrow e^+ + e^-$ d) $p + e^- \rightarrow n + \nu_e$

ANS. a) not allowed: violates angular momentum, baryon number
     b) not allowed: violates angular momentum, baryon number
     c) not allowed: violates angular momentum, charge, muon family number
     d) allowed

CHAPTER 33

ASTROPHYSICS AND COSMOLOGY

KEY TERMS AND PHRASES

astrophysics is a branch of astronomy that applies the techniques and theories of physics to the study of celestial objects.

cosmology is the study of the organization and structure of the universe and its evolution.

parallax is a method for measuring the distance to nearby stars. Parallax employs the apparent shift of position of a star against the background of more distant stars as the Earth moves about the Sun.

parsec is a measure of stellar distance. The distance to a star which changes its apparent position in the sky through an angle of 1″ (1 second of arc) in the course of the year is one parallax second or parsec. 1 parsec = 3.26 light years.

redshift refers to the displacement of a star's spectra toward the red end of the spectrum. Based on the Doppler effect, the red shift indicates that the universe is expanding.

A **galaxy** contains millions to hundreds of thousands of millions of stars. The Milky Way Galaxy contains about 10^{11} stars.

absolute luminosity refers to the total power radiated in watts by a star or galaxy.

apparent brightness is the power crossing per unit area perpendicular to the path of the light at the Earth.

Hertzsprung-Russell (H-R) diagram is a graph of absolute magnitude of stars versus temperature with stars represented by a point on the diagram. Most stars fall along the diagonal band called the **main sequence**.

black hole is the residual mass of a star which is so dense that no matter or light can escape from its gravitational field.

Schwarzchild radius represents the event horizon of a black hole. This is the surface beyond which no signals can ever reach us.

Hubble's law states that galaxies are moving away from one another at speeds proportional to the distance between them.

Big Bang theory suggests that the expansion of the universe is due to an explosion which probably occurred 10 to 15 billion years ago.

standard cosmological model gives an explanation of how the universe evolved after the Big Bang.

SUMMARY OF MATHEMATICAL FORMULAS

redshift	$\lambda' = \lambda[(1 + v/c)/(1 - v/c)]$	The observed wavelength (λ') depends on the true wavelength (λ) as well as the velocity of recession (v) and the speed of light (c).
apparent brightness	$\ell = L/4\pi d^2$	The apparent brightness (ℓ) is the total power (L) crossing per unit area perpendicular to the path of the light at the Earth. d is the distance from the Earth to the star or galaxy and the total area of a sphere of radius d is $4\pi d^2$. Note: the total power radiated in watts by a star or galaxy is called the absolute luminosity (L).
apparent magnitude	$\ell_1/\ell_2 = 100(m_2 - m_1)/5$ $m_2 - m_1 = 2.5 \log_{10}(\ell_1/\ell_2)$	The apparent magnitude (m) is based on a logarithmic scale and is defined so that for two objects whose luminosity ℓ_1 and ℓ_2 differ by a factor of 100, their apparent magnitudes m_1 and m_2 differ by 5.
absolute magnitude	$M = m - 5 \log_{10}(d/10 \text{ parsecs})$	The absolute magnitude (M) of an object is defined as the apparent magnitude that the object would have at a distance of 10 parsecs. d is the distance from the Earth to the object.

Schwarzchild radius	R = 2 G M/c²	Schwarzchild radius (R) represents the radius of the event horizon of a black hole. M is the object's mass, G is the gravitational constant, and c is the speed of light in a vacuum.

CONCEPT SUMMARY

Astrophysics is a branch of astronomy that applies the techniques and theories of physics to the study of celestial objects. **Cosmology** is the study of the organization and structure of the universe and its evolution.

Stellar Distances

One method for measuring the distance to nearby stars is **parallax**. Parallax employs the apparent shift of position of a star against the background of more distant stars as the Earth moves about the Sun. The distance to a star which changes its apparent position in the sky through an angle of 1″ (1 second of arc) in the course of the year is one parallax second or parsec. 1 parsec = 3.26 light years where 1 light year is the distance that light travels through a vacuum in one year (approximately 10^{13} km or 6 trillion miles).

As shown in the diagram , the sighting angle of a star relative to the plane of the Earth's orbit (θ) is measured at different times of the year. The distance from the Earth to the Sun (d) is known and trigonometry can be used to determine the distance (D) to the star.

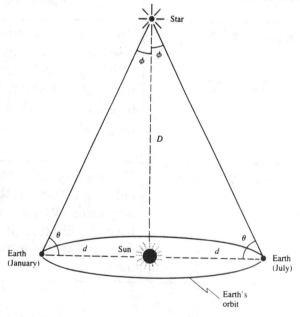

Parallax is useful for stars up to 100 light years (about 30 parsecs) from the Earth. The apparent brightness of more distant stars can give an approximate measurement of distance. Also useful is analysis of the shift of a star's spectra, the so-called redshift.

Redshift

Analysis of the spectra of stars and galaxies indicates a shift toward the red end of the spectrum. The Doppler effect suggests that this shift toward the red or **redshift** of light indicates that the star or galaxy is receding from the Earth and that the universe is expanding in size.

Based on equations derived from the Doppler effect, the observed wavelength (λ') is related to the true wavelength (λ) by the equation

$$\lambda' = \lambda[(1 + v/c)/(1 - v/c)]$$

where v = velocity of recession and c = speed of light.

The shift in the wavelength ($\Delta\lambda$) can be determined from the equation $\Delta\lambda = \lambda' - \lambda$.

Stars and Galaxies

Our Sun is a single star in a **galaxy** known as the Milky Way Galaxy. A typical **galaxy** contains millions to hundreds of thousands of millions of stars. The Milky Way Galaxy contains about 10^{11} stars. It has a diameter of 100,000 light years and a thickness of about 6000 light years. Our Sun is located about halfway from the center to the edge, about 28,000 light years from the center. The Sun orbits the galactic center approximately once every 200,000 years at a speed of 250 km/s relative to the center of the galaxy.

Galaxies tend to be grouped in **galaxy clusters** with the clusters organized into clusters of clusters called **super-clusters**.

Luminosity and Magnitude

The total power radiated in watts by a star or galaxy is called the **absolute luminosity (L)**. The **apparent brightness** (ℓ) is the power crossing per unit area perpendicular to the path of the light at the Earth. Ignoring any absorption of the light as it travels through space,

$\ell = L/(4 \pi d^2)$ where d is the distance from the earth to the star or galaxy and the total area of area of a sphere of radius d is $4 \pi d^2$

The **apparent magnitude (m)** is based on a logarithmic scale and is defined so that for two objects whose luminosity ℓ_1 and ℓ_2 differ by a factor of 100, their apparent magnitudes m_1 and m_2 differ by 5; that is,

$$\ell_1/\ell_2 = 100(m_2 - m_1)/5 \quad \text{and} \quad m_2 - m_1 = 2.5 \log_{10}(\ell_1/\ell_2)$$

Visible stars, other than the Sun, range in apparent magnitude from approximately -1 (the brightest) to m = 6 (just visible to the naked eye).

The **absolute magnitude (M)** of an object is defined as the apparent magnitude that the object would have at a distance of 10 parsecs:

$$M = m - 5 \log_{10}(d/10 \text{ parsecs})$$ where d is the distance from the Earth to the object

Hertzsprung-Russell (H-R) Diagram

For most stars, the color is related to its absolute luminosity and therefore to its mass. This relationship is represented on a diagram called the Hertzsprung-Russell or H-R diagram. On the H-R diagram, shown on the next page, one axis represents the absolute magnitude while the other represents the temperature, and each star is represented by a point on the diagram.

Most stars fall along the diagonal band called the **main sequence**. The coolest stars, reddish in color, are located on the lower right. These stars are the least luminous and therefore must be of low mass. Stars on the upper left, bluish in color, are much more massive and have a higher luminosity. Stars that are not main-sequence type stars include red giants and white dwarfs.

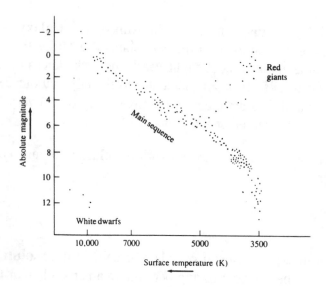

Stellar Evolution

Current theories suggest that **protostars** are formed from collapsing masses of hydrogen gas which exist in great clouds in space. During collapse, gravitational potential energy is transformed into kinetic energy, causing a heating of the gas. When the temperature reaches 10 million degrees kelvin, nuclear fusion begins. The dominant early reaction taking place in the core is the proton-proton cycle, in which four protons fuse to form a helium nucleus. After a time, the protostar becomes stable and can be placed on the main sequence of the H-R diagram. For a star with the mass of our Sun, it takes about 30 million years to reach the main sequence and it will remain there for approximately 10 billion years.

When the supply of hydrogen in the core is sufficiently depleted, the core contracts and the decrease in gravitational potential energy results in higher temperatures. This results in an increase in nuclear reactions, some of which result in the formation of more massive nuclei such as carbon. The outer part of the star expands and cools and the star's position on the H-R diagram moves from the main sequence to become a **red giant**. In about 5 billion years our Sun should become a red giant and its size could increase beyond the orbit of some if not all of the inner planets.

Stars with residual mass less than 1.4 solar masses will cool and become **white dwarfs**. The white dwarf will continue radiating energy until it eventually becomes a black dwarf, a dark, cold piece of ash.

More massive stars can continue to contract, due to their greater gravitational fields, and will eventually approach nuclear densities. Under sufficient pressure, the electrons combine with protons to form neutrons and the result may be a **neutron star**. Neutron stars are thought to have diameters in the order of ten kilometers. It is thought that **supernovae** are the result of the energy released in the final contraction of a neutron star.

It is theorized that an explosion in a supernovae can result in a rotating neutron star known as a **pulsar**. Pulsars are thought to be rotating neutron stars with extremely high magnetic fields which result in intense emission of radio waves in narrow beams. The result is similar to a lighthouse with a beam of light emitted at regular intervals as the light rotates.

If the star's residual mass is greater than two or three solar masses, it may contract still further and form a **black hole**, which is so dense that no matter or light can escape from its gravitational field.

General Relativity: Gravity and the Curvature of Space

Einstein's **principle of equivalence** states that "no experiment can be performed in a small region of space that could distinguish between a gravitational field and an equivalent uniform acceleration." As a result, there is an equivalence between gravitational mass and inertial mass.

Gravity is considered to be curvature in space-time. The curvature is greater near massive bodies. The theory requires the use of non-Euclidian geometries where the shortest distance between two points is not a straight line but possibly a curve called a geodesic. In such a space the sum of the angles of a triangle is not equal to 180°. The theory predicts that light rays passing near a massive object will be deflected. Confirmation came in 1919 when starlight passing the edge of the Sun during a solar eclipse was deflected by an amount consistent with the theory. In addition, the universe as a whole may be curved. If there is sufficient mass, the curvature of the universe is positive and the universe is closed and finite. Otherwise, the curvature is negative and the universe is open and infinite.

A black hole could produce extreme curvature of space-time. A black hole is the residual mass of a star which is so dense that no matter or light can escape from it. The **Schwarzchild**

radius represents the **event horizon** of a black hole. This is the surface beyond which no signals can ever reach us. Inside the event horizon, at r = 0, there is, classically, an infinitely dense **singularity**. Whether such singularities can form is still unknown. Schwarzchild radius (R) is given by the equation R = 2 G M/c² where G is the gravitational constant and c is the speed of light in a vacuum.

The Expanding Universe

Distant galaxies display a redshift of spectral lines. The Doppler effect is used to interpret the redshift, the interpretation indicates that the universe is expanding. The galaxies are moving away from one another at speeds (v) proportional to the distance (d) between them. **Hubble's law** describes the relationship as follows:

v = H d   where   H is Hubble's constant

Hubble's constant is not precisely known but generally estimated to be approximately 50 km/s/Mpc, i.e., 50 kilometers per second per million parsecs of distance.

The expansion of the universe is thought to be due to an explosive origin of the universe called the **Big Bang** which probably occurred 10 to 15 billion years ago.

There is a class of objects called **quasars** which do not seem to conform to Hubble's law. Quasars are quasi-stellar objects which are as bright as nearby stars but display very large redshifts. The large redshift indicates that they are very far away and because of their brightness they must be thousands of times brighter than normal galaxies.

The Big Bang and the Cosmic Microwave Background

In 1964, evidence to support the Big Bang theory was discovered. The discovery came in the form of radiation of wavelength 7.35 cm, which is in the microwave region of the spectrum. The intensity of this radiation was found to be independent of the time of day and came from all directions in the universe with equal intensity. The intensity of this radiation at λ = 7.35 cm corresponds to a blackbody radiation temperature of 2.7 ± 0.1 K.

The **standard cosmological model** gives an explanation of how the universe evolved after the Big Bang. According to the model, starting at 10^{-43} s after the Big Bang, there was a series of phase transitions during which previously unified forces of nature condensed out one by one. The inflationary scenario assumes that during one of these phase transitions, the universe underwent a brief but rapid exponential expansion. Until about 10^{-35} s there was no distinction between quarks and leptons. Shortly thereafter the universe entered the hadron era. About 10^{-6} s after the Big Bang, the majority of hadrons disappeared, introducing the lepton era.

By the time the universe was 10 s old, the electrons too had mostly disappeared and the universe became radiation dominated. As the universe continued to expand, the radiation density decreased faster than the matter density. This eventually resulted in a matter-dominated universe. After about 500,000 years the universe was at a temperature of 3000 K and was cool enough for electrons to combine with nuclei and form atoms. After this time, the universe was cool enough for formation of stars and galaxies. In the 10 to 15 billion years since the Big Bang, the radiation has cooled to a temperature of 2.7 K which produces the 7.35 cm background radiation discovered in 1964.

The Future of the Universe

Whether the universe is open or closed depends on the matter density of the universe. If the average density is above a critical value known as the **critical density**, about 10^{-26} kg/m^3, then gravity will eventually stop the expansion and the universe will collapse back into a big crunch. If the average density is less than this value, the universe will continue to expand forever.

SELECTED TEXT QUESTIONS WITH ANSWERS

QUESTION 7. Does the H-R diagram reveal anything about the core of a star?

ANSWER: The H-R diagram relates only the surface temperature of a star to the absolute luminosity of the surface of the star. It does not directly reveal anything about the core. Theories on stellar evolution provide ideas on the core of a star and its place on the main sequence of the H-R diagram during its life cycle. Therefore, knowledge of a star's position on the H-R diagram might give some indirect knowledge of the stellar interior.

QUESTION 14. What is the difference between the Hubble age of the universe and the actual age? Which is greater?

ANSWER: The Hubble age of the universe is approximately 18 billion years. However, calculations using Hubble's law ignore the effect of gravitation. The matter contained in the universe exerts a gravitational effect which slows the expansion as time passes. Because of this, the commonly accepted age is approximately 15 billion years.

QUESTION 19. Why were atoms unable to exist until hundreds of thousands of years after the Big Bang?

ANSWER: As the universe expanded, the energy spread out over an increasingly larger volume and the temperature dropped. At temperatures higher than 3000 K the energy of the electrons and nuclei was too high for atoms to exist. It was necessary for the temperature of the universe to drop to approximately 3000 K before atoms could form.

PROBLEM SOLVING SKILLS

For problems involving the parallax angle of a star and the stellar distance expressed in parsecs:

1. The parallax angle ϕ can be determined from the equation $\tan \phi = d/D$. d is the radius of the

Earth's orbit about the Sun in km and D is the distance to the star in km.
Note: 1 Ly = 9.46 x 10^{12} km
2. The parallax angle is small and therefore tan $\phi \approx \phi$ when ϕ is expressed in radians.
3. The stellar distance in parsecs = $1/\phi$ where ϕ is expressed in seconds of arc. An alternative method to use is the conversion 1 pc = 3.26 Ly.

For problems involving the absolute magnitude of a star:

1. If necessary, convert stellar distances to parsecs.
2. The absolute magnitude (M) is related to the apparent magnitude (m) by the equation M = m - 5 $\log_{10}$ d/10 where d is the distance to the star expressed in parsecs.

For problems involving the Schwarzchild radius:

1. Determine the mass of the object in kg.
2. The Schwarzchild radius (R) is related to the object's mass by the equation R = G M/c^2.

For problems involving Hubble's law:

1. Hubble's law states that the velocity (v) of a star/galaxy relative to the Earth is related to the distance to the star/galaxy by the equation v = H d. Express Hubble's constant in the appropriate units and solve the problem.

For problems involving the redshift of light from a star or galaxy receding from the Earth:

1. Express the velocity (v) of the galaxy as a fraction of the speed of light.
2. The observed wavelength (λ') is related to the true wavelength (λ) by the equation:
   $\lambda' = \lambda[(1 + v/c)/(1 - v/c)]$
3. The shift in the wavelength ($\Delta\lambda$) can be determined from the equation $\Delta\lambda = \lambda' - \lambda$.

PROGRAMMED PROBLEMS

PROBLEM 1. The brightest star in the night sky is Sirius, a star in the constellation Canis Major. Sirius is 8.6 light years from the Earth. Determine the a) parallax angle for Sirius, and b) distance to Sirius in parsecs.

a. 1 Determine the parallax in radians.	Solution: Section 33-1 Use the figure on page 33-3, note that $\tan \phi = d/D$. d is the angle radius of the Earth's orbit about the Sun; $d = 1.5 \times 10^8$ km. D is the distance from the Earth to Sirius. $D = (8.6 \text{ Ly})(9.46 \times 10^{12} \text{ km/Ly})$ $D = 8.14 \times 10^{13}$ km $\tan \phi = (1.5 \times 10^8 \text{ km})/(8.14 \times 10^{13} \text{ km})$ $\tan \phi = 1.8 \times 10^{-6}$ The angle is small; therefore, $\tan \phi \approx \phi$. $\phi = 1.8 \times 10^{-6}$ radians
a. 2 Express the angle in seconds of arc.	$\phi = (1.8 \times 10^{-6} \text{ rad})(360°/2\pi \text{ rad})(3600''/1°)$ $\phi = 0.377''$ or 0.377 seconds of arc
b. 1 Determine the distance to Sirius in parsecs.	1 parsec (pc) $= 1/\phi$ where ϕ is expressed in seconds of arc distance in parsecs $= 1/(0.377'') = 2.63$ pc alternative solution: $(8.6 \text{ Ly})(1 \text{ pc}/3.26 \text{ Ly}) = 2.63$ pc

PROBLEM 2. The apparent magnitude of Sirius is -1.47. Determine its absolute magnitude based on your answer to problem 1, part b.

a. 1 Determine the absolute magnitude.	Solution: Section 33-2 The absolute magnitude (M) is related to the apparent magnitude (m) by the equation $M = m - 5 \log_{10}(d/10)$ where d is the distance to Sirius in parsecs.

$$M = -1.47 - 5 \log_{10} (2.63/10)$$

$$= -1.47 - 5 \log_{10} 0.263$$

$$= -1.47 - 5 (-0.579)$$

$$M = -1.47 + 2.90 = + 1.43$$

PROBLEM 3. The largest planet in our solar system is Jupiter. The mass of Jupiter is 318 times the mass of the Earth. Determine the Schwarzchild radius for Jupiter.

a. 1	Solution: Section 33-3
Determine the mass of Jupiter in kg.	$M = 318 \, M_e$
	$\quad = 318 \times 6.0 \times 10^{24}$ kg
	$M = 1.91 \times 10^{27}$ kg
a. 2	$R = G \, M/c^2$
Determine the Schwarzchild radius for Jupiter.	$\quad = (6.67 \times 10^{-11} \text{ N m}^2/\text{kg}^2)(1.91 \times 10^{27} \text{ kg})/(3 \times 10^8 \text{ m/s})^2$
	$R = 1.41$ m
	This compares to a Schwarzchild radius of 2.95 km for the Sun and 8.9 mm for the Earth.

PROBLEM 4. a) Use Hubble's law to estimate the recessional velocity of a galaxy 1.0 billion light years from the Earth. Note: let Hubble's constant H = 15 km/s per million light years of distance. b) Express the answer to part a as a fraction of the speed of light.

a. 1	Solution: Section 33-4
Determine the recessional velocity of the galaxy.	$v = H \, d$
	$v = (15 \text{ km/s})/(1 \times 10^6 \text{ Ly}) \times 1.0 \times 10^9 \text{ Ly} = 15000$ km/s
b. 1	$15000 \text{ km/s} \times 1000 \text{ m}/1 \text{ km} = 1.5 \times 10^7$ m/s
Express the answer to part a as a fraction of the speed of light.	$(1.5 \times 10^7 \text{ m/s})/(3 \times 10^8 \text{ m/s}) = 0.5 \times 10^{-1} \, c = 0.05 \, c$

PROBLEM 5. a) Calculate the observed wavelength for the 656 nm line in the Balmer series for hydrogen in the spectrum of a galaxy which is receding at 0.1c from the Earth. b) Determine the change in the wavelength.

a. 1	Solution: Section 33-4
Calculate the observed wave-length.	$\lambda' = \lambda[(1 + v/c)/(1 - v/c)]$
	$\lambda' = 656$ nm $[(1 + 0.1c/c)/(1 - 0.1c/c)]$
	$\lambda' = 656$ nm $[(1 + 0.1)/(1 - 0.1)]$
	$\lambda' = 656$ nm $(1.1/0.9) = (656$ nm$)(1.22)$
	$\lambda' = 802$ nm
b. 1	$\Delta\lambda = \lambda' - \lambda$
Determine the shift in the wavelength.	$= 802$ nm $- 656$ nm
	$\Delta\lambda = 146$ nm (redshift)

PRACTICE PROBLEMS

PROBLEM 1. With the exception of the Sun, the star Alpha Centuri is closest to the Earth. If Alpha Centuri is 4.3 light years from the Earth, a) estimate this distance in parsecs and b) determine the parallax angle.

ANS. a) 1.3 pc, b) 0.77″

PROBLEM 2. The apparent magnitude of Alpha Centuri A is -0.01. Using the information given in practice problem 1, determine the star's absolute magnitude.

ANS. +4.4

PROBLEM 3. Determine the Schwarzchild radius for a planet the size of Saturn. The mass of Saturn is 94 times that of the Earth.

ANS. 0.42 m or 42 cm

PROBLEM 4. a) Estimate the observed wavelength for 656 nm line in the Balmer series for hydrogen in the spectrum of a galaxy which is receding from the Earth at 0.25c. b) Determine the change in the wavelength.

ANS. a) 1090 nm, b) 437 nm